MW00906318

Cyanobacterial Toxins of Drinking Water Supplies

Cyanobacterial Toxins of Drinking Water Supplies

Cylindrospermopsins and Microcystins

Ian Robert Falconer

CRC PRESS

Boca Raton London New York Washington, D.C.

Library of Congress Cataloging-in-Publication Data

Falconer, Ian R.
Cyanobacterial toxins of drinking water supplies : cylindrospermopsins and microcystins
Ian R. Falconer.
p. cm.
Includes bibliographical references and index.
ISBN 0-415-31879-3 (alk. paper)
1. Cyanobacterial toxins. 2. Microcystins. 3. Cyanobacteria. 4. Freshwater microbiology. 5. Bacterial pollution of water. I. Title

QP632.C87F34 2004
615.9'52939—dc22

2004054551

Direct all inquiries to CRC Press, 2000 N.W. Corporate Blvd., Boca Raton, Florida 33431.

Visit the CRC Press Web site at www.crcpress.com

International Standard Book Number 0-415-31879-3
Library of Congress Card Number 2004054551
Printed in the United States of America 1 2 3 4 5 6 7 8 9 0
Printed on acid-free paper

Preface

The importance of cyanobacterial toxins in drinking water sources has been highlighted by the adoption of a provisional drinking water "Guideline Value" for microcystin-LR, one of the most abundant toxins, by the World Health Organization (WHO). A number of nations have now legislated a guideline for microcystins into their drinking water regulations, with the consequent need for monitoring and analytical techniques. The Chemical Safety Committee of the WHO also has under consideration a Guideline Value for cylindrospermopsin, the other most damaging cyanobacterial toxin.

The need for careful study of the cyanobacterial toxins, their sources, and their removal from water supplies was emphasized by the substantial death toll among dialysis patients in Brazil who were accidentally treated with water containing these toxins. Consumers of treated drinking water have also suffered injury due to microcystins and cylindrospermopsin in the water supply, as have people exposed through recreational activities.

Two aspects of cyanobacterial toxicity that require substantial attention are the possible long-term effects on the population of exposure to low doses of the toxins and intermittent exposure to higher doses. In addition, there is increasing experimental evidence of tumor promotion and carcinogenesis in rodents due to the toxins.

This book assesses the present knowledge of toxic species of cyanobacteria and their ecology, the chemistry and toxicology of the most relevant toxins, safe concentrations in drinking and recreational water, monitoring of organisms and toxins, mitigation of reservoir problems, and water treatment technologies. Each of these areas is the subject of considerable recent research, with North America, Europe, Japan, and Australia contributing substantially. This volume is intended to be useful to environmental and public health agencies, water supply utilities, and managers of drinking and recreational water, as well as to researchers in this field.

Acknowledgments

I would like to express my gratitude to my wife, Mary, who has supported my teaching, research, and writing throughout my career and without whose help little would have been achieved. My scientific colleagues Maria Runnegar and Andrew Humpage have, over 30 years, contributed greatly to this research, from its infancy to what is now a rapidly expanding worldwide investigation. Our research collaborators and postgraduate students deserve recognition for their systematic contributions to the field, which are evident from the coauthorship of the many papers quoted in this volume.

I would like to thank the following publishers, organizations, and individuals for permission to reproduce or modify copyrighted material used in this text: the American Chemical Society for Figures 3.2, 3.3, and 10.2; Australian government for Figure 8.4; Australasian Medical Publishing Company for Figure 5.1; Peter Baker, CRC for Water Quality and Treatment, Salisbury, South Australia, for Figure 4.6; Cyanosite Image Gallery — Dr. Roger Burke, University of California, Riverside, and Dr. Mark Schneegurt, Wichita State University — for Figure 2.3; Dr. Bernard Ernst, University of Konstanz, for Figure 2.4d; Dr. Larelle Fabbro for Figure 4.4; Gladstone Area Water Board for Figures 4.2 and 4.5; *Journal of the Association of Official Analytical Chemists* for Figure 10.1; Royal Society of Chemisty for Figure 8.2; Taylor & Francis Journals for Figure 7.3; Wiley for Figures 2.4, 6.1, and 7.1; and the World Health Organization for Figures 4.1 and 4.3.

Table of Contents

1 Introduction

In many parts of the world, surface waters are used for the drinking water supply. The quality of these surface waters is very variable both within and between countries. In developed countries, this water is treated through purification processes; in less developed areas, many people have to rely on untreated water. Toxic cyanobacteria are a normal part of the phytoplankton of surface waters and therefore can present a hazard to consumers if they are present in sufficient numbers. The toxins from cyanobacteria are resistant to boiling and can also pass through conventional water treatment plants. The understanding of cyanobacteria and their toxins and measures for the control of both has expanded greatly in recent years. This volume aims to provide a current account of present knowledge of the two potentially most damaging cyanobacterial toxins in drinking water: cylindrospermopsin and microcystin.

Cyanobacteria are generally distributed in the biosphere, with many species in freshwaters (Whitton and Potts 2000). In clean (oligotrophic — few nutrients) lakes and rivers, their cell concentration is low and there is a range of species at any given time. The size of the cyanobacterial population under these circumstances will be limited by lack of nutrients, particularly available phosphate. The organisms are photosynthetic and capable of growth under low light intensities, so that they can be found at depth in the water of clear lakes. Some species have specialized nitrogen-fixing cells, called heterocysts, that allow them to grow when very little inorganic nitrogen is available in the water. These cells provide a morphological feature that assists in their identification (Chapter 2). As nutrient availability in lakes and rivers increases through human activity, whether from intensification of agriculture or human waste disposal, the size of the cyanobacterial population rises. The final condition of freshwater is termed eutrophic (good nutrients) when the nutrients are sufficient to support high populations of phytoplankton.

These eutrophic waters may have dominant green algae or diatoms, or they may be seasonally subject to water blooms of cyanobacteria. The circumstances of cyanobacterial dominance are discussed in Chapter 4 and occur relatively frequently in reservoirs and weir pools in slow-flowing rivers. One of the factors that may contribute to a single species of cyanobacterium becoming the dominant organism in a water body is the ability to produce toxins. Phytoplankton, including cyanobacteria, are the primary food source for a diversity of consuming organisms in freshwater. The presence of toxins in particular species of cyanobacteria may provide a competitive advantage by suppressing consumption, allowing the toxic organisms to outgrow nontoxic phytoplankton.

The cyanobacterial toxins include a range of chemical compounds, with those currently identified being predominantly alkaloids and peptides (Chapter 3). The

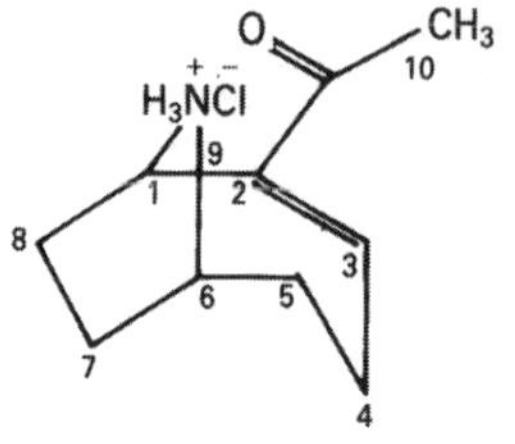

Anatoxin–a hydrochloride

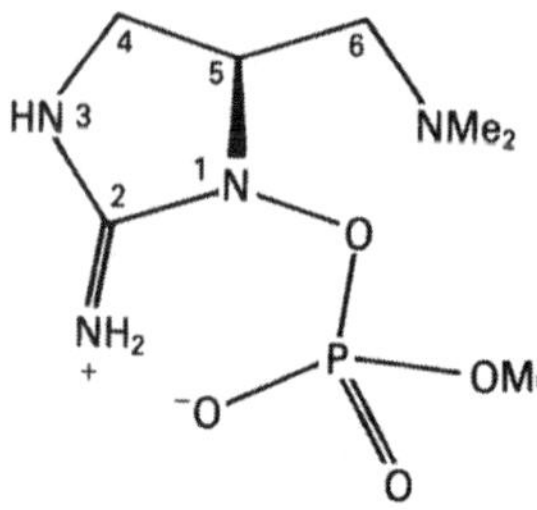

Anatoxin–a(s)

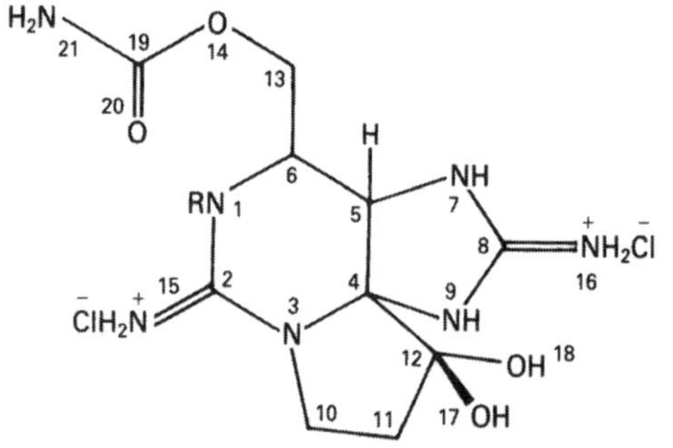

R = H; Saxitoxin dihydrochloride
R = OH; Neosaxitoxin dihydrochloride

FIGURE 1.1 Alkaloid neurotoxins from cyanobacteria.

toxins are formed as secondary metabolites, not as parts of metabolic pathways leading to other compounds. The toxins remain with the cells to a variable extent, with the alkaloid toxins more likely to be present in the free water solution than the peptide toxins, which are liberated from the cells only on damage or death. If water containing toxic cells is consumed, the toxins will be liberated in the gastrointestinal tract. In sufficient concentration, the toxins will cause clinical injury or even death. Many cases of death among livestock have been reported after they drank water containing cells from a cyanobacterial bloom (Carmichael and Falconer 1993).

Three types of neurotoxic alkaloid have been isolated from cyanobacteria, as illustrated in Figure 1.1. All three types of toxin have been identified after livestock poisonings due to the consumption of cyanobacteria in drinking water. The anatoxins, initially isolated from *Anabaena*, comprise anatoxin-a, which is a neuromuscular junction blocking agent (Carmichael, Biggs et al. 1979) and anatoxin-a(s), which resembles an organophosphate anticholinesterase, with effects exerted through inhibition of the breakdown of acetylcholine at the nerve synapse (Mahmood and Carmichael 1987).

The third type of neurotoxic alkaloid from cyanobacteria is the saxitoxin group of tricyclic guanidinium molecules, which are commonly called paralytic shellfish poisons. These were originally isolated from shellfish after human poisoning

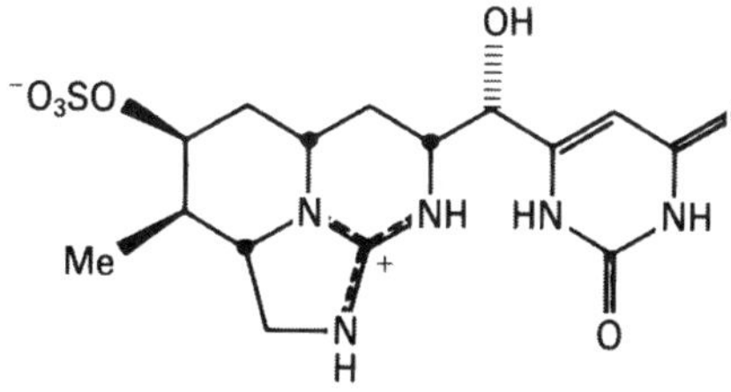

Cylindrospermopsin

FIGURE 1.2 Alkaloid cytotoxin from cyanobacteria.

episodes, in which cases the toxins originated in marine dinoflagellates consumed by the shellfish (Steidinger 1993). These toxins are also produced by several genera of cyanobacteria. The most extensive poisoning event caused by saxitoxins from freshwater cyanobacteria was the water bloom of *Anabaena circinalis*, which covered about 1000 km of the Darling River in Australia in the summer of 1990. This caused the deaths of more than a thousand livestock and also contaminated the drinking water supply of several towns (Humpage, Rositano et al. 1993, 1994).

Another tricyclic guanidinium alkaloid, of quite different structure and toxicity, is cylindrospermopsin (Figure 1.2). This was named from *Cylindrospermopsis raciborskii*, the first organism from which it was isolated. In this case the cyanobacteria in a drinking water reservoir caused a major human poisoning event, which led to the investigation of the cause of the poisoning and the nature of the toxin involved. This is discussed in Chapter 5 and Chapter 6, as cylindrospermopsin is a major potential source of human injury through the drinking water supply. The toxin is not specific to liver but causes damage in a range of organs, as it appears to enter cells readily. As toxins that are consumed first reach the intestinal lining and then are taken up by the liver, followed by other tissues, gastroenteritis and liver injury are the initial symptoms of this poisoning (Chapter 5).

The peptide toxins that have received the majority of investigation to the present time are the microcystins and nodularins, named from *Microcystis* and *Nodularia*, the first two genera of toxic cyanobacteria from which the toxins were isolated. Both of these groups of compounds are cyclic peptides, containing D-amino acids in the ring. Both contain a unique β-linked amino acid bearing a phenyl residue (Figure 1.3). These compounds are selectively hepatotoxic because they require a transporter mechanism to enter cells. This transporter occurs in hepatocytes and also in the cells of the gastrointestinal lining. Toxic cyanobacteria such as *Microcystis*, which contain the toxin within the cells, leak toxin into the gut contents as a result of attack by the digestive system, from which it is transported across the intestinal lining to the hepatic portal vein. On entering the liver, the toxin is actively concentrated into the hepatocytes, with damaging effects (Chapter 7). The earliest scientific report of livestock poisoning from toxic cyanobacteria was based on animals that drank water containing *Nodularia* at the edge of a large estuarine lake in South Australia (Francis 1878). The cyanobacteria still occur in the lake, which is used as

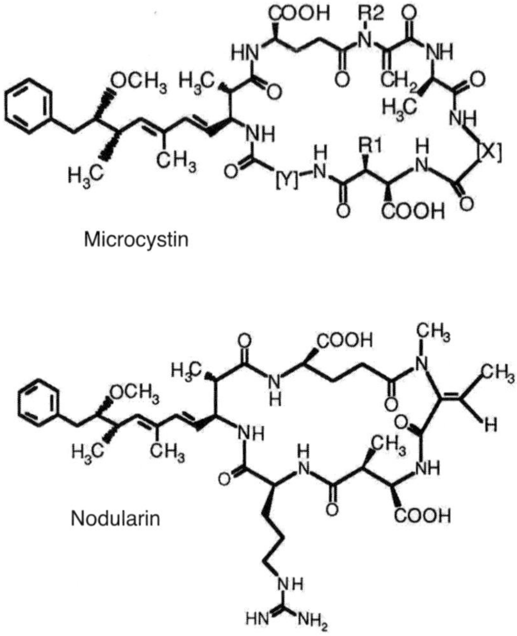

Microcystin

Nodularin

FIGURE 1.3 Cyclic peptide hepatotoxins from cyanobacteria.

a human drinking water source for several small towns and is carefully monitored for cell numbers of the organism.

This volume focuses its attention only on the two cyanobacterial toxins that are most likely to cause long-term adverse health effects in the human population. Both cylindrospermopsin and microcystins have appeared in potentially dangerous concentrations in "finished" drinking water supplies. Hence, there is a risk to consumers that the early, measurable, adverse effects of the toxins on individuals may be followed by more potentially damaging long-term consequences. Cylindrospermopsin has pronounced genotoxic effects on rodent and human cells. An initial study of carcinogenesis following cylindrospermopsin dosing in mice gave evidence of tumor initiation and growth (Falconer and Humpage 2001). Much more research is needed to determine the magnitude of the consequences of cylindrospermopsin poisoning, especially studies of human cancer epidemiology. In the interim, an assessment of the "safe" concentration of toxin in domestic water supplies has been carried out (Chapter 8).

Microcystins have been examined in detail by an expert group under the auspices of the World Health Organization (WHO), which has determined a provisional "Guideline Value" for microcystin-LR in drinking water of 1 μg/L (Chorus and Bartram 1999). This was based on subchronic toxicity to rodents and pigs in the absence of adequate data for carcinogenesis and teratogenesis. Microcystin has clearly been shown to promote the growth of tumor precursors, particularly in the liver, and there is some evidence from southern China that it may be a factor in human liver carcinoma (Yu 1995); (see Chapter 8). The WHO Guideline Value has been used as a basis for revised drinking water legislation in a number of countries, which now require monitoring of the toxin in supplies that may be at risk.

Considerable attention is now being paid to effective monitoring techniques for toxic cyanobacteria and for dissolved toxins in drinking water supplies in order to meet the WHO guidelines. Advances in both areas are reported, with increasing attention to genetic approaches to identifying toxic organisms. Much of the uncertainty in identification of species and toxicity is removed if it can be demonstrated that organisms in a water supply do or do not possess toxin-synthesizing genes (Chapter 9). Increasing availability of enzyme-linked immunosorbent assay techniques for the specific toxins allows sensitive toxin detection, which then needs to be fully quantitated by chemical analytical methods. The alternative techniques for toxin quantitation are currently being validated across laboratories and countries. The European Commission is active in supporting the standardization of methods (Chapter 9).

As more water bodies become eutrophic and these waters are increasingly used as drinking water sources, the problem of reservoir management becomes a major issue. If cyanobacterial blooms can be prevented by management techniques, then much of the health risk associated with toxic cyanobacteria is removed. Cyanobacterial mitigation techniques have proved difficult and long-term, as nutrient reduction is the most ecologically sound and the most difficult to achieve (Chapter 11). Destratification of reservoirs in summer has had limited success. The most ecologically damaging method of reservoir management is also the most effective, which is killing cyanobacteria by adding a sufficient concentration of copper to the water. This will clear a reservoir of cyanobacteria within a few days, but the cell lysis that then occurs releases toxin into the water, which is then more difficult to remove by water treatment. Hence frequent additions of copper are needed before actual bloom conditions occur in order to minimize the risk of dissolving high concentrations of cyanobacterial toxins into the water.

Water treatment to remove these toxins has been extensively investigated, and new technologies are under development (Chapter 12). The effectiveness of conventional water treatment for toxin removal depends on the conditions of the toxic bloom and the operation and design of the treatment plant. If cells containing toxin can be removed intact from the water flow and the toxic cells taken out of the treatment system before they lyse, then a substantial reduction in toxicity results (Chapter 12). Modern, sophisticated treatment technology can remove a wide range of potentially harmful organic chemicals, including cyanobacterial toxins. New technologies with catalytic oxidation or membrane filtration are showing potential for toxin removal. Cyanobacterial blooms continue to present problems in water treatment because of the large organic load in the water combined with the presence of toxins.

The magnitude of the problem of toxic cyanobacteria in drinking water sources relates to two major world issues (Chapter 13). These are population growth and global warming. Population growth results in increased demand for drinking water, coupled with an increased likelihood of eutrophication of previously clean water supplies. As the population of cities grows, the intensity of land use in the water catchments supplying the cities rises, with nutrients from agriculture and human waste increasing in the reservoirs. One of the early consequences of this population increase is seasonal blooms of toxic cyanobacteria in drinking water reservoirs. As eutrophication increases, so does the cell concentration of cyanobacteria.

Global warming may be the cause of the observed migration of the warm-water toxic species *C. raciborskii* northward in the Northern Hemisphere. This species is a characteristically tropical organism, first identified in Indonesia, which is now appearing in Europe and the northern U.S. At present it has been recorded during warm summers in Europe and can be expected to become more frequent and at higher cell concentrations with rising ambient temperatures (Padisak 1997). As a toxic contaminant of drinking water supplies in subtropical regions with potentially severe adverse health effects, the spread of the species into temperate climates presents a new health risk. The species is currently abundant in Florida, where it has been recorded in drinking water reservoirs, and it is likely to be found increasingly in the central and northern U.S. as the ambient temperature increases (Chapter 13).

It is apparent that the public health implications of cyanobacterial toxins in drinking water will result in attention being focused on the risk to the population and ways to minimize that risk. It is hoped that this volume will assist in the evaluation of the field and the development of strategies to protect the public health from the damaging effects of cyanobacterial toxins.

REFERENCES

Carmichael, W. W., D. F. Biggs, et al. (1979). Pharmacology of anatoxin-a produced by the freshwater cyanophyte *Anabaena flos-aquae* NRC-44-1. *Toxicon* 17: 229–236.

Carmichael, W. W. and I. R. Falconer (1993). Diseases related to freshwater blue-green algal toxins, and control measures. *Algal Toxins in Seafood and Drinking Water.* I. R. Falconer, ed. London, Academic Press: 187–209.

Chorus, I. and J. Bartram (1999). *Toxic Cyanobacteria in Water: A Guide to Their Public Health Consequences, Monitoring and Management.* London, E & FN Spon (on behalf of WHO).

Falconer, I. R. and A. R. Humpage (2001). Preliminary evidence for *in-vivo* tumour initiation by oral administration of extracts of the blue-green alga *Cylindrospermopsis raciborskii* containing the toxin cylindrospermopsin. *Environmental Toxicology* 16(2): 192–195.

Francis, G. (1878). Poisonous Australian lake. *Nature (London)* 2 May: 11–12.

Humpage, A. R., J. Rositano, et al. (1993). Paralytic shellfish poisons from freshwater blue-green algae. *Medical Journal of Australia* 159: 423.

Humpage, A. R., J. Rositano, et al. (1994). Paralytic shellfish poisons from Australian cyanobacterial blooms. *Australian Journal of Marine and Freshwater Research* 45: 761–771.

Mahmood, N. A. and W. W. Carmichael (1987). Anatoxin-a(s), an anticholinesterase from the cyanobacterium *Anabaena flos-aquae* NRC-525-17. *Toxicon* 25: 1221–1227.

Padisak, J. (1997). *Cylindrospermopsis raciborskii* (Woloszynska) Seenayya et Subba Raju, an expanding, highly adaptive cyanobacterium: worldwide distribution and review of its ecology. *Archiv für Hydrobiologie* 107 (suppl): 563–593.

Steidinger, K. A. (1993). Some taxonomic and biologic aspects of toxic dinoflagellates. *Algal Toxins in Seafood and Drinking Water*. I. R. Falconer, ed. London, Academic Press: 1–28.

Whitton, B. A. and M. Potts (2000). *The Ecology of Cyanobacteria: Their Diversity in Time and Space*. Dordrecht, Kluwer Academic Publishers.

Yu, S. Z. (1995). Primary prevention of hepatocellular carcinoma. *Journal of Gastroenterology and Hepatology* 10(6): 674–682.

2 Toxic Cyanobacteria and Their Identification

2.1 THE ORIGINS OF CYANOBACTERIA

The cyanobacteria are exceedingly ancient organisms, identifiable in rocks dating from the first thousand million years of the earth's history. As cyanobacterial colonies occur in shallow water, they appear in the fossil record in sedimentary rocks deposited in shallow seas and lakes. The older rocks containing cyanobacteria are the cherts, generated from silt, sand, and mud by heat and pressure over the large extent of geological time. The cyanobacterial colonies called stromatolites appear in rocks as fossilized mushroom shapes and sheets in widely distributed locations around the world. One of the best-known stromatolite formations is the Gunflint chert of the Lake Erie region of North America, which dates from 2.09 billion years before the present. The oldest described in detail are the Apex cherts of Western Australia, dated to approximately 3.5 billion years before the present. As the earth's crust dates to approximately 4.5 billion years before the present, cyanobacteria are among the very earliest life forms (Thorpe, Hickman et al. 1992; Schopf 2000). These rocks have been shown to contain fossil evidence of a wide range of both filamentous and spherical organisms, many identical in size and shape to current cyanobacteria (Schopf 2000). Isotopic ratio data from carbon within these and other cherts show evidence of photosynthetic activity, as living organisms incorporate carbon 12 preferentially to carbon 13 and residues of the organic carbon from the organisms remain in the rocks, providing a ratio of the isotopes characteristic of photosynthetic life (Strauss, Des Marais et al. 1992).

Geologically adjacent iron-rich rocks show fine banding of ferric iron, indicative of oxygen presence in local areas and demonstrating photosynthesis in an otherwise anaerobic atmosphere (Klein and Buekes 1992).

Stromatolites have been described in geological strata that date from these earliest examples to the modern day, through the Precambrian period and into the recent rocks. Good examples of living stromatolites can be seen in the Caribbean and in Shark Bay, Western Australia (Figure 2.1). Less well known occurrences are in salt lakes and hypersaline lagoons (Figure 2.2). The laminated appearance of sections through stromatolites is due to layers containing more cyanobacterial cells alternating with layers of calcareous deposition or trapped sand/silt. A freshly broken stromatolite shows a clear green band of cyanobacteria under the hard surface, with successive less green bands below. Recent use of genetic analysis on DNA from present-day stromatolites showed only a single cyanobacterial strain in each sample, and successfully examined internal core samples at least 10 years old (Neilan, Burns et al. 2002).

FIGURE 2.1 (See color insert following page 146.) Stromatolites exposed at low tide in a hypersaline bay, Shark Bay, Western Australia.

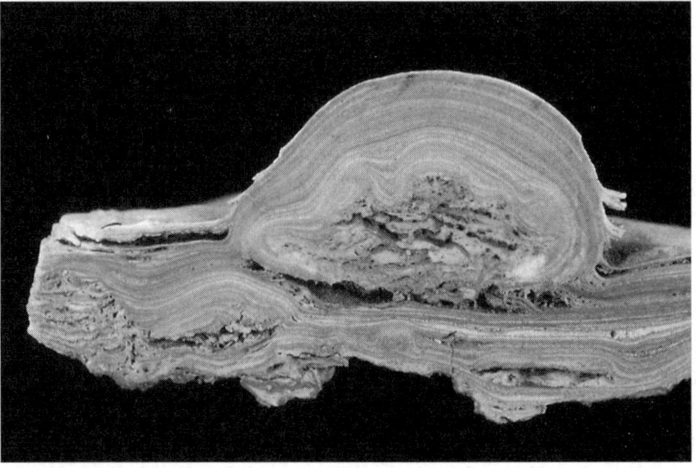

FIGURE 2.2 (See color insert.) Section of stromatolite from a saline lake in Innes National Park, South Australia, showing cyanobacterial layers.

2.2 CYANOBACTERIAL ORGANISMS

Cyanobacteria are photosynthetic prokaryotes, part of the bacterial domain, with no structured nucleus. They possess a single circular chromosome, which has been completely sequenced in several species (Kaneko, Sato et al. 1996). Some also carry plasmids, small circular strands of DNA, which do not appear to have a role in toxicity (Schwabe, Weihe et al. 1988). Their photosynthetic membranes contain chlorophyll-*a* and the pigment phycocyanin, which provides the characteristic blue-green color of many species (Whitton and Potts 2000). Other pigments may also be present, particularly carotenoids and phycoerythrin, which give a strong red color to some species. The protein-synthesizing organelles of cyanobacteria, the ribosomes, are of the bacterial type (Bryant 1994). They are not therefore eukaryotic cells, despite the common name blue-green algae, and are not directly related to the algae. It is possible that cyanobacteria were the precursors of the plant chloroplasts. Like the algae, cyanobacteria are predominantly oxygen-releasing photosynthetic cells, using water as the electron source and releasing oxygen gas.

Nitrogen fixation is an important feature of some species of cyanobacteria. The specialist nitrogen-fixing cells are called heterocysts, have a thickened cell wall, and do not possess photosynthetic membranes. In appearance under the light microscope they are larger, clear, highly refractive cells. They may occur within the filament of photosynthetic cells or terminally on a filament. Because of the differences in size, shape, and location of the heterocysts, they form a significant component in species identification. Within the heterocysts the enzyme nitrogenase reduces molecular nitrogen to ammonia, which is incorporated into the amido group of glutamine (Bryant 1994). The thickened cell wall enables molecular oxygen entry to the cell to be reduced, thus helping to maintain a highly reducing environment within the cell, necessary for nitrogen reduction. Some species of cyanobacteria appear to be able to fix atmospheric nitrogen without visible heterocysts, which may relate to the anaerobic conditions in which the organisms can survive.

The other very characteristic cell type found in some filamentous cyanobacterial genera is the akinete, a very large spherical to oval-shaped cell with granular contents. Akinetes form resting cells when the filament dies, regenerating a new filament when the environmental conditions are favorable (Adams and Duggan 1999). Both heterocysts and akinetes are illustrated in Figure 2.3. A good color illustration of *Cylindrospermopsis raciborskii* with a heterocyst and an akinete is found at www.unc.edu/~moisande/image3.html. The size, shape, location on the filament, and frequency of heterocysts and akinetes are major taxonomic features identifying genera and species among the cyanobacterial orders Nostocales and Stigonematales.

2.3 CLASSIFICATION AND NOMENCLATURE

The systematic nomenclature of the cyanobacteria has been a subject of disagreement and revision due to the early application of botanical nomenclature to organisms that are not related to plants. As with plant classification, the structure of the organisms and their colonies has formed the present basis of classification and identification. Several recent books and reports on cyanobacterial identification have been published, which are most useful in identification to genus level. In the field, classification to genera can often be achieved, but species identification may be exceptionally difficult and is a specialist preserve. In the U.K. a computer-based system of identification has been developed, which includes 320 species found in the British Isles (Whitton, Robinson et al. 2000). Komarek in Hungary has published (in German) a consolidated account of the spherical-celled colonial Chroococcales, which are among the most difficult to identify (Komarek and Anagnostides 1999).

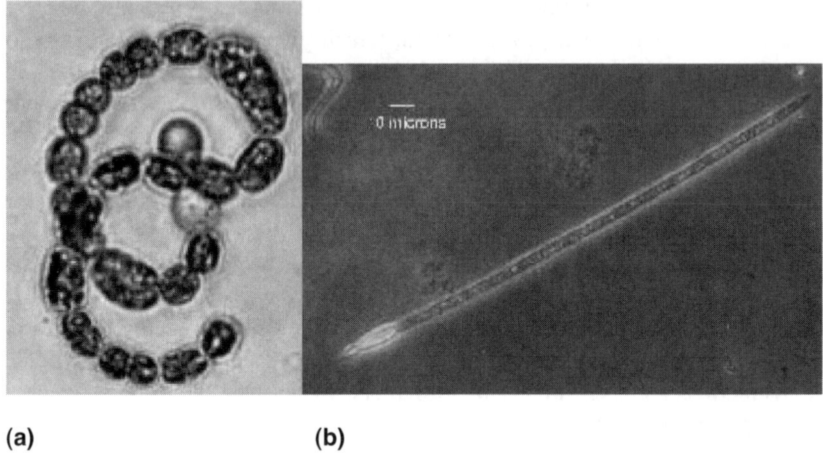

FIGURE 2.3 (See color insert.) (a) *Anabaena circinalis* showing akinetes (large dense oval cells) and heterocysts (translucent spherical cells); **(b)** *Cylindrospermopsis raciborskii* showing akinete (large oval cell) and terminal heterocyst. (Images from Roger Burks, University of California at Riverside; Mark Schneegurt, Wichita State University; and Cyanosite, www.cyanosite.bio.purdue.edu. With permission.)

These cyanobacteria form colonies in a mucilaginous gel matrix, which in field samples is characteristic of the species. However in culture they change to unicellular suspensions of cells, which makes species identification from a cultured strain almost impossible. The Urban Water Research Association of Australia published *Identification of Common Noxious Cyanobacteria: Part 1 — Nostocales* in 1991 and *Part 2 — Chroococcales and Oscillatoriales* in 1992, illustrated with photographs and line drawings (Baker 1991, 1992). These are useful guides for field identification of species with morphometry as well as appearance. A more recent guide was published by the Australian Cooperative Research Centre for Freshwater Ecology in 2002 (Baker and Fabbro 2002).

Some of the most abundant toxic cyanobacteria are illustrated in Figure 2.4 to help readers to identify them in field samples. Table 2.1 gives a botanical description of the main cyanobacterial orders, which contain the toxic species as well as many species in which no toxicity has been recorded up to now. Examples of genera that include toxic species are listed under the appropriate order. Table 2.2 lists most of the planktonic (free-floating) freshwater species presently identified as toxic, but this list extends continually and cannot be regarded as complete. The references to the toxic species are chosen to be illustrative rather than comprehensive and to assist in further reading.

In particular, the benthic (growing on rocks or sediment) species have not been extensively tested for toxicity, as they only infrequently contaminate drinking water supplies. In two cases, after poisoning incidents with domestic animals, benthic species have been tested and found toxic. In a third case the organisms dislodged naturally from the sediments in a drinking water holding reservoir and were tested to evaluate the safety of the supply. Table 2.3 lists these few benthic cyanobacteria

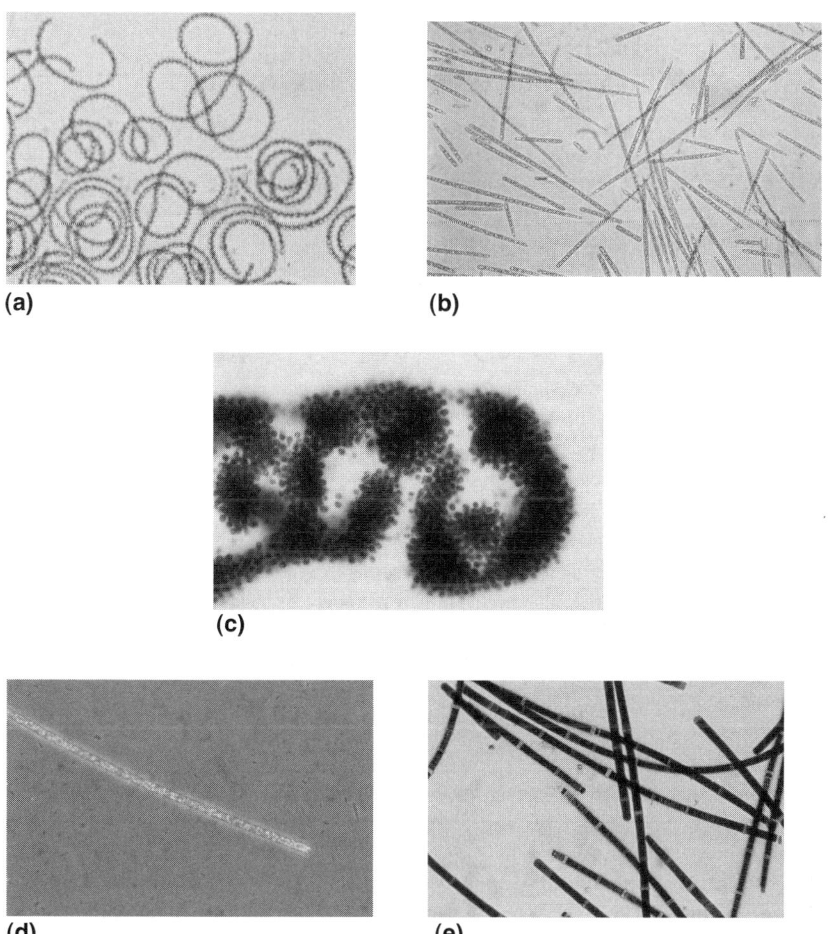

FIGURE 2.4 (See color insert.) Photomicrographs of toxic species of cyanobacteria: **(a)** *Anabaena circinalis*; **(b)** *Cylindrospermopsis raciborskii*; **(c)** *Microcystis aeruginosa*; **(d)** *Planktothrix* sp.; **(e)** *Nodularia spumigena*. (Images (b), (c), and (e) from Cyanobacteria-toxins in drinking water, Ian R. Falconer, *Encyclopedia of Microbiology*, p. 985. With permission from Wiley. Image (d) from Dr. B. Ernst. With permission.)

reported to contain toxins. It can be expected that, when more species are tested, species of benthic cyanobacteria will be found to be toxic in equal proportion to planktonic species.

2.4 MOLECULAR TAXONOMY

As a consequence of the great advances in the molecular characterization of living organisms, attention is increasingly being paid to use of both proteins and DNA in identifying cyanobacteria. Alloenzyme determination has been used in differentiating species within the genus *Anabaena*, which has a large number of similar species

TABLE 2.1
Orders of Cyanobacteria with Examples of Toxic Genera

Filamentous		**Toxic Genera**
Order Oscillatoriales	Unbranched filaments (may have false branches); cells reproduce by binary fission; no heterocysts; no recorded akinetes.	*Planktothrix* *Phormidium* *Lyngbya*
Order Nostocales	Growth similar to Oscillatoriales; form heterocysts; some species have akinetes.	*Anabaena* *Aphanizomenon* *Cylindrospermopsis* *Nodularia*
Order Stigonematales	Growth similar to Oscillatoriales but branched filaments; form heterocysts; some species have akinetes.	*Haphalosiphon* *Umezakia*
Unicellular Aggregates		
Order Chroococcales	Held together by outer wall or gel matrix; binary division in one, two, or three planes, symmetrically or asymmetrically; or may reproduce by budding; akinetes rare.	*Microcystis* *Snowella*
Order Pleurocapsales	Held together by outer wall or gel matrix; cells reproduce by internal multiple divisions with production of smaller daughter cells, or by this method plus binary fission; akinetes rare.	Yet to be characterized for toxicity.

From Castenholz and Waterbury 1989, modified from Whitton and Potts 2000.

tending to grow to different dimensions under differing conditions of nutrition (Tatsumi, Watanabe et al. 1991). The presence and quantity of cyanobacteria, as against most other life forms, can be determined by analysis of water samples for phycocyanin pigment, as this photosynthetic component is highly conserved (de Lorimer, Bryant et al. 1984).

More precise analysis for elements of the cyanobacterial genome coding for phycocyanin will differentiate cyanobacteria from other phycocyanin-containing organisms, and also provide taxonomic information. The phycocyanin operon (functional genetic unit) contains genes coding for two bilin subunits (α and β) and three linking polypeptides. The intergenic spacing element between the bilin coding regions demonstrated a highly variable region, containing enough sequence differences to assist in taxonomic determination (Neilan, Jacobs et al. 1995; Baker, Neilan et al. 2001). Two approaches have been successful. Both used the polymerase chain reaction (PCR) to amplify the cyanobacterial DNA in the intergenic spacer by selection of primers from sequences beyond each end of the intergenic spacer. These are spacer-flanking sequences within the DNA coding for the two bilin subunit proteins, selected because their sequences are completely conserved in the phycocyanin genome

TABLE 2.2
Planktonic Cyanobacterial Species Shown to Contain Toxins

Species	Toxin	Sample Location	References
Anabaena bergii	Cylindrospermopsins	Australia	Fergusson and Saint 2003
Anabaena circinalis	Microcystins	France	Vezie, Brient et al. 1998
Anabaena circinalis	Saxitoxins	Australia	Humpage, Rositano et al. 1994
Anabaena flos-aquae	Anatoxin-a	Canada Germany	Carmichael, Biggs et al. 1975; Carmichael and Gorham 1978 Bumke-Vogt, Mailahn et al. 1999
Anabaena flos-aquae	Anatoxin-a(s)	Canada	Mahmood and Carmichael 1986
Anabaena flos-aquae	Microcystins	Canada Norway	Khrishnamurthy, Szafraniec et al. 1989; Sivonen, Namikoshi et al. 1992
Anabaena lemmermannii	Anatoxin-a(s)	Denmark	Henriksen, Carmichael et al. 1997
Anabaena lemmermannii	Microcystins	Norway	Skulberg 1996
Anabaena planktonica	Anatoxin-a	Italy	Bruno, Barbini et al. 1994
Anabaenopsis millerii	Microcystins	Greece	
Aphanizomenon flos-aquae	Saxitoxins	U.S.	Jackim and Gentile 1968; Ikawa, Wegener et al. 1982
Aphanizomenon ovalisporum	Cylindrospermopsins	Israel Australia	Banker, Carmeli et al. 1997; Shaw, Sukenik et al. 1999
Aphanizomenon sp.	Anatoxin-a	Finland Germany	Sivonen, Himberg et al. 1989; Bumke-Vogt, Mailahn et al. 1999
Cylindrospermum sp.	Anatoxin-a	Finland	Sivonen, Himberg et al. 1989
Cylindrospermopsis raciborskii	Cylindrospermopsins	Australia	Hawkins, Runnegar et al. 1985 Hawkins, Chandrasena et al. 1997
		Thailand U.S.	Li, Carmichael et al. 2001a Williams, Burns et al. 2001
Cylindrospermopsis raciborskii	Saxitoxins	Brazil	Lagos, Onodera et al. 1999
Cylindrospermopsis raciborskii	Toxin(s) not related to cylindrospermopsin or saxitoxin	France Germany Portugal	Bernard, Harvey et al. 2003 Fastner, Heinze et al. 2003 Saker, Nogueira et al. 2003
Lyngbya wollei	Saxitoxins	U.S.	Carmichael, Evans et al. 1997

(continued)

TABLE 2.2 (CONTINUED)
Planktonic Cyanobacterial Species Shown to Contain Toxins

Species	Toxin	Sample Location	References
Microcytis aeruginosa	Microcystins, examples only, worldwide distribution	South Africa Australia Japan U.K. U.S.	Botes, Viljoen et al. 1982; Botes, Wessels et al. 1985 Harada, Ogawa et al. 1991 Codd and Carmichael 1982; Codd, Brooks et al. 1989 Rinehart, Namikoshi et al. 1994
Microcystis botrys	Microcystins	Denmark	Henriksen 1996
Microcystis ichthyoblabe	Microcystins	Czech Republic	Marsalek, Blaha et al. 2001
Microcystis viridis	Microcystins	Japan	Kusumi, Ooi et al. 1987 Watanabe 1996
Nodularia spumigena	Nodularins	Baltic Sea Australia	Sivonen, Kononen et al. 1989 Baker and Humpage 1994
Nostoc sp.	Microcystins	Finland U.K.	Sivonen, Niemela et al. 1990 Beattie, Kaya et al. 1998
Planktothrix agardhii	Microcystins	Finland China	Sivonen, Niemela et al. 1990 Ueno, Nagata et al. 1996
Planktothrix formosa	Homoanatoxin-a	Norway	Skulberg, Carmichael et al. 1992
Planktothrix mougeotii	Microcystins	Denmark	Henriksen 1996
Planktothrix rubescens	Microcystins	Norway Germany	Skulberg 1996 Fastner, Erhard et al. 2001
Raphidiopsis curvata	Cylindrospermopsin	China	Li, Carmichael et al. 2001b
Snowella lacustris	Microcystins	Norway	Skulberg 1996
Umezakia natans	Cylindrospermopsin	Japan	Harada, Ohtani et al. 1994
Woronichinia naegeliana	Microcystins	Denmark	Henriksen 2001

(Neilan, Jacobs et al. 1995). Using these primer sequences to generate DNA amplification fragments the first approach demonstrated that cyanobacteria could be clearly distinguished from eukaryotic algae, red algae (rhodophytes), and cryptophytes, but species could not be assigned.

However in the second approach, these fragments were then digested with restriction endonuclease enzymes cleaving the DNA at known locations to yield a "DNA fingerprint" — or restriction fragment length polymorphism (RFLP) — from which both species and genetic relationships could be assigned (Neilan, Jacobs et al. 1995). Three different approaches were employed to analyze the data, based on phenetic and cladistic methods. All three trees of strain relationships were identical, and as far as genus level largely consistent with the existing morphological classifications. Two main groupings emerged, one consisting of strains from the genera

TABLE 2.3
Benthic Cyanobacterial Genera and Species Shown to Contain Toxins

Genus or species	Toxin	Sample Location	Reference
Haphalosiphon hibernicus	Microcystins	U.S.	Prinsep, Caplan et al. 1992
Oscillatoria limnosa	Microcystins	Switzerland	Mez, Beattie et al. 1997
Oscillatoria sp.	Anatoxin-a	Scotland	Edwards, Beattie et al. 1992
Phormidium aff. *formosum*	Not yet known	Australia	Baker, Steffensen et al. 2001

Anabaena, *Aphanizomenon*, *Cylindrospermopsis*, and *Nodularia*, all morphologically located in the order Nostocales. The other was genetically more diverse and appeared to contain at least three genetic lineages, one comprising *Planktothrix/Oscillatoria* and an *Anabaena* species, the second several *Microcystis* species showing great genetic diversity with no clear relationship between species designation and genetic fingerprint, and the third *Microcystis aeruginosa* strains genetically distinct from the others. This grouping thus contained representatives of three orders: Oscillatoriales, Nostocales, and Chroococcales (Neilan, Jacobs et al. 1995).

Further genetic characterization using this approach examined 19 strains of cyanobacteria morphologically identified as *Anabaena circinalis*, *M. aeruginosa*, and *Nodularia spumigena* (Bolch, Blackburn et al. 1996). The *Microcystis* strains of the same morphological species gave RFLP patterns which were quite different, whereas the *Anabaena* and *Nodularia* strains were much less variable. This research strengthens the potential for cyanobacterial classification on a genetic basis.

Another study using the phycocyanin intergenic spacer for cyanobacterial identification employed three levels of discrimination, including DNA sequencing (Baker, Neilan et al. 2001). This study investigated water-bloom material and mixed species from cultures to ascertain that the techniques had field application for species identification. The sequences of the spacer region were determined for strains of *Aphanizomenon* and *Cylindrospermopsis* as well as the genera previously investigated by Neilan et al. (1995) and Bolch et al. (1996). The main feature shown in this study is the very highly conserved DNA sequence within a genus but substantial differences between genera. As the database extends through ongoing research, the genetic analysis of this region of cyanobacterial DNA will cast increasing light on cyanobacterial systematics, particularly in the Chroococcales, where considerable genetic divergence is seen.

Other regions of the cyanobacterial chromosome have also been investigated for use in genus and species identification, including the DNA coding for the 16S ribosomal subunit. This genetic component has been widely used in bacterial identification and was assessed for use in establishing the evolutionary relationships among the genus *Microcystis*. A number of species within the genus have been named, but they are most difficult cyanobacterial species to identify from morphology

(Komarek and Anagnostides 1999), and molecular phylogeny is likely to result in some revision. DNA sequences have been determined for 16S ribosomal RNA in a range of strains of *Microcystis*, showing some disparity between morphological species identification and genetic linkage (Neilan, Jacobs et al. 1997).

Use of base composition of DNA has also been applied to taxonomic differentiation of *Anabaena* species, which morphologically are difficult to characterize. It was shown that strains of a single species could be separated on this basis (Li and Watanabe 2002).

Most recently these techniques have been applied to *C. raciborskii*, which appears worldwide but shows a variety of different toxicities in different locations; see Table 2.2. Using 16S rRNA sequencing, cultures of this species from Europe, the U.S., Brazil, and Australia were examined. A sequence similarity of 99.1% was found, indicating that the morphological species identification was accurate (Neilan, Saker et al. 2003). Sequence differences showed three groupings, the North and South American group, the European group, and the Australian group. In comparison with *Cylindrospermopsis*, sequence assessment of 16S rRNA from the nostocalean genera *Cylindrospermum* sp., *Nostoc* sp., *Anabaena (bergii)*, and *Anabaenopsis* sp. showed considerable similarities of 93.7, 93.7, 93.3, and 93.2%, respectively. *Umezakia natans*, from the order Stigonematales, which also produces the toxin cylindrospermopsin, had only 84.6% similarity with *Cylindrospermopsis* (Neilan, Saker et al. 2003).

A second approach by Neilan, Saker et al. (2003) used a short tandem repeat sequence specific to cyanobacteria to evaluate genetic differences, which had previously been shown to be effective for phylogenetic assessment of *Anabaena* (Smith, Parry et al. 1998). This approach also supported a phylogenetic tree that grouped geographical origins of isolates and showed the greatest divergence between the Australian and Brazilian isolates. The European isolates from Germany, Hungary, and Portugal were closer to the Australian organisms than to the American group (Neilan, Saker et al. 2003). In parallel, investigation of a nitrogen-fixing gene component (nifH), and the phycocyanin intergenic spacer region of strains of *C. raciborskii* showed separation of American, European, and Australian strains, with the European strain closer to the Australian than to the American, confirming the consistency of the approach (Dyble, Paerl et al. 2002).

A concerted investigation of *Nodularia* strains at the University of Helsinki has further strengthened the value of genetic approaches to the study of cyanobacterial taxonomy. As a major toxic cyanobacterium in the Baltic Sea and associated brackish water lakes, *Nodularia* has public health significance for water supply, recreation, and potential food contamination. In particular, it is necessary to be able to distinguish toxic from nontoxic species or strains. Eighteen *Nodularia* strains were examined from the Baltic region and from Australia. Morphologically they classified into four species as well as unclassified strains. A range of genetic assessments were employed, including RFLP of 16S rRNA genes, sequencing of 16S rRNA genes, and several intergenic spacer methodologies, one of which was the phycocyanin intergenic spacer described previously (Lehtimaki, Lyra et al. 2000; Laamanen, Gugger et al. 2001). The three planktonic *Nodularia* species identified from morphology—*N. spumigena*, *N. baltica*, and *N. litorea*—were genetically indistinguishable species: all were planktonic, had gas vacuoles, and produced the toxin nodularin. However, the benthic strains classified as *N. harveyana* and *N. sphaerocarpa* genetically differed from each other and from the planktonic strains. The conclusion was that there is only one planktonic *Nodularia* species in the Baltic Sea, and it is toxic (Laamanen, Gugger et al. 2001).

It is apparent from the discussion in this chapter that the systematic identification and nomenclature of cyanobacteria, formerly entirely based on the morphology of the organisms, is under rapid revision. The range of genetic tools for exploring the genome of the cyanobacteria is impressive, from well-understood 16S rRNA sequences to variable intergenetic spacer regions. These approaches have already been utilized to revise phylogenetic trees for cyanobacterial orders and amend species designations. With the advances in genetic analysis of the cyanobacterial toxin genes themselves, discussed in the next chapter, approaches to the understanding of cyanobacteria through molecular biology have proved enormously productive.

REFERENCES

Adams, D. G. and P. S. Duggan (1999). Heterocyst and akinete differentiation in cyanobacteria. *New Phytology* 144: 3–33.

Baker, J. A., B. A. Neilan, et al. (2001). Identification of cyanobacteria in environmental samples by rapid molecular analysis. *Environmental Toxicology* 16(6, special issue): 472–482.

Baker, P. (1991). *Identification of Common Noxious Cyanobacteria: Part 1 — Nostocales*. Melbourne, Urban Water Research Association of Australia.

Baker, P. (1992). *Identification of Common Noxious Cyanobacteria: Part 2 — Chroococcales, Oscillatoriales*. Melbourne, Urban Water Research Association of Australia.

Baker, P. and L. D. Fabbro (2002). *A Guide to the Identification of Common Blue-Green Algae (Cyanoprokaryotes) in Australian Freshwaters*. Thurgoona, Australia, Cooperative Research Centre for Freshwater Ecology.

Baker, P. D. and A. R. Humpage (1994). Toxicity associated with commonly occurring cyanobacteria in surface waters of the Murray-Darling Basin, Australia. *Australian Journal of Marine and Freshwater Research* 45: 773–786.

Baker, P. D., D. A. Steffensen, et al. (2001). Preliminary evidence of toxicity associated with the benthic cyanobacterium *Phormidium* in South Australia. *Environmental Toxicology* 16(6, special issue): 506–511.

Banker, P., S. Carmeli, et al. (1997). Identification of cylindrospermopsin in *Aphanizomenon ovalisporum* (Cyanophyceae) isolated from Lake Kinneret, Israel. *Journal of Applied Phycology* 33: 613–616.

Beattie, K. A., K. Kaya, et al. (1998). Three dehydrobutyrine containing microcystins from *Nostoc*. *Phytochemistry* 47(7): 1289–1292.

Bernard, C., M. Harvey, et al. (2003). Toxicological comparison of diverse *Cylindrospermopsis raciborskii* strains: Evidence of liver damage caused by a French *C. raciborskii* strain. *Environmental Toxicology* 18: 176–186.

Bolch, C. J., S. I. Blackburn, et al. (1996). Genetic characterization of strains of cyanobacteria using PCR-RFLP of the cpcBA intergenic spacer and flanking regions. *Journal of Phycology* 32(3): 445–451.

Botes, D. P., C. C. Viljoen, et al. (1982). Structure of toxins of the blue-green alga *Microcystis aeruginosa*. *South African Journal of Science* 78: 378–379.

Botes, D. P., P. L. Wessels, et al. (1985). Structural studies on cyanoginosins-LR, -YR, -YA, and -YM, peptide toxins *Microcystis aeruginosa*. *Journal of the Chemical Society, Perkin Transactions* 1: 2747–2748.

Bruno, M., D. A. Barbini, et al. (1994). Anatoxin-a and a previously unknown toxin in *Anabaena planctonica* from blooms found in Lake Mulargia (Italy). *Toxicon* 32(3): 369–373.

Bryant, D. A., ed. (1994). *The Molecular Biology of Cyanobacteria.* Dordrecht, Kluwer Academic Publishers.

Bumke-Vogt, C., W. Mailahn, et al. (1999). Anatoxin-a and neurotoxic cyanobacteria in German lakes and reservoirs. *Environmental Toxicology* 14(1): 117–125.

Carmichael, W. W., D. F. Biggs, et al. (1975). Toxicology and pharmacological action of *Anabaena flos-aquae* toxin. *Science* 187: 542–544.

Carmichael, W. W., W. R. Evans, et al. (1997). Evidence for paralytic shellfish poisons in the freshwater cyanobacterium *Lyngbya wollei* (Farlow ex Gomont) comb. nov. *Applied and Environmental Microbiology* 63(8): 3104–3110.

Carmichael, W. W. and P. R. Gorham (1978). Anatoxins from clones of *Anabaena flos-aquae* isolated from lakes in western Canada. *Mitteilungen — Internationale Vereinigung fur Theoretische und Angewandte Limnologie* 21: 285–295.

Castenholz, R. and J. Waterbury (1989). Group 1, Cyanobacteria. Preface. *Bergey's Manual of Systematic Bacteriology.* J. Staley, M. Bryant, N. Pfennig, and J. Holt, eds. Baltimore, Williams & Wilkins. 3: 1710–1727.

Codd, G. A., W. P. Brooks, et al. (1989). Production, detection and quantification of cyanobacterial toxins. *Toxicity Assessment* 4: 499–511.

Codd, G. A. and W. W. Carmichael (1982). Toxicity of a clonal isolate of the cyanobacterium *Microcystis aeruginosa* from Great Britain. *FEMS Microbiology Letters* 13: 409–411.

de Lorimer, R., D. A. Bryant, et al. (1984). Genes for the a and b subunits of phycocyanin. *Proceedings of the National Academy of Sciences of the United States of America* 81: 7946–7950.

Dyble, J., H. W. Paerl, et al. (2002). Genetic characterization of *Cylindrospermopsis raciborskii* (cyanobacteria) isolates from diverse geographic origin based on nifH and cpcBA-IGS nucleotide sequence analysis. *Applied Environmental Microbiology* 68(5): 2567–2571.

Edwards, C., K. A. Beattie, et al. (1992). Identification of anatoxin-a in benthic cyanobacteria (blue-green algae) and in associated dog poisonings at Loch Insh, Scotland. *Toxicon* 30(10): 1165–1175.

Fastner, J., M. Erhard, et al. (2001). Microcystin variants in *Microcystis* and *Plantothrix* dominated field samples. *Cyanotoxins: Occurrence, Causes, Consequences.* I. Chorus, ed. Berlin, Springer-Verlag: 148–152.

Fastner, J., R. Heinze, et al. (2003). Cylindrospermopsin occurrence in two German lakes and preliminary assessment of toxicity and toxin production of *Cylindrospermopsis raciborskii* (Cyanobacteria) isolates. *Toxicon* 42(3): 313–321.

Fergusson, K. M. and C. P. Saint (2003). Multiplex PCR assay for *Cylindrospermopsis raciborskii* and cylindrospermopsin-producing cyanobacteria. *Environmental Toxicology* 18(2): 120–125.

Harada, K. I., K. Ogawa, et al. (1991). Microcystins from *Anabaena flos-aquae* NRC 525-17. *Chemical Research in Toxicology* 4: 535–540.

Harada, K., I. Ohtani, et al. (1994). Isolation of cylindrospermopsin from a cyanobacterium *Umezakia natans* and its screening method. *Toxicon* 32: 73–84.

Hawkins, P. R., N. R. Chandrasena, et al. (1997). Isolation and toxicity of *Cylindrospermopsis raciborskii* from an ornamental lake. *Toxicon* 35(3): 341–346.

Hawkins, P. R., M. T. C. Runnegar, et al. (1985). Severe hepatoxicity caused by the tropical cyanobacterium (blue-green alga) *Cylindrospermopsis raciborskii* (Woloszynska) Seenaya and Subba Raju isolated from a domestic supply reservoir. *Applied and Environmental Microbiology* 50(5): 1292–1295.

Henriksen, P. (1996). Toxic cyanobacteria/blue-green algae in Danish fresh waters. Department of Phycology. Copenhagen, University of Copenhagen.

Henriksen, P. (2001). Toxic freshwater cyanobacteria in Denmark. *Cyanotoxins: Occurrence, Causes, Consequences.* I. Chorus, ed. Berlin, Springer-Verlag: 49–56.

Henriksen, P., W. Carmichael, et al. (1997). Detection of an anatoxin-a(s)-like anticholinesterase in natural blooms and cultures of cyanobacteria/blue-green algae from Danish lakes and in the stomach contents of poisoned birds. *Toxicon* 35: 901–913.

Humpage, A. R., J. Rositano, et al. (1994). Paralytic shellfish poisons from Australian cyanobacterial blooms. *Australian Journal of Marine and Freshwater Research* 45: 761–771.

Ikawa, M., K. Wegener, et al. (1982). Comparisons of the toxins of the blue-green alga *Aphanizomenon flos-aquae* with the *Gonyaulax* toxins. *Toxicon* 20, 4: 747–752.

Jackim, E. and J. Gentile (1968). Toxins of a blue-green alga: similarity to a saxitoxin. *Science* 162: 915–916.

Kaneko, T., S. Sato, et al. (1996). Sequence analysis of the genome of the unicellular cyanobacterium *Synechocystis* sp. strain PCC66803: II. Sequence determination of the entire genome and assignment of potential protein coding regions. *DNA Research* 3: 109–136.

Khrishnamurthy, T., L. Szafraniec, et al. (1989). Structural characterization of toxic cyclic peptides from blue-green algae by tandem mass spectrometry. *Proceedings of the National Academy of Sciences of the United States of America* 86: 770–774.

Klein, C. and N. J. Buekes (1992). Time distribution, stratigraphy, sedimentologic setting, and geochemistry of precambrian iron formations. *The Proterozoic Biosphere: A Multidisciplinary Study.* J. W. Schopf and C. Klein, eds. New York, Cambridge University Press: 139–146.

Komarek, J. and K. Anagnostides (1999). *Cyanoprokaryota. 1. Teil Chroococcales. Subwasserflora von Mitteleuropa.* Munich, Gustav Fischer Verlag.

Kusumi, T., T. Ooi, et al. (1987). Cyanoviridin-RR, a toxin from the cyanobacterium (blue-green alga) *Microcystis aeruginosa. Tetrahedron Letters* 28: 4695–4698.

Laamanen, M. J., M. Gugger, et al. (2001). Diversity of toxic and nontoxic *Nodularia* isolates (cyanobacteria) and filaments from the Baltic Sea. *Applied Environmental Microbiology* 67: 4638–4647.

Lagos, N., H. Onodera, et al. (1999). The first evidence of paralytic shellfish toxins in the freshwater cyanobacterium *Cylindrospermopsis raciborskii*, isolated from Brazil. *Toxicon* 37: 1359–1373.

Lehtimaki, J., C. Lyra, et al. (2000). Characterization of *Nodularia* strains, cyanobacteria from brackish waters, by genotypic and phenotypic methods. *International Journal of Systematic Evolutionary Microbiology* 50: 1043–1053.

Li, R., W. W. Carmichael, et al. (2001a). Isolation and identification of the cyanotoxin cylindrospermopsin and deoxy-cylindrospermopsin from a Thailand strain of *Cylindrospermopsis raciborskii* (cyanobacteria). *Toxicon* 39: 973–980.

Li, R. H., W. W. Carmichael, et al. (2001b). The first report of the cyanotoxins cylindrospermopsin and deoxycylindrospermopsin from *Raphidiopsis curvata* (cyanobacteria). *Journal of Phycology* 37(6): 1121–1126.

Li, R. H. and M. M. Watanabe (2002). DNA base composition of planktonic species of *Anabaena* (cyanobacteria) and its taxonomic value. *Journal of General Microbiology* 48: 77–82.

Mahmood, N. A. and W. W. Carmichael (1986). The pharmacology of anatoxin-a(s), a neurotoxin produced by the freshwater cyanobacterium *Anabaena flos-aquae* NRC 525-17. *Toxicon* 24(5): 425–434.

Marsalek, B., L. Blaha, et al. (2001). Microcystin-LR and total microcystins in cyanobacterial blooms in the Czech Republic 1993–1998. *Cyanotoxins: Occurrence, Causes, Consequences* I. Chorus, ed. Berlin, Springer-Verlag: 56–62.

Mez, K., K. Beattie, et al. (1997). Identification of a microcystin in benthic cyanobacteria linked to cattle deaths on alpine pastures in Switzerland. *European Journal of Phycology* 32(2): 111–117.

Neilan, B. A., B. P. Burns, et al. (2002). Molecular identification of cyanobacteria associated with stromatolites from distinct geographical locations. *Astrobiology* 2(3): 271–280.

Neilan, B. A., D. Jacobs, et al. (1995). Genetic diversity and phylogeny of toxic cyanobacteria determined by DNA polymorphisms within the phycocyanin locus. *Applied and Environmental Microbiology* 61: 3875–3883.

Neilan, B. A., D. Jacobs, et al. (1997). rRNA sequences and evolutionary relationships among toxic and nontoxic cyanobacteria of the genus *Microcystis*. *International Journal of Systematic Bacteriology* 47(3): 693–697.

Neilan, B. A., M. L. Saker, et al. (2003). Phylogeography of the invasive cyanobacterium *Cylindrospermopsis raciborskii*. *Molecular Ecology* 12: 133–140.

Prinsep, M. R., F. R. Caplan, et al. (1992). Microcystin-LR from a blue-green alga belonging to the stigonematales. *Phytochemistry* 31(4): 1247–1248.

Rinehart, K. L., M. Namikoshi, et al. (1994). Structure and biosynthesis of toxins from blue-green-algae (cyanobacteria). *Journal of Applied Phycology* 6(2): 159–176.

Saker, M. L., I. C. G. Nogueira, et al. (2003). First report and toxicological assessment of the cyanobacterium *Cylindrospermoposis raciborskii* from Portuguese freshwaters. *Ecotoxicology and Environmental Safety* 55: 243–250.

Schopf, J. W. (2000). The fossil record: tracing the roots of the cyanobacterial lineage. *The Ecology of Cyanobacteria*. B. A. Whitton and M. Potts, eds. Dordrecht, Kluwer: 13–35.

Schwabe, W., A. Weihe, et al. (1988). Plasmids in toxic and nontoxic strains of the cyanobacterium *Microcystis aeruginosa*. *Current Microbiology* 17: 133–137.

Shaw, G. R., A. Sukenik, et al. (1999). Blooms of the cylindrospermopsin containing cyanobacterium, *Aphanizomenon ovalisporum* (Forti), in newly constructed lakes, Queensland, Australia. *Environmental Toxicology* 14(1): 167–177.

Sivonen, K., K. Himberg, et al. (1989). Preliminary characterization of neurotoxic cyanobacteria blooms and strains from Finland. *Toxicity Assessment* 4: 339–352.

Sivonen, K., K. Kononen, et al. (1989). Toxicity and isolation of the cyanobacterium *Nodularia spumigena* from the southern Baltic Sea in 1986. *Hydrobiologia* 185: 3–8.

Sivonen, K., M. Namikoshi, et al. (1992). Isolation and structures of five microcystins from a Russian *Microcystis aeruginosa* strain calu 972. *Toxicon* 30(11): 1481–1485.

Sivonen, K., S. I. Niemela, et al. (1990). Toxic cyanobacteria (blue-green algae) in Finnish fresh and coastal waters. *Hydrobiologia* 190: 267–275.

Skulberg, O. M. (1996). Toxins produced by cyanophytes in Norwegian inland waters — health and environment. *Chemical Data as a Basis for Geomedical Investigations*. J. Lag, ed. Oslo, The Norwegian Academy of Science and Letters: 197–216.

Skulberg, O. M., W. W. Carmichael, et al. (1992). Investigations of a neurotoxic oscillatorian strain (cyanophyceae) and its toxin. Isolation and characterization of homoanatoxin-a. *Environmental Toxicology and Chemistry* 11: 321–329.

Smith, J. K., J. D. Parry, et al. (1998). A PCR technique based on the Hip 1 interspersed repetitive sequence distinguishes cyanobacterial species and strains. *Microbiology* 144: 2791–2801.

Strauss, H., D. J. Des Marais, et al. (1992). The carbon isotopic record. *The Proterozoic Biosphere, A Multidisciplinary Study.* J. W. Schopf and C. Klein, eds. New York, Cambridge University Press: 117–127.

Tatsumi, K., M. F. Watanabe, et al. (1991). Alloenzyme divergence in *Anabaena* (Cyanophyceae) and its taxonomic inference. *Archives of Hydrobiology* 92(suppl): 129–140.

Thorpe, R. I., A. H. Hickman, et al. (1992). U-Pb zircon geochronology of Archaean felsic units in the Marble Bar region Pilbara Craton, Western Australia. *Precambrian Research* 56: 169–189.

Ueno, Y., S. Nagata, et al. (1996). Detection of microcystins, a blue-green algal hepatotoxin, in drinking water sampled in Haimen and Fusui, endemic areas of primary liver cancer in China, by highly sensitive immunoassay. *Carcinogenesis* 17: 1317–1321.

Vezie, C., L. Brient, et al. (1998). Variation of microcystin content of cyanobacterial blooms and isolated strains in lake Gand-lieu (France). *Microbial Ecology* 35(2): 126–135.

Watanabe, M. (1996). Isolation, cultivation and classification of bloom-forming *Microcystis* in Japan. *Toxic Microcystis.* M. F. Watanabe, K.-I. Harada, W. W. Carmichael, and H. Fujiki, eds. Boca Raton, FL, CRC Press: 13–34.

Whitton, B. A. and M. Potts (2000). *The Ecology of Cyanobacteria: Their Diversity in Time and Space.* Dordrecht, Kluwer Academic Publishers.

Whitton, B. A., P. Robinson, et al. (2000). *Key to Blue-Green Algae of the British Isles.* Environment Agency (England and Wales) and University of Durham, Department of Biological Sciences, Durham, England.

Williams, C. D., J. Burns, et al. (2001). Assessment of cyanotoxins in Florida's lakes, reservoirs, and rivers. *Cyanobacteria Survey Project.* Harmful Algal Bloom Task Force, St. John's River Water Management District, Palatka, FL.

3 Toxin Chemistry and Biosynthesis

The cyanobacterial toxins are secondary metabolites synthesized within the cells of some species from at least four of the five orders of cyanobacteria. The toxins show great diversity, ranging from simple alkaloids to complex polycyclic compounds and cyclic peptides. In all probability, the characterized toxins illustrate only a small proportion of the total toxins of cyanobacteria, as most cyanobacterial species have not yet been examined for toxicity. The best-understood peptide toxin group, the microcystins, originally isolated from the genus *Microcystis*, have more than 60 molecular variants identified at the present time. Similarly the saxitoxin-related alkaloids in cyanobacteria show a family of compounds of differing toxicity, some of which are different from those of marine dinoflagellates (Onodera, Satake et al. 1997). By comparison with toxins identified from marine dinoflagellates (Baden and Trainer 1993), cyanobacteria have not yet been shown to possess polyether neurotoxins, though there have been numerous reports of uncharacterized neurotoxicity from cyanobacteria (Hawser, Capone et al. 1991; Baker et al. 2001).

From the viewpoint of safety of drinking water supplies, the major area of concern is the water-soluble cyanobacterial toxins. Lipid-soluble toxins are bound to cells or particulate fragments that will be removed by coagulation and sedimentation in standard water treatment (see Chapter 12). Lipid-soluble toxins are, however, of medical significance in food, especially shellfish and fish exposed to dinoflagellate blooms, and have caused widespread illness and death in human populations (Falconer 1993). They are also relevant in recreational exposure to the cyanobacteria. The best characterized example is *Lyngbya majuscula* in tropical coastal waters, which causes severe skin burns to people bathing and fishing (Banner 1959). This benthic cyanobacterium, which grows on rocks, seagrass, and marine macroalgae, contains several tumor-promoting irritant toxins that readily penetrate the skin and gastrointestinal tract as a consequence of their lipophilic nature (Cardelina, Marner et al. 1979; Ito, Satake et al. 2002). Their molecular structure is illustrated in Figure 3.1. The mechanism of toxicity has been identified for these compounds, which activate the enzyme protein kinase C in a similar manner to the phorbol esters of plant origin (Basu, Kozikowski et al. 1992).

The water-soluble toxins from cyanobacteria of greatest significance for the safety of the drinking water supply are the cylindrospermopsins and the microcystins. These toxins damage the liver in particular and have carcinogenic or tumor-promoting properties. Their toxicity is discussed in Chapter 5 and Chapter 6 and their chemistry and biosynthesis in this chapter.

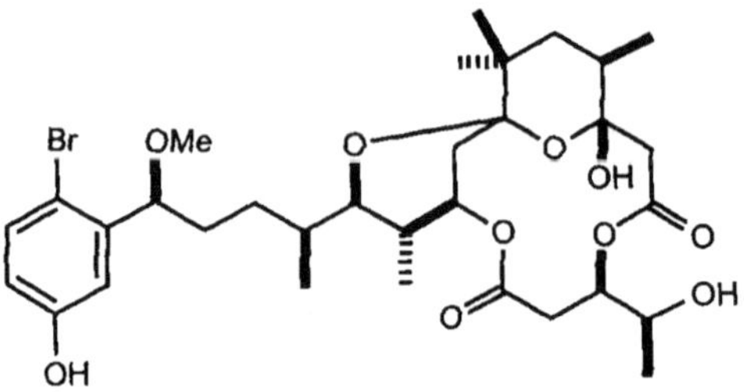

Debromoaplysiatoxin

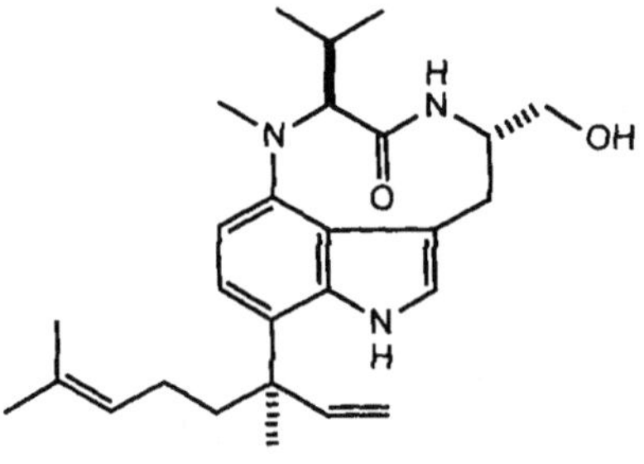

Lyngbyatoxin A

FIGURE 3.1 Structures of debromoaplysiatoxin and lyngbyatoxin A.

3.1 CHEMISTRY OF CYLINDROSPERMOPSINS

The toxic alkaloid cylindrospermopsin was isolated from a culture of the cyanobacterium *Cylindrospermopsis raciborskii* originating in a water supply reservoir on Palm Island, off the tropical coast of Queensland, Australia (Ohtani, Moore et al. 1992). The organism, and its toxicity, came to attention as a result of a severe gastroenteritis outbreak among children who were drinking water from that supply (Byth 1980; Hawkins, Runnegar et al. 1985). This event is discussed in detail in Chapter 5.

Cylindrospermopsin was isolated from an ultrasonicated, freeze-dried culture of *C. raciborskii* extracted in 0.9% NaCl, with a 0.5% yield of the alkaloid as a white crystalline powder. Purification was done by repeated gel filtration on Toyopearl WH40F in 1:1 methanol:water, followed by reversed-phase high-performance liquid chromatography (HPLC) purification using a C-18 column eluted with 5% methanol. Positive ion mass spectroscopy by high-resolution fast-atom bombardment mass spectroscopy (HRFABMS) yielded a protonated (MH^+) ion of mass (*m/z*) 416.1236, and fragmentation evidence of uracil and sulfate groups. Detailed analysis of the 500-MHz ^1H and 125-MHz ^{13}C nuclear magnetic resonance (NMR) spectra in D_2O, together with homonuclear and heteronuclear correlation techniques (COSY, HMQC, and HMBC; see Bax and Subramanian 1986), provided the information for the structure shown in Figure 3.2 (Ohtani, Moore et al. 1992).

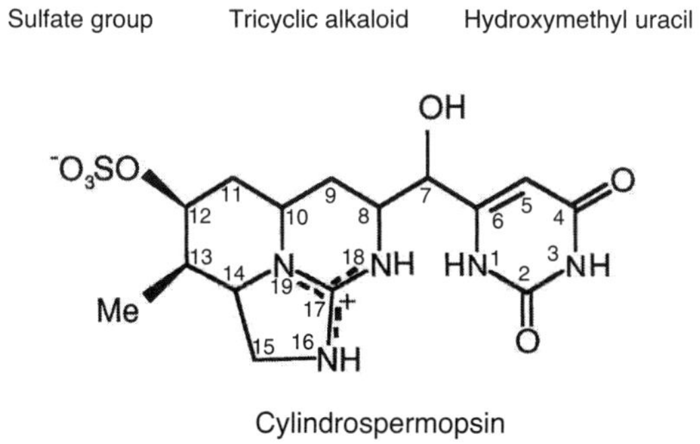

Sulfate group Tricyclic alkaloid Hydroxymethyl uracil

Cylindrospermopsin

FIGURE 3.2 Cylindrospermopsin molecular shape. (From Ohtani, Moore et al. 1992. With permission.)

Figure 3.2 illustrates cylindrospermopsin, an alkaloid with a tricyclic ring structure containing a guanido group. At position C-12 a sulfate group is attached, and at position C-13 a methyl group. Hydroxymethyl uracil is linked to the ring structure at C-8, with the bridging hydroxymethyl at C-7, linking to C-6 of the uracil pyrimidine ring. The uracil showed shifts between keto and enol forms at pH 7, as occurs in nucleotides. Cylindrospermopsin is a zwitterion, carrying a double (positive and negative) charge, and as a consequence is very water-soluble (Ohtani, Moore et al. 1992). The UV spectrum in water has a strong peak at 262 nm, with an extinction coefficient of 5800. The tricyclic structure is essentially flat, with rotational bonds at the hydroxymethyl at C-7 linking to the tricyclic ring and the uracil ring. It is therefore possible to build an accurate model of the whole cylindrospermopsin molecule that is flat and has the potential capacity for intercalating into the DNA double helix. This has relevance for the observed capacity of the molecule to cause chromosome breaks in replicating DNA and also for the mechanism of protein synthesis inhibition by the toxin, both of which are discussed in Chapter 6.

Investigation of *C. raciborskii* for unexplained *in vivo* toxicity resulted in the identification of deoxycylindrospermopsin, which occurred in freeze-dried samples of the cyanobacterium at 10 to 50% of the quantity of cylindrospermopsin (Norris, Eaglesham et al. 1999). The protonated molecule on HPLC tandem mass spectroscopy (MS/MS) was of mass 400 *m/z*, with NMR spectral evidence that the oxygen at C-7 in cylindrospermopsin was absent. Initial toxicity assessment indicated that no toxicity could be identified (Norris, Eaglesham et al. 1999). However, recent chemical synthesis of 7-deoxycylindrospermopsin showed that the compound inhibited protein synthesis *in vitro* using the reticulocyte lysate protein synthesis system and also inhibited protein synthesis in isolated hepatocytes. Inhibition was approximately 10-fold lower than that shown by cylindrospermopsin (Runnegar, M.T.C. personal communication). The issue of noncylindrospermopsin toxicity in *C. raciborskii* isolates is yet to be resolved.

A further variant on cylindrospermopsin was identified by Banker, Teltsch et al. (2000) from the cylindrospermopsin-containing cyanobacterium *Aphanizomenon ovalisporum*, isolated from Lake Kinneret, Israel. This molecule differed only by the hydroxyl group at C-7 being in the epimer position compared to the cylindrospermopsin structure. 7-Epicylindrospermopsin occurred as only a minor proportion of the cylindrospermopsin content but was of equal toxicity (Banker, Carmeli et al. 2001).

Chlorination of the cylindrospermopsin molecule yielded 5-chlorocylindrospermopsin, with the chlorine atom attached to position 5 of the uracil ring. The compound was tested and found to be nontoxic at doses up to 50 times higher than the lethal dose killing 50% of the organisms (LD_{50}) of the native molecule (Banker, Carmeli et al. 2001). Another product of chlorine treatment was formed by cleavage of the cylindrospermopsin molecule at C-6 by oxidation to a carboxylic acid group, thus displacing the uracil residue. This too was found to be nontoxic. It is therefore apparent that the uracil residue is essential for toxicity, and that a chlorine atom at carbon 5 of uracil is sufficient to block the toxic mechanism. Use of synthetic cylindrospermopsin and structural analogues of the molecule provided further information on the structure–activity relationship. The molecule without the sulfate group (cylindrospermopsin diol) showed protein synthesis inhibition in both cell-free reactions and in cultured rat hepatocytes at similar concentration to natural toxin, indicating that the group was not required for cell entry or for toxicity. A simple cylindrospermopsin model (AB-Model) (Xie, Runnegar et al. 2000), which lacked the sulfate, the methyl group at C-13, and the guanido C-ring, was shown to inhibit protein synthesis *in vitro* and in hepatocytes but required an approximately 1000 times higher concentration (Runnegar, Xie et al. 2002).

3.2 SYNTHESIS OF CYLINDROSPERMOPSIN

The total synthesis of racemic cylindrospermopsin was first achieved by Xie, Runnegar et al. (2000) in 20 steps commencing from 4-methoxy-3-methylpyridine. The synthesis was a very difficult task, as the molecule has six chiral centers with three functional groups—the guanine in the tricyclic ring, the sulfate, and uracil. Several partial syntheses had been published earlier (Heintzelman, Weintreb et al. 1996; Murphy and Thomas 2001), showing the progress of the synthesis. Because of the need to verify the structure by synthesis and also to have a method that can potentially produce quantities of toxin for experimental use, much effort has been expended in the synthetic pathway. The synthesis of Xie, Runnegar et al. (2000) achieved an overall yield of 3.5%.

Recently a different approach requiring about 30 steps was undertaken by Heintzelman, Fang et al. (2002). A stereoselective synthesis was used that produced 7-epicylindrospermopsin, and resulted in a revision of the stereochemical assignments of the 7-hydroxyl position in the naturally occurring toxin. It is proposed that the described structure for cylindrospermopsin (Ohtani, Moore et al. 1992) is actually that for 7-epicylindrospermopsin, and vice versa. These are illustrated in Figure 3.3. As both epimers are of equal toxicity and both occur naturally in cyanobacteria, the stereochemical revision is not likely to result in any change in the biochemical or toxicological research in progress.

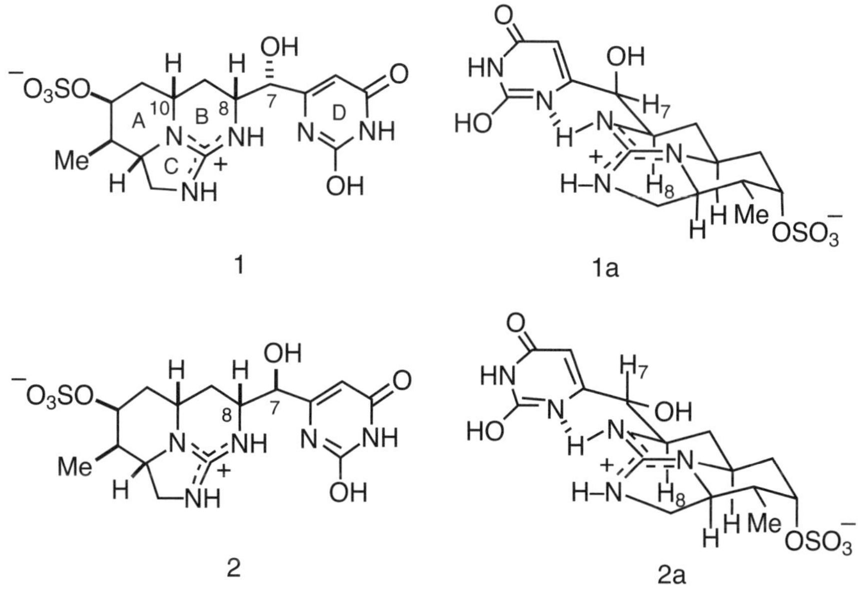

FIGURE 3.3 Stereospecific assignment of the 7-hydroxyl group of cylindrospermopsin. 7-epicylindrospermopsin, molecular structure 1; cylindrospermopsin structure 2. (From Heintzelman, Fang et al. 2002. With permission.)

3.3 BIOSYNTHESIS OF CYLINDROSPERMOPSIN

The biosynthesis of cylindrospermopsin has been investigated from two different directions. The traditional approach to establishing a biosynthetic pathway is to supply radioactive precursor molecules and to identify the products forming the intermediates along the pathway. This method has been used with spectacular success in the past — for example, in the discovery of the carbon fixation pathway for photosynthesis in green plants (Bassham and Calvin 1957). With the advance in capability of NMR techniques, isotope-enriched precursors can be used, which allow the location of the precursor atom to be identified in the product molecule. This approach is described below (Burgoyne, Hemscheidt et al. 2000).

A new and independent approach is to undertake genetic analysis of likely DNA regions in the organism, looking for nucleic acid sequences for enzymes that may be part of the biosynthetic pathway. As the gene sequences for increasing numbers of enzymes are being reported and listed in the computer databases and sequence comparison programs are widely available, the enzymes involved in biosynthesis of new compounds can be identified from sequence data alone. A further advantage of the genetic approach is that the biosynthesis of secondary metabolites often follows an initially common pathway, followed by relatively small changes to common types of enzyme reactions to produce the specialized products. This method has been employed with success in clarifying the biosynthesis of cylindrospermopsin, as discussed later.

The understanding of the genes responsible for cylindrospermopsin biosynthesis in *A. ovalisporum* is progressing, with the identification of an amidinotransferase gene that is likely to code for the enzyme forming guanidinoacetic acid, the first step in the biosynthetic pathway to cylindrospermopsin (Shalev-Alon, Sukenik et al. 2002). This gene is located in the region carrying the polyketide synthase and peptide synthetase genes, supporting a role in cylindrospermopsin biosynthesis. A recent poster by Shalev-Alon, Sukenik et al. (2004) illustrated their concept of the gene group responsible for this biosynthesis; it comprised a gene sequence of a dehydrogenase, an acyl transferase, and a β-ketoacyl synthetase all reading left, with a linking amidinotransferase reading right followed by an AMP-binding domain, a phosphopantotheine-binding domain, a β-keto acyl synthase, and an acyl transferase.

Using the isotope label approach, cultures of *C. raciborskii* were grown with the simple precursor molecules, for example, acetate and glycine labeled with the stable isotopes ^{2}H, ^{13}C, ^{15}N, and ^{18}O. The biosynthesized cylindrospermopsin was extracted from the cells, and NMR used to locate the position in the toxin molecule of the labeled atoms. When the possible precursor was not incorporated, no labeled atoms appeared in the product. All of the carbon atoms from C-4 to C-13 of cylindrospermopsin were labeled by feeding the culture with uniformly labeled ^{13}C sodium acetate, demonstrating that the carbon "backbone" was a polymer of five acetate units (see Figure 3.2 for numbering of atoms). Carbon atoms 14 and 15 and the associated nitrogen atom 16 came from glycine, as demonstrated by feeding ^{13}C, ^{15}N glycine to the culture. The methyl group at C-13 came from the single carbon pool of the cell (Burgoyne, Hemscheidt et al. 2000). This left the problems of the initial starter molecule, the origins of carbon atom 17 and nitrogen atoms 18 and 19 of the guanido group, and uracil atoms 1, 2, and 3.

Guanidinoacetic acid was synthesized with four ^{13}C and three ^{15}N labels and fed to the culture. The resulting cylindrospermopsin was labeled at C-17 and adjacent nitrogen atoms, showing that the guanidino group had been incorporated. It was concluded that guanidinoacetic acid was the starter group onto which the successive acetate groups were added. The source of the atoms in the N-1, C-2, and N-3 portions of the uracil group is currently unknown (Burgoyne, Hemscheidt et al. 2000).

The successive additions of acetate groups occur commonly in biosynthetic pathways, the enzymes responsible in many cases being very large complex multifunctional proteins. The best-studied example is fatty acid biosynthesis, in which acetyl groups are cyclically linked and reduced to form an elongating hydrocarbon chain. The enzyme complex includes an acyltransferase, which accepts and transfers acyl groups to a carrier protein, a ketosynthase, which condenses the existing acyl or starter group with an incoming carboxy-acyl group with decarboxylation, a reductase for conversion of keto groups to hydroxyl groups, a dehydratase that extracts water leaving double-bonded carbon atoms and a reductase that inserts hydrogen to form the saturated chain. A thioesterase finally cleaves the acyl carrier protein from the fatty acid. A very similar multienzyme complex is polyketide synthase (type 1), which occurs in cyanobacteria and other life forms (Hutchinson 1999; Moffitt and Neilan 2003). This enzyme complex produces a range of secondary metabolites, including several antibiotics and toxins of pharmaceutical interest (Hutchinson 1999).

The genetic approach to cylindrospermopsin biosynthesis was based on the earlier exploration of polyketide antibiotic synthesis, which demonstrated conserved sequences of amino acids within the peptides of the enzyme complex (Schembri, Neilan et al. 2001). Knowledge of these sequences enabled suitable DNA sequences to be identified for use as primers for selective polymerase chain reaction amplification of cyanobacterial DNA coding for polyketide biosynthesis. These primers for polyketide gene sequences have been used with success to isolate genes likely to code for enzymes which carry out cylindrospermopsin biosynthesis. Examination of a series of strains of *C. raciborskii* for DNA fragments amplified using these specific primers demonstrated that the presence of the characteristic DNA fragments was coincident with the presence of cylindrospermopsin in the cells (Schembri, Neilan et al. 2001). A second enzyme coding region was also identified for the prokaryotic nonribosomal peptide synthetase, which will be discussed later as it is a key component of the biosynthesis of microcystins.

This technique for identifying the polyketide synthase gene has been extended to examination of a wide range of cyanobacterial species and strains associated with cylindrospermopsin production, and also *C. raciborskii* strains that have not been shown to produce the toxin. In all cases where the toxin has been found in the tested strain of cyanobacterium, including the species *Anabaena bergii* and *A. ovalisporum*, the gene has also been found. When the toxin was absent — as in strains of *C. raciborskii* from Germany and Brazil and in species in which the toxin has not been found, such as *Anabaena circinalis* or *Microcystis aeruginosa* — the gene was also absent (Fergusson and Saint 2003). The definitive proof of the polyketide synthase gene being responsible for cylindrospermopsin biosynthesis requires a "knockout" mutant of the gene, which has not yet been achieved. However the evidence is compelling that this genetic region of the chromosome is required for the toxin production. Detailed genetic analysis of the region is not yet available for cyanobacteria, though it has been established for fungal polyketide synthase (Hutchinson 1999).

The structural information on cylindrospermopsin, the isotopic feeding experiments examining the biosynthetic pathway, and the genetic exploration of polyketide synthase in cyanobacteria provide a clear general picture of the mechanism of biosynthesis. The detailed enzymology has yet to be explored. At the time of writing the enzyme responsible for addition of the sulfate group has not been examined, and the process of ring closure remains speculative. The ecological and nutritional influences on toxin production are considered later in Chapter 4.

3.4 CHEMISTRY OF MICROCYSTINS

Toxic water blooms of *Microcystis* have been reported widely across the world, associated with livestock, pet, and wildlife deaths (see Carmichael and Falconer 1993). They have also been implicated in human injury, both through drinking water (Falconer, Beresford et al. 1983) and dialysis fluid (Jochimsen, Carmichael et al. 1998; Pouria, de Andrade et al. 1998). The first identification that the toxin was peptide in nature was made by Bishop, Anet et al. (1959), who isolated the "fast-death factor" from *M. aeruginosa* in culture. Later, electrophoretically purified toxin was obtained from *M. aeruginosa* collected from a natural water bloom and shown to have a very simple amino acid composition, including the amino acids alanine and glutamic acid in D configuration, erythro β-methyl aspartic acid, and tyrosine and methionine in L configuration (Elleman, Falconer et al. 1978). The final structural determination was carried out on toxin samples from South Africa and Australia, using FABMS and NMR techniques at Cambridge University in the U.K. The toxins were purified by ammonium bicarbonate extraction of cell homogenates, followed by multistep column fractionation using Sephadex G-50 and DEAE cellulose. Final purification of the toxins from dam samples was done by high-voltage paper electrophoresis (Botes, Tuinman et al. 1984; Botes, Wessels et al. 1985).

The structure of microcystin is a cyclic heptapeptide, the first structure published having the L-amino acids leucine and alanine, together with five other unusual amino acids. The sequence in the peptide ring is γ-linked D-glutamic acid, N-methyldehydroalanine, D-alanine, L-alanine, β-linked erythro-β-methylaspartic acid, L-leucine, and a completely novel β-amino acid, abbreviated to ADDA (3-amino-9-methoxy-10-phenyl-2,6,8,-trimethyldeca-4,6-dienoic acid). The molecular weight is 909 Da. and the structure is illustrated in Figure 3.4. Soon after the first structure was published, the structures of a further four microcystin variants were published (Botes, Wessels et al. 1985). These five published toxins were obtained from toxic *M. aeruginosa* collected or cultured from reservoirs in South Africa and from a *M. aeruginosa* water bloom in an Australian drinking water reservoir (Botes, Viljoen et al. 1982; Botes, Wessels et al. 1985). All the microcystin variants had a characteristic UV absorption spectrum, with a strong peak at 238 nm due to the conjugated diene of the ADDA residue (Botes, Viljoen et al. 1982).

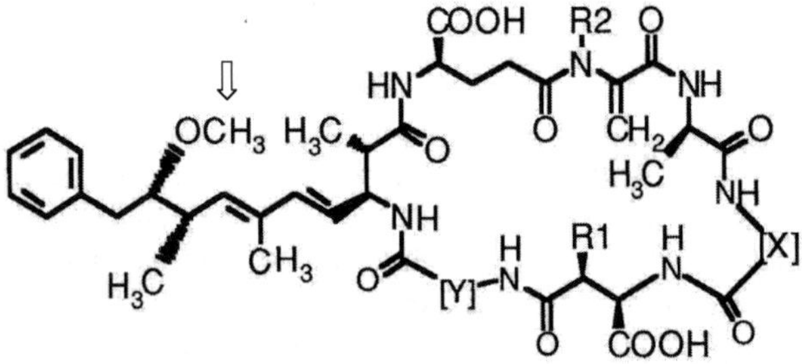

Microcystin

FIGURE 3.4 General structure of the cyclic heptapeptide toxin microcystin. X = L-leucine, Y = L-alamine in microcystin-LA, the first toxin variant totally structurally identified. (From Botes et al. 1984.) R1 and R2 are methyl groups. In other microcystin variants, positions X and Y may be substituted by a range of other L-amino acids and the methyl groups in R1 and R2 may be substituted by hydrogen in desmethyl variants. The methoxy groups at carbon-9 (⇓) in ADDA may be substituted by a hydrogen or an acetoxy group.

In these initial analyses only the two L-amino acids showed changes with the different samples. Using the amino acid abbreviations for the L-acids, the microcystin variants were as follows:

Microcystin-LR: X = leucine; Y = arginine; MW 994; South African
Microcystin-YR: X = tyrosine; Y = arginine; MW 1044; South African
Microcystin-YA: X = tyrosine; Y = alanine; MW 959; South African
Microcystin-YM: X = tyrosine; Y = methionine; MW 1019; Australian

Since these structures were determined, some 60 different microcystin variants have been described, and the number continues to increase (Harada 1996; Sivonen and Jones 1999). The majority of L-amino acid variants of microcystin have hydrophobic amino acids at position X and hydrophilic amino acids at position Y. The most frequent amino acids are leucine at X and arginine at Y, though tyrosine, phenylalanine, methionine, tryptophan, arginine, and other rarer amino acids are also found at X and alanine, methionine, tyrosine, and other acids occur at position Y. The methyl groups at position 3 in the methylaspartic acid and position 7 in the methyldehydroalanine may also be absent. Some variations in the ADDA molecule also occur, with the methoxy group at carbon 9 being replaced with a hydroxy or acetoxy group.

The different toxicities of the variants of microcystins provide some insight into the key elements of the toxic effect. The most toxic of the microcystins are those with the more hydrophobic L-amino acids, for example, microcystins-LA, -LR, -YR, -YM, with the least toxic those with more hydrophilic amino acids, for example, microcystin-RR. The difference is six- to tenfold. Loss of the methyl group from β-methyl aspartic acid or from methyldehydroalanine reduces toxicity roughly by half (see summary by Sivonen and Jones 1999). The ADDA group appears to be crucial for toxicity, as removal or saturation of the group greatly reduced toxicity (Dahlem 1989). Isomers of microcystin-LR and -RR, which were isolated from field samples of *Microcystis viridis*, differed from the toxic peptides only by isomerization of the ADDA diene at C-6; C-7 from 6(*E*) to 6(*Z*), which effectively abolished toxicity (Harada, Matsuura et al. 1990; Harada, Ogawa et al. 1990). Together these results demonstrate the essential nature of the ADDA residue and its stereochemical configuration for the toxicity of the microcystin molecule. The methyoxy group at C-9 of ADDA however seems less significant, as no major differences in toxicity appear when comparing the methoxy, acetoxy, and hydroxy forms of the otherwise identical microcystin (Sivonen and Jones 1999).

Computation of the three-dimensional shape of the microcystin molecule in solution has shown a saddle- or boat-shaped peptide ring with flexibility in the large ADDA side chain, which is an essential part of the molecule for toxicity (Rudolph-Bohner, Mierke et al. 1994; Bagu, Sonnichsen et al. 1995; Trogen, Annila et al. 1996; Trogen, Edlund et al. 1998). The arginine residue in microcystin-LR also projects out from the ring, which allows some movement of the terminal guanidinium group. While this is a prominent part of the molecule, its function in toxicity is minor as microcystin with alanine or methionine at that position is equally toxic

(Sivonen and Jones 1999). The information on solution and crystal structure of microcystin has immediate relevance to the mechanism of toxicity, which is discussed extensively in Chapter 7.

The microcystins are very stable molecules, resistant to boiling at neutral pH or 40°C at pH 1 (Harada, Tsuji et al. 1996). They are not attacked by the hydrolytic enzymes of the gut, such as trypsin or chymotrypsin, or the bacterial enzymes subtilisin, thermolysin, and *Staphylococcus aureus* protease due to the presence of D-amino acids (Botes, Viljoen et al. 1982). Natural degradation of microcystins in lakes by enzymes from specific bacteria is discussed in Chapter 7.

3.5 SYNTHESIS OF MICROCYSTINS

The first total synthesis of the ADDA β-amino acid component of microcystins was carried out in 1989 (Namikoshi, Rinehart et al. 1989). Synthesis proceeded in three stages: the synthesis of the aromatic portion C-7 to C-10 with the terminal benzene ring; addition of C-5 and C-6; and finally synthesis and addition of the β-amino acid portion C-1 to C-4. Other routes of ADDA synthesis have since been published (Humphrey, Aggen et al. 1996; Sin and Kallmerten 1996; Candy, Donohue et al. 1999). As this amino acid is essential for biological activity, knowledge of the stereochemistry and the ability to synthesize and alter the molecule are valuable for understanding the mechanism of action. The mechanism of action is discussed in detail in Chapter 7. More recently, synthetic ADDA has been used to raise antibodies, which provide a general reactivity to microcystins independent of the variant (Fischer, Garthwaite et al. 2001). This is discussed in Chapter 9 and Chapter 10, as it offers a monitoring approach to the toxin with wide future potential.

The amino acid sequence of microcystins has also been synthesized, and the ring closed to form the complete molecule. Solid-phase peptide synthesis of Ac-D-γ-Glu-[N-Me-Δ Ala]-D-Ala-Leu amide was followed by synthesis of N-methyldehydroalanine (Zetterstrom, Trogen et al. 1995). Synthesis of the 3-methylaspartic acid was described by Echavarren and Castano (1995). Solid-phase synthesis was also used to synthesize a range of peptide rings modeled on microcystin and a related cyclic heptapeptide, nodularin (Taylor, Quinn et al. 1996).

Total synthesis of microcystin-LA was achieved by Humphrey, Aggen et al. (1996), using a new (at that time) route to ADDA synthesis and solution-phase amino acid coupling. There is a continuing interest in synthesis of microcystin analogues as the mechanism of action of the toxin (discussed in Chapter 7) involves inhibition of an important set of phosphatase enzymes with pharmacological implications (Mehrotra, Webster et al. 1997; Aggen, Humphrey et al. 1999; Gulledge, Aggen et al. 2002, 2003a, 2003b).

Exploration of the structure–activity relationships of synthetic or modified microcystins by the toxicity of the compounds and the inhibition of phosphatase enzymes has shown that the dehydroalanine can be saturated without loss of activity (Mehrotra, Webster et al. 1997), but alteration of the ADDA results inactivation. Harada demonstrated that the geometric isomer at C-7 diene was nontoxic (Harada,

Matsuura et al. 1990; Harada, Ogawa et al. 1990; Mehrotra, Webster et al. 1997) and Dahlem that saturation of the dienes has the same effect (Dahlem 1989). The free glutamic acid group also appears essential for activity (Stotts, Namikoshi et al. 1993).

3.6 BIOSYNTHESIS OF MICROCYSTINS: BIOCHEMICAL APPROACHES

The first research on the biosynthesis of the microcystin molecule was carried out by Moore, Chen et al. (1991) using precursor molecules labeled with ^{13}C. Cultures of *M. aeruginosa* were grown in a range of likely precursors, and the biosynthesized microcystin molecules subsequently analyzed by NMR spectroscopy to locate the labeled atoms within the microcystin molecule.

Attention was particularly paid to the precursor atoms of the ADDA portion of the microcystin. The methyl groups on C-6 and C-8 of ADDA were shown to be derived from methionine by feeding with L-[methyl-^{13}C] methionine. The methyl group on C2 appeared to be derived from methionine if acetate was available for biosynthesis, but in the absence of acetate, propionate may have been the precursor. Acetate was shown to be the main precursor of the linear portion of the ADDA providing carbon atoms 1 through 8, whereas [U-^{13}C]-L-phenylalanine was incorporated directly into the terminal phenyl unit of the molecule.

The precursor molecules for the cyclic amino acid structure were also investigated. Acetate was incorporated into C-4 and C-5 of the γ-linked D-glutamic acid residue and C-1 and C-2 of the β-methyl aspartic acid. [U-^{13}C]-Pyruvate supplied the precursor for C-3 and C-4 and the methyl on C-3 of the β-methyl aspartic acid. Pyruvate also supplied the carbon atoms for D-alanine.

[1,2-^{13}C]-L-glutamic acid feeding provided labeled C-1 and C-2 of the γ-linked glutamic acid of microcystin, demonstrating the direct incorporation of this amino acid when available. L-Glutamate was also shown to be the precursor of the L-arginine residue.

These studies demonstrated that the general processes of L-amino acid metabolism occurred in the biosynthesis of the amino acids incorporated into microcystin rather than unique pathways for the production of D-amino acids. The implication from this is that the racemization of the amino acids from L- to D- occurs during biosynthesis of the peptide ring.

Studies of the biosynthesis of the closely related cyclic pentapeptide nodularin, the toxin from *Nodularia spumigena*, demonstrated a similar pathway to that identified for microcystin (Rinehart, Namikoshi et al. 1994). Nodularin differs from microcystin by the absence of D-alanine and one L-amino acid and the substitution of N-methyldehydrobutyrin [2-(methylamino)-2-dehydrobutyric acid] for *N*-methyl dehydroalanine. It was suggested that threonine was the precursor in nodularin of the dehydro acid whereas serine was the precursor in microcystin (Rinehart, Namikoshi et al. 1994).

3.7 MOLECULAR GENETIC APPROACHES

Following this highly successful research by chemical and biochemical methodology, the next major advance occurred by the application of molecular biological methods to explore the genetic basis of microcystin biosynthesis. The biosynthesis of several peptide antibiotics, some cyclic and containing D-amino acids, had earlier been investigated, demonstrating a nonribosomal peptide synthesis pathway (Kleinkauf and von Dohren 1990). This pathway utilizes large multifunctional enzyme complexes that involve amino-acyl adenylates as the activated donor molecules via a thioester carrier to synthesize sequential peptide bonds. Use of conserved sequences found in the DNA of bacteria and fungi that synthesize nonribosomal peptides provided an approach to identifying potential peptide synthetase genes in *Microcystis aeruginosa* (Borchert, Patil et al. 1992). This technique allowed identification of fragments of genomic DNA from *Microcystis* possessing homologies to the amino acid adenylate–forming regions of peptide synthetases from other organisms. Sequencing of overlapping fragments of *Microcystis* DNA allowed the identification of a region of 2982 bp showing considerable homology to previously characterized genes encoding peptide synthetases, which hybridized only with DNA from toxic strains (Meisner, Dittmann et al. 1996). Further study identified three modular regions (subsequently named *mcy*A, *mcy*B, and *mcy*C) corresponding to components of nonribosomal peptide synthetases (Dittman, Meissner et al. 1996). This research provided both an analytical approach to the genetics of microcystin synthesis and clear evidence of the nonribosomal pathway to synthesis in cyanobacteria.

The necessity of a functional nonribosomal peptide synthetase gene in *Microcystis* for the synthesis of microcystin was demonstrated by insertional mutagenesis (Dittman, Neilan et al. 1997). A hepatotoxic strain of *Microcystis* was transformed by insertion of a chloramphenicol resistance gene flanked by peptide synthetase sequences. This resulted in replacement of the original "wild-type" DNA sequence with the antibiotic resistance sequence and inactivated the peptide synthetase, resulting in loss of microcystin synthesizing ability. The insertion occurred in the *mcy*B region of the gene, demonstrating that it is essential in the synthesis of the functional toxin.

Further research into the genetic components of microcystin synthesis revealed a set of seven more independent modules within the total gene cluster. These are read in the opposite direction to the earlier identified *mcy*A, *mcy*B, *mcy*C, which are responsible for the activation and incorporation of five amino acid constituents of microcystin-LR (Nishizawa, Asayama et al. 1999; Nishizawa, Ueda et al. 2000). The other modules are *mcy*D, which contains two polyketide synthase modules; *mcy*E, consisting of a polyketide synthase module and a peptide synthetase module; *mcy*F, which resembles an epimerase/racemase enzyme; *mcy*G, containing both a polyketide synthase and peptide synthetase (Nishizawa, Ueda et al. 2000) and *mcy*H, which appears to be related to transporter function; *mcy*I, likely to be a component of N-methyldehydroalanine synthesis; and *mcy*J, the last open reading frame in the cluster, which has a similarity to a previously described *O*-methyltransferase (Tillett, Dittmann et al. 2000). Transcriptional analysis demonstrated the microcystin synthetase to be transcribed as two polycistronic operons *mcy*ABC and *mcy*DEFGHIJ,

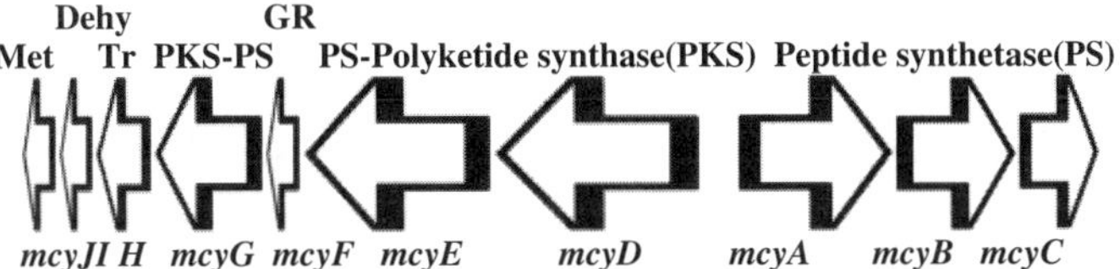

FIGURE 3.5 Organization of the genes responsible for the biosynthesis of the cyanobacterial toxin microcystin in *Microcystis aeruginosa.* Two operons of total length approximately 55 kb, comprising 10 open reading frames, transcribed in two groups in opposite directions. Arrow size relates to module length. Gene module functions: Met, *O*-methyltransferase; Dehy, *N*-methyldehydroalanine synthesis; Tr, toxin transport; PKS, polyketide synthase; PS, nonribosomal peptide synthetase; some modules also contain epimerase; GR, glutamate racemase. (From Tillett, Dittmann et al. 2000. With permission.)

beginning a central bidirectional promoter between *mcy*A and *mcy*D (Kaebernick, Dittmann et al. 2002).

Figure 3.5 illustrates the organization of the two operons concerned with microcystin biosynthesis — their component genes and biochemical functions. This genetic analysis has resolved the majority of the components of the biosynthetic pathway and relates well to the earlier studies in which the biosynthetic precursors were identified (Moore, Chen et al. 1991). For the biosynthesis of the ADDA components, both Nishizawa, Ueda et al. (2000) and Tillett, Dittmann et al. (2000) propose similar pathways commencing with the phenylalanine residue. This is proposed to be activated via a thioester bond to an acyl carrier protein, in a comparable manner to the biosynthesis of fatty acids. Additional acetate groups are proposed to be sourced from malonyl-coenzyme A, with decarboxylative condensation similar to that found in other polyketide and fatty acid biosynthesis. Methylation of the oxygen of the phenylacetate and of the acetate units is carried out by the *O*-methyl and *C*-methyltransferases using *S*-adenosyl methionine as the donor. The origin of the β-amino group of ADDA, and the mechanism of condensation of the γ-carboxyic acid group on glutamate to ADDA are not resolved. It is also unclear whether the amino acids are epimerized from L- to D- prior to or following addition to the growing peptide chain. My own preferred assumption is that they are epimerized after activation by an adenylation process prior to peptide bond formation.

Nishizawa et al. (2000) consider that genes *mcy*A, *mcy*B, and *mcy*D are respectively responsible for the addition of L-serine, which is metabolized to *N*-methyldehydroalanine, D-alanine, L-leucine, D-methylaspartic, and L-arginine in the biosynthesis of microcystin-LR.

Racemization of the L-glutamic acid to D-glutamic acid appears to be coded a specific gene locus, *mcy*F, which was demonstrated to carry out this activity when transferred into a D-glutamate-requiring strain of *Escherichia coli* (Nishizawa, Asayama et al. 2001). Similarly, an aspartate racemase has been identified within the microcystin biosynthesis gene cluster (Sielaff, Dittmann et al. 2003).

Further analysis of the microcystin biosynthesizing genes from different strains of *Microcystis*, *Anabaena*, and *Planktothrix* producing different L-amino acid variants of the toxin are beginning to resolve the genetic basis of the toxin variants. A

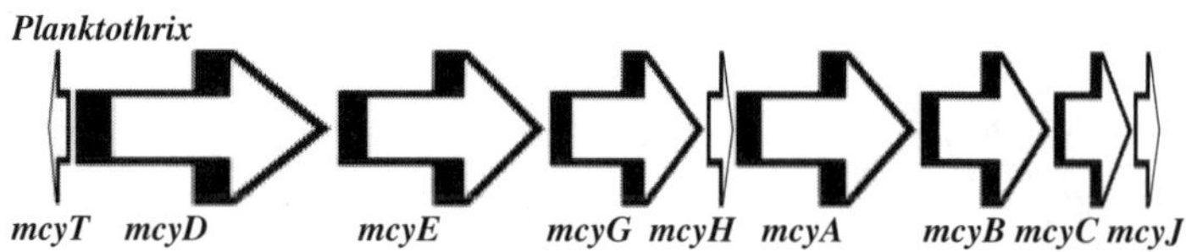

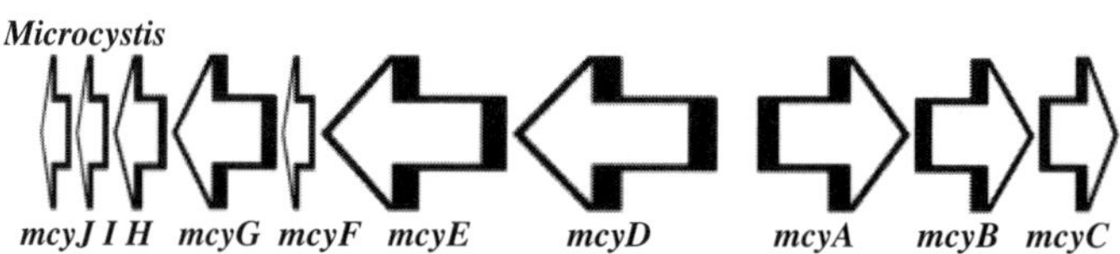

FIGURE 3.6 Comparison of the genetic organization responsible for the biosynthesis of the cyanobacterial toxin microcystin in *Planktothrix agardhii* with *Microcystis aeruginosa*. Overall length approximately 55 kb in both cases. Arrow length approximates to DNA sequence length. *Mcy*T is proposed to be a thioesterase. (From Christiansen, Fastner et al. 2003. With permission.)

study of the microcystin synthetase gene cluster in the toxic microcystin-producing species *Planktothrix agardhii* demonstrated similar modular units to those seen in *Microcystis* but differently organized and all reading in the same direction (Figure 3.6) (Christiansen, Fastner et al. 2003). The DNA sequence variation in the same modules between the two species ranged from 47 to 88% identity, indicating a substantial flexibility in the amino acid sequences of the functional proteins of the complex. *P. agardhii* is a motile, filamentous cyanobacterium, which moves towards light, as compared with the unicellular *Microcystis*, which forms colonies in a gelatinous matrix.

A detailed study of the microcystin synthetase genes in *Anabaena* shows considerable differences in the gene order and the sequence match compared to *Microcystis* and *Planktothrix*, indicating divergence or independent origins for these genes in *Anabaena* (Rouhiainen, Vakkilainen et al. 2004). These species are well separated in any phylogenetic analysis (Neilan, Jacobs et al. 1995), indicating that microcystin synthesis may have evolved early in cyanobacterial evolution. It has been postulated that the microcystin synthetase complex arose from the comparable gene complex synthesizing nodularin, the toxic cyclic pentapeptide from *N. spumigena*. This peptide lacks two amino acids of microcystin, D-alanine and the adjacent variable L-amino acid designated X in the microcystin molecule. The corresponding gene modules are the second module of *mcy*A and the first module of *mcy*B, which show the lowest sequence identity between *Microcystis* and *Planktothrix* genes. Thus it is proposed that the *Microcystis* and *Planktothrix* gene clusters arose independently from the precursor of the nonribosomal peptide synthetase of *Nodularia* DNA by addition of the C-terminal module of *mcy*A, which incorporates D-alanine in most microcystins, and the N-terminal module of *mcy*B, which incorporates a range of different L-amino acids into microcystin (Christiansen, Fastner et al. 2003). An alternative explanation for the origin of nodularin is that it is a recent derivation from the genes encoding microcystin synthetase (Rantala, Fewer et al. 2004).

Examination of the genetic basis of the many variants of microcystin, particularly at the locations of the two L-amino acids, points to genetic variation in the adenylate activating regions. This essential step activates the amino acid prior to condensation into the growing peptide and determines which amino acid is incorporated. DNA sequence analysis of the adenylation domain in *mcy*B1 was interpreted to indicate that recombination between *mcy*B1 and *mcy*C could give rise to the two groups of toxin variants observed (Mikalsen, Boison et al. 2003). The authors conclude that recombination between imperfect repeats, gene loss, and horizontal gene transfer can explain the variation in toxin chemistry between *Microcystis* strains. Recent phylogenetic analysis, however, indicated a coevolution of "housekeeping" genes and microcystin synthetase genes over the entire evolutionary history of the toxin. This implies that horizontal gene transfer between genera is unlikely (Rantala, Fewer et al. 2004).

The evolution of eukaryotic cells is envisaged as commencing about 1.5 billion years ago (Heckman, Geiser et al. 2001), whereas geological evidence for cyanobacteria dates from at least 1.5 to 2 billion years ago (see Chapter 2). It therefore seems likely that the microcystin synthesis genes evolved prior to commencement of eukaryotic life for purposes other than toxicity to eukaryote organisms.

Recent studies of the phylogenetic evidence for the evolution of the microcystin biosynthesis genes have resolved the problem of whether the genes entered unrelated cyanobacterial genera by horizontal gene transfer or convergent evolution or are of ancient origin. A comparison of genes involved in primary metabolism (housekeeping genes) and in biosynthesis of microcystins and nodularin has showed evidence for coevolution of both sets of genes over the whole evolutionary history of cyanobacteria. It is therefore suggested that the current microcystin synthesis genes evolved from an ancestral gene set and that the later-evolved genera lacking this capability lost it during their evolution. The evidence also suggests that the current nodularin synthesis genes evolved recently from the same set of ancient microcystin synthesis genes (Rantala, Fewer et al. 2004).

Control of the microcystin gene cluster has been investigated in cultured *M. aeruginosa*, which showed an activation of gene transcription of *mcy*B and *mcy*D at high light intensities and at red light intensity equivalent to the red component of the high-intensity illumination. As *mcy*B codes for peptide synthetase and *mcy*D for polyketide synthase, components read in opposite directions in the genome, this indicated a general activation of biosynthesis by light. However, analysis of the microcystin content of the cells did not show an increase, so further research will be required to clarify the regulation of microcystin biosynthesis (Kaebernick, Neilan et al. 2000).

REFERENCES

Aggen, J. B., J. M. Humphrey, et al. (1999). The design, synthesis, and biological evaluation of analogues of the serine-threonine protein phosphatase 1 and 2A selective inhibitor microcystin-LA: Rational modifications imparting PPI selectivity. *Bioorganic and Medicinal Chemistry* 7(3): 543–564.

Baden, D. G. and V. L. Trainer (1993). Mode of action of toxins of seafood poisoning. *Algal Toxins in Seafood and Drinking Water.* I. R. Falconer, ed. London, Academic Press: 49–74.

Bagu, J. R., F. D. Sonnichsen, et al. (1995). Comparison of the solution structures of microcystin-LR and motuporin [letter]. *Nature: Structural Biology* 2(2): 114–116.

Baker, P. D., D. A. Steffensen, et al. (2001). Preliminary evidence of toxicity associated with the benthic cyanobacterium *Phormidium* in South Australia. *Environmental Toxicology* 16(6): 506–511.

Banker, R., S. Carmeli, et al. (2001). Uracil moiety is required for toxicity of the cyanobacterial hepatotoxin cylindrospermopsin. *Journal of Toxicology and Environmental Health A* 62(4): 281–288.

Banker, R., B. Teltsch, et al. (2000). 7-Epicylindrospermopsin, a toxic minor metabolite of the cyanobacterium *Aphanizomenon ovalisporum* from Lake Kinneret, Israel. *Journal of Natural Products* 63(3): 387–389.

Banner, A. H. (1959). A dermatitis-producing alga in Hawaii. *Hawaii Medical Journal* 19, 1: 35–36.

Bassham, J. A. and M. Calvin (1957). *The Path of Carbon in Photosynthesis.* Engelwood Cliffs, NJ, Prentice Hall.

Basu, A., A. P. Kozikowski, et al. (1992). Structural requirements of lyngbyatoxin a for activation and downregulation of protein kinase C. *Biochemistry* 31: 3824–3830.

Bishop, C. T., E. Anet, et al. (1959). Isolation and identification of the fast-death factor in *Microcystis aeruginosa* NRC-1. *Canadian Journal of Biochemistry and Physiology* 37: 453–471.

Borchert, S., S. S. Patil, et al. (1992). Identification of putative multifunctional peptide synthetase genes using highly conserved oligonucleotide sequences derived from known synthetases. *FEMS Microbiology Letters* 82: 175–180.

Botes, D. P., A. A. Tuinman, et al. (1984). The structure of cyanoginosin-LA, a cyclic heptapeptide toxin from the cyanobacterium *Microcystis aeruginosa. Journal of the Chemical Society, Perkin Transactions* 1: 2311–2318.

Botes, D. P., C. C. Viljoen, et al. (1982). Structure of toxins of the blue-green alga *Microcystis aeruginosa. South African Journal of Science* 78: 378–379.

Botes, D. P., P. L. Wessels, et al. (1985). Structural studies on cyanoginosins-LR, -YR, -YA, and -YM, peptide toxins *Microcystis aeruginosa. Journal of the Chemical Society, Perkin Transactions* 1: 2747–2748.

Burgoyne, D. L., T. K. Hemscheidt, et al. (2000). Biosynthesis of cylindrospermopsin. *Journal of Organic Chemistry* 65(1): 152–156.

Byth, S. (1980). Palm Island mystery disease. *Medical Journal of Australia* 2: 40–42.

Candy, D. J., A. C. Donohue, et al. (1999). An assymetric synthesis of ADDA and ADDA-glycine dipeptide using the beta-lactam synthon method. *Journal of the Chemical Society, Perkin Transactions* 1(5): 559–567.

Cardelina, J. H., F. J. Marner, et al. (1979). Structure and absolute configuration of malyngolide, an antibiotic from the marine blue-green alga. *Journal of Organic Chemistry* 44: 4039–4042.

Carmichael, W. W. and I. R. Falconer (1993). Diseases related to freshwater blue-green algal toxins, and control measures. *Algal Toxins in Seafood and Drinking Water.* I. R. Falconer, ed. London, Academic Press: 187–209.

Christiansen, G., J. Fastner, et al. (2003). Microcystin biosynthesis in *Planktothrix*: Genes, evolution, and manipulation. *Journal of Bacteriology* 185(2): 564–572.

Dahlem, A. M. (1989). *Structure/Toxicity Relationships and Fate of Low Molecular Weight Peptide Toxins from Cyanobacteria.* Urbana-Champaign, University of Illinois: 135.

Dittman, E., K. Meissner, et al. (1996). Conserved sequences of peptide synthetase genes in the cyanobacterium *Microcystis aeruginosa. Phycologia* 35(6 suppl): 62–67.

Dittman, E., B. Neilan, et al. (1997). Insertional mutagenesis of a peptide synthetase gene which is responsible for hepatotoxin production in the cyanobacterium *Microcystis aeruginosa* PCC 7806. *Molecular Microbiology* 26: 779–787.

Echavarren, A. M. and A. M. Castano (1995). Synthesis of 3-methylaspartic acids by ring-contraction of a nickelacycle derived from glutamic anhydride. *Tetrahedron* 51: 2369–2378.

Elleman, T. C., I. R. Falconer, et al. (1978). Isolation, characterization and pathology of the toxin from a *Microcystis aeruginosa* (*Anacystis cyanea*) bloom. *Australian Journal of Biological Science* 31: 209–218.

Falconer, I. R. (1993). *Algal Toxins in Seafood and Drinking Water.* London, Academic Press.

Falconer, I. R., A. M. Beresford, et al. (1983). Evidence of liver damage by toxin from a bloom of the blue-green alga, *Microcystis aeruginosa. Medical Journal of Australia* 1(11): 511–514.

Fergusson, K. M. and C. P. Saint (2003). Multiplex PCR assay for *Cylindrospermopsis raciborskii* and cylindrospermopsin-producing cyanobacteria. *Environmental Toxicology* 18(2): 120–125.

Fischer, W. J., I. Garthwaite, et al. (2001). Congener-independent immunoassay for microcystins and nodularins. *Environmental Science and Technology* 35(24): 4849–4856.

Gulledge, B. M., J. B. Aggen, et al. (2002). The microcystins and nodularins: Cyclic polypeptide inhibitors of PP1 and PP2A. *Current Medicinal Chemistry* 9(22): 1991–2003.

Gulledge, B. M., J. B. Aggen, et al. (2003a). Linearized and truncated microcystin analogues inhibitors of protein phosphatases 1 and 2a. *Bioorganic and Medicinal Chemistry Letters* 13(17): 2907–2911.

Gulledge, B. M., J. B. Aggen, et al. (2003b). Microcystin analogues comprised only of adda and a single additional amino acid retain moderate activity as PP1/PP2A inhibitors. *Bioorganic and Medicinal Chemistry Letters* 13(17): 2907–2911.

Harada, K., K. Matsuura, et al. (1990). Isolation and characterization of the minor components associated with microcystins LR and RR in the cyanobacterium (blue-green algae). *Toxicon* 28: 55–64.

Harada, K., K. Ogawa, et al. (1990). Structural determination of geometrical isomers of microcystins-LR and -RR from the cyanobacteria by two-dimensional NMR spectroscopic techniques. *Chemical Research in Toxicology* 3: 473–481.

Harada, K.-I. (1996). Chemistry and detection of microcystins. *Toxic Microcystis.* M. F. Watanabe, K.-I. Harada, W. W. Carmichael, and H. Fujiki, eds. Boca Raton, FL, CRC Press: 103–148.

Harada, K. I., K. Tsuji, et al. (1996). Stability of microcystins from cyanobacteria-III. Effect of pH and temperature. *Phycologia* 35: 83–88.

Hawkins, P. R., M. T. C. Runnegar, et al. (1985). Severe hepatotoxicity caused by the tropical cyanobacterium (blue-green alga) *Cylindrospermopsis raciborskii* (Woloszynska) Seenaya and Subba Raju isolated from a domestic supply reservoir. *Applied and Environmental Microbiology* 50(5): 1292–1295.

Hawser, S. P., D. G. Capone, et al. (1991). A neurotoxic factor associated with the bloom-forming cyanobacteria *Trichodesmium. Toxicon* 29: 277–278.

Heckman, D. S., D. M. Geiser, et al. (2001). Molecular evidence for the early colonization of land by fungi and plants. *Science* 293: 1129–1133.

Heintzelman, G. R., W. K. Fang, et al. (2002). Stereoselective total syntheses and reassignment of stereochemistry of the freshwater cyanobacterial hepatotoxins cylindrospermopsin and 7-epicylindrospermopsin. *Journal of the American Chemical Society* 124(15): 3939–3945.

Heintzelman, G. R., S. M. Weinreb, et al. (1996). Imino Diels-Alder-based construction of a piperidine A-ring unit for total synthesis of the marine hepatotoxin cylindrospermopsin. *Journal of Organic Chemistry* 61(14): 4594–4599.

Humphrey, J. M., J. B. Aggen, et al. (1996). Total synthesis of the serine-threonine phosphatase inhibitor microcystin-LA. *Journal of the American Chemical Society* 118(47): 11759–11770.

Hutchinson, C. R. (1999). Microbial polyketide synthases: More and more prolific. *Proceedings of the National Academy of Sciences of the United States of America* 96(7): 3336–3338.

Ito, E., M. Satake, et al. (2002). Pathological effects of lyngbyatoxin A upon mice. *Toxicon* 40(5): 551–556.

Jochimsen, E. M., W. W. Carmichael, et al. (1998). Liver failure and death after exposure to microcystins at a hemodialysis center in Brazil. *New England Journal of Medicine* 338(13): 873–878.

Kaebernick, M., E. Dittmann, et al. (2002). Multiple alternate transcripts direct the biosynthesis of microcystin, a cyanobacterial nonribosomal peptide. *Applied Environmental Microbiology* 68(2): 449–455.

Kaebernick, M., B. A. Neilan, et al. (2000). Light and the transcriptional response of the microcystin biosynthesis gene cluster. *Applied and Environmental Microbiology* 66(8): 3387–3392.

Kleinkauf, H. and H. von Dohren (1990). Nonribosomal synthesis of peptide antibiotics. *European Journal of Biochemistry* 1892: 1–15.

Mehrotra, A. P., K. L. Webster, et al. (1997). Design and synthesis of serine-threonine protein phosphatase inhibitors based upon the nodularin and microcystin toxin structure. 1. Evaluation of key inhibitory features and synthesis of a rationally stripped down molecule. *Journal of the Chemical Society, Perkin Transactions* 1(17): 2495–2511.

Meisner, K., E. Dittmann, et al. (1996). Toxic and non-toxic strains of the cyanobacterium *Microcystis aeruginosa* contain sequences homologous to peptide synthetase genes. *FEMS Microbiology Letters* 135: 295–303.

Mikalsen, B., G. Boison, et al. (2003). Natural variation in the microcystin synthetase operon mcyABC and impact on microcystin production in *Microcystis* strains. *Bacteriology* 185(9): 2774–2785.

Moffitt, M. C. and B. A. Neilan (2003). Evolutionary affiliations within the superfamily of ketosynthases reflect complex pathway associations. *Journal of Molecular Evolution* 56(4): 446–457.

Moore, R. E., J. L. Chen, et al. (1991). Biosynthesis of microcystin-LR. Origin of the carbons in the adda and masp units. *Journal of the American Chemical Society* 113: 5083–5084.

Murphy, P. J. and C. W. Thomas (2001). The synthesis and biological activity of the marine metabolite cylindrospermopsin. *Chemical Society Reviews* 30(5): 303–312.

Namikoshi, M., K. L. Rinehart, et al. (1989). Total synthesis of adda, the unique C 20 amino acid of cyanobacterial hepatotoxins. *Tetrahedron Letters* 33: 4349–4352.

Neilan, B. A., D. Jacobs, et al. (1995). Genetic diversity and phylogeny of toxic cyanobacteria determined by DNA polymorphisms within the phycocyanin locus. *Applied and Environmental Microbiology* 61: 3875–3883.

Nishizawa, T., M. Asayama, et al. (1999). Genetic analysis of the peptide synthetase genes for a cyclic heptapeptide microcystin in *Microcystis* spp. *Journal of Biochemistry (Tokyo)* 126: 520–529.

Nishizawa, T., M. Asayama, et al. (2001). Cyclic heptapeptide microcystin biosynthesis requires the glutamate racemase gene. *Microbiology* 147: 1235–1241.

Nishizawa, T., A. Ueda, et al. (2000). Polyketide synthase gene coupled to the peptide synthetase module involved in the biosynthesis of the cyclic heptapeptide microcystin. *Journal of Biochemistry* 127(5): 779–789.

Norris, R. L., G. K. Eaglesham, et al. (1999). Deoxycylindrospermopsin, an analog of cylindrospermopsin from *Cylindrospermopsis raciborskii. Environmental Toxicology* 14(1): 163–165.

Ohtani, I., R. E. Moore, et al. (1992). Cylindrospermopsin: A potent hepatotoxin from the blue-green alga *Cylindrospermopsis raciborskii. Journal of the American Chemical Society* 114: 7941–7942.

Onodera, H., M. Satake, et al. (1997). New saxitoxin analogues from the freshwater filamentous cyanobacterium *Lyngbya wollei. Natural Toxins* 5: 146–151.

Pouria, S., A. de Andrade, et al. (1998). Fatal microcystin intoxication in haemodialysis unit in Caruaru, Brazil. *Lancet* 352: 21–26.

Rantala, A., D. P. Fewer, et al. (2004). Phylogenetic evidence for the early evolution of microcystin synthesis. *Proceedings of the National Academy of Sciences of the United States of America* 101(2): 568–573.

Rinehart, K. L., M. Namikoshi, et al. (1994). Structure and biosynthesis of toxins from blue-green-algae (cyanobacteria). *Journal of Applied Phycology* 6(2): 159–176.

Rouhiainen, L., T. Vakkilainen, et al. (2004). Genes coding for hepatotoxic heptapeptides (microcystins) in the cyanobacterium *Anabaena* strain 90. *Applied Environmental Microbiology* 70(2): 686–692.

Rudolph-Bohner, S., D. F. Mierke, et al. (1994). Molecular structure of the cyanobacterial tumor-promoting microcystins. *FEBS Letters* 349(3): 319–323.

Runnegar, M. T., C. Xie, et al. (2002). In vitro hepatotoxicity of the cyanobacterial alkaloid cylindrospermopsin and related synthetic analogues. *Toxicological Science* 67(1): 81–7.

Schembri, M. A., B. A. Neilan, et al. (2001). Identification of genes implicated in toxin production in the cyanobacterium *Cylindrospermopsis raciborskii. Environmental Toxicology* 16(5): 413–421.

Shalev-Alon, G., A. Sukenik, et al. (2002). A novel gene encoding amidinotransferase in the cylindrospermopsin producing cyanobacterium *Aphanizomenon ovalisporum. FEMS Microbiology Letters* 209(2002): 87–91.

Shalev-Alon, G., A. Sukenik, et al. (2004). Regulation of the expression of genes encoding amidinotransferase and polyketide synthase probably involved in the biosynthesis of cylindrospermosin in *Aphanizomenon ovalisporum*. Poster presentation at the Sixth International Conference on Toxic Cyanobacteria, Bergen, Norway.

Sielaff, H., E. Dittmann, et al. (2003). The mcyF gene of the microcystin biosynthetic gene cluster from *Microcystis aeruginosa* encodes an aspartate racemase. *Biochemical Journal* 373: 909–916.

Sin, N. and J. Kallmerten (1996). Synthesis of (2S,3S,8S,9S)-ADDA from D-glucose. *Tetrahedron Letters* 37(32): 5645–5648.

Sivonen, K. and G. Jones (1999). Cyanobacterial toxins. *Toxic Cyanobacteria in Water. A Guide to Their Public Health Consequences, Monitoring and Management.* I. Chorus and J. Bartram, eds. London, E & FN Spon (on behalf of WHO): 41–111.

Stotts, R. R., M. Namikoshi, et al. (1993). Structural modifications imparting reduced toxicity in microcystins from *Microcystis* spp. *Toxicon* 31: 783–789.

Taylor, C., R. J. Quinn, et al. (1996). Synthesis of cyclic peptides modelled on the microcystin and nodularin rings. *Bioorganic and Medicinal Chemistry Letters* 6(17): 2107–2112.

Tillett, D., E. Dittmann, et al. (2000). Structural organization of microcystin biosynthesis in *Microcystis aeruginosa* PCC7806: an integrated peptide-polyketide synthetase system. *Chemistry and Biology* 2000(7): 753–764.

Trogen, G., A. Annila, et al. (1996). Conformational studies of microcystin-LR using NMR spectroscopy and molecular dynamics calculations. *Biochemistry* 35(10): 3197–3205.

Trogen, G.-B., U. Edlund, et al. (1998). The solution NMR structure of a blue-green algae hepatotoxin, microcystin-RR — A comparison with the structure of microcystin-LR. *European Journal of Biochemistry* 258: 301–312.

Xie, C., M. T. C. Runnegar, et al. (2000). Total synthesis of (+/–)-cylindrospermopsin. *Journal of the American Chemical Society* 122(21): 517–524.

Zetterstrom, M., L. Trogen, et al. (1995). Synthesis of an n-methyldehydroalanine-containing fragment of microcystin by combination of solid phase peptide synthesis and beta-elimination in solution. *Acta Chemica Scandinavica* 49: 696–700.

4 Cyanobacterial Ecology

Cyanobacteria are found throughout the great variety of ecosystems on the surface of the planet. They are abundant colonizers of the most dessicated environments, ranging from the cold deserts of the Antarctic continent to the stones and sand of the world's hot deserts. They form a predominant component of surface crusts in the Taylor Valley of Antarctica, where the mean July temperature was –32.2°C averaged over 3 years. They also occur under rocks in stony desert, with surface temperatures reaching 60°C and above (Wynn-Williams 2000).

The major locations of cyanobacteria are, however, in moist or aquatic environments. But even in these more favorable environments, cyanobacteria survive in the most extreme conditions. Much research has investigated cyanobacteria in hot springs, where spectacular mats of cyanobacteria occur, changing their appearance as the temperature falls away from the geothermal source. The genera *Synechococcus*, *Phormidium*, and *Oscillatoria* are found in hot springs at temperatures from 74 to 55°C (Ward and Castenholz 2000).

Cyanobacterial species are also capable of abundant growth over a very wide range of salinity, hence osmotic pressure regimes. While the major focus of this volume is freshwater environments, saline and hypersaline aquatic environments also provide favorable growth conditions. An example of this is the stromatolites of Shark Bay in Western Australia, found in a shallow hypersaline bay in which the normal grazing organisms are suppressed by the salinity (Logan 1961). Cyanobacteria are common in salt lakes and can survive in brine solutions of 3- to 4-M concentration (Reed, Chudek et al. 1984). Cyanobacterial species can also survive and opportunistically proliferate in aquatic environments of widely oscillating salinity, as in the Peel-Harvey estuary in Western Australia, where winter rainfall results in almost freshwater conditions, followed by evaporation and increased salinity. *Nodularia spumigena* flourishes there until the water becomes hypersaline in late summer, when the filaments die (Huber 1986; Lukatelich and McComb 1986).

In the open ocean, cyanobacteria play a major role in carbon and nitrogen fixation, particularly in low-nutrient areas (oligotrophic), where they may occupy more than 50% of the biomass (Paerl 2000). The filamentous cyanobacterium *Trichodesmium* forms red tides in open and coastal oceans, in which the appearance of the water is reddish-brown and filled with sawdust-like aggregates of filaments (Gallon, Jones et al. 1996).

4.1 CYANOBACTERIA IN FRESHWATER

The ecology of freshwater cyanobacteria has been extensively studied, both from the viewpoint of their contribution to the energy and nutrient dynamics of aquatic biological systems and from the viewpoint of their growth to nuisance or even health hazard proportions in lakes and rivers. In this chapter ecological conditions favoring cyanobacterial proliferation are discussed, as the most relevant to drinking and recreational water issues.

Cyanobacterial cell numbers in water bodies vary seasonally, as a consequence of changes in water temperature and irradiance as well as meteorological conditions and nutrient supply. Peak cyanobacterial concentrations in lakes and rivers usually occur in mid to late summer, in both subtropical and temperate latitudes, when water temperatures in the surface layers reach a maximum. A sequence of dominant organisms is frequently observed in deeper lakes and reservoirs, commencing with diatoms in spring. These are followed by green algae if nutrient concentrations are sufficiently high, then, as nutrient concentrations fall and surface temperature rises, cyanobacteria become dominant (Reynolds 1984; Oliver and Ganf 2000). The major underlying process is that of stratification of the water body, as the upper layers become warmer and a distinct temperature transition develops with depth, where temperature and oxygen saturation fall sharply. The warmer water above this level is termed the *epilimnion*; the lower, colder water the *hypolimnion*. The partially mixed layer between is termed the *metalimnion*.

The larger the temperature gradient between upper and lower levels, the more stable the layers and the less mixing. A temperature profile of a deep stratified lake in northern Europe in summer is shown in Figure 4.1.The large temperature difference between the epilimnion at above 17°C down to 6 m depth and the hypolimnion at 10°C from 14 to 20 m depth shows a high level of stable stratification.

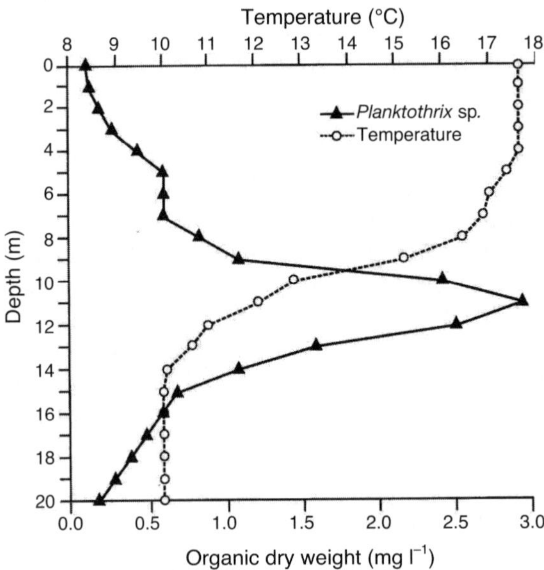

FIGURE 4.1 Vertical distribution of *Planktothrix* sp. in a deep, thermally stratified meso-oligotrophic lake during bloom conditions. (From Mur, Skulberg et al. 1999. With permission.)

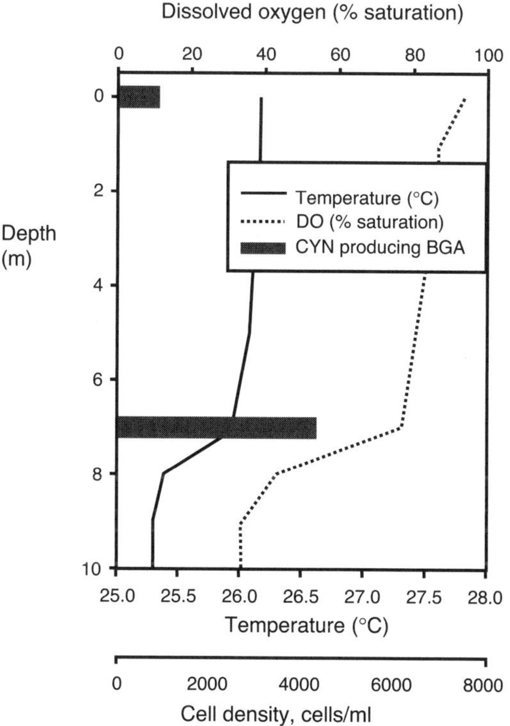

FIGURE 4.2 Temperature, oxygen saturation, and cylindrospermopsin (CYN)-producing cyanobacterial (BGA) cell concentration (cell/ml). (From Fabbro and Andersen 2003. With permission.)

As a result of oxygen consumption by microorganisms in the organically enriched layers of the sediment, the hypolimnion becomes oxygen-depleted and can become anaerobic in deeper lakes and in organically enriched shallow stratified lakes, as shown in Figure 4.2 (Fabbro and Andersen 2003). This anaerobic environment results in mobilization of nutrients from sediments, which diffuse into the adjacent water, providing opportunity for cyanobacterial growth, as discussed later.

As the plankton in the epilimnion deplete the nutrients and the water stability allows the heavier plankton to sediment down away from light, cyanobacteria are progressively advantaged.

4.2 LIGHT

Light availability and its converse turbidity have a strong influence on the species of cyanobacteria that predominate and the depth at which they occur. In clear cool lakes in northern Europe and Scandinavia, for example, the toxic filamentous cyanobacterium *Planktothrix agardhii* can form dense bands of filaments at the metalimnion, where there is sufficient light intensity for growth and also nutrient enrichment from the deeper layer. These bands can form at depths of 12 m in a sufficiently clear lake in summer, occupying the full depth of the metalimnion, as illustrated in Figure 4.1. The capability of cyanobacteria to grow at depth is determined by the turbidity or clarity of the water; this is quantified by the term *euphotic zone*. This is the depth at which photosynthesis can occur and is arbitrarily defined as the depth at which 1% of the surface light intensity can be detected (Mur, Skulberg et al. 1999). In very clear lakes, such as that illustrated in Figure 4.1, the euphotic zone extends to at least 12 m, allowing effective photosynthesis to occur at considerable depth. The species illustrated, *P. agardhii*, is especially well adapted to low light conditions and hence can grow at depths below those occupied by other cyanobacteria or green algae. As nutrients are commonly at higher concentrations in the hypolimnion, the ability to grow at depth is a substantial advantage when the epilimnion is nutrient depleted.

In deeper rivers, lakes, and reservoirs in subtropical environments, which also show marked summer stratification, the warmer-water cyanobacterium *Cylindrospermopsis raciborskii* similarly forms dense layers of filaments at depths down to the bottom of the euphotic zone, as seen in Figure 4.2. Both of these species tolerate low light intensities and may be associated with other finely filamentous cyanobacteria, such as *Limnothrix redekei*, *Pseudanabaena limnetica*, and *Planktolyngbya subtilis* with considerable shade tolerance (McGregor and Fabbro 2000).

Other cyanobacteria predominate at higher light intensities and in shallow, mixed lakes. *Microcystis aeruginosa* occurs commonly in both stratified and in mixed lakes, with the most rapid growth under stratified conditions. After turbulent mixing, cyanobacterial photosynthesis was temporarily inhibited at the surface but recovered within two or more calm days (Kohler 1992). Adaptation to light intensity occurs as a result of changes in the photosystem pigments, including carotene content. The proportion of chlorophyll-a to the phycobiliproteins, the size of the phycobilisome (Wyman and Fay 1986), as well as the number of photosynthetic units (Falkowski and LaRoche 1991) vary with light intensity. Phycobilisomes are rounded assemblies of accessory pigments of photosynthesis that project from the thylakoid membranes in cyanobacteria. They contain phycocyanin, allophycocyanin (both blue-green), and phycoerythrin (red) and act as light trapping systems. Because they can utilize green light and operate at low light intensities, they provide an advantage to cyanobacteria in eutrophic lakes containing high concentrations of eukaryotic phytoplankton, in which the light at depth is green (Oliver and Ganf 2000).

In eutrophic lakes with high biological productivity and with lower light penetration, other species of cyanobacteria will form bands just below the surface; for example, the filamentous nitrogen-fixing genus *Anabaena*. This prefers higher light intensity than *Planktothrix* and forms dense bands in the water of stratified lakes at shallow depths, as shown in Figure 4.3. Under these conditions, the euphotic depth will be shallower and will often be above the metalimnion, so that normal mixing processes in the upper layers of the water will carry phytoplankton below the depth at which they can photosynthesize (Mur, Skulberg et al. 1999). This can advantage cyanobacteria, many species of which can position themselves in the water column by variations in buoyancy.

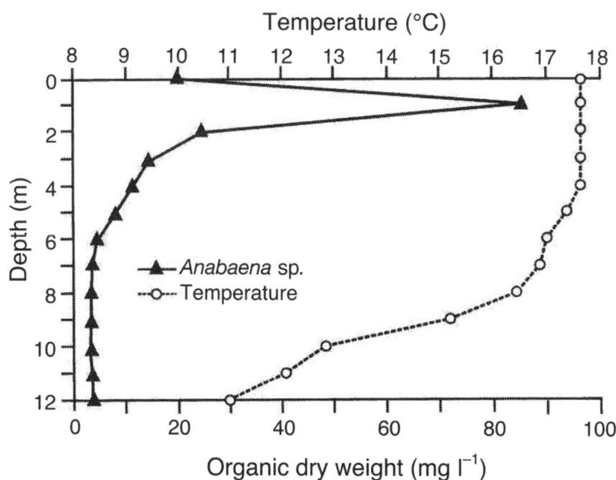

FIGURE 4.3 Vertical distribution of *Anabaena* sp. in a thermally stratified eutrophic lake during bloom conditions. (From Mur, Skulberg et al. 1999. With permission.)

4.3 BUOYANCY

The capacity of cyanobacteria to form bands within the water column reflects their variable buoyancy. This is accomplished by the presence of gas vacuoles and variable cell density. It confers a substantial ecological advantage to the organisms, as they can congregate at favorable levels in the water of stratified lakes and also move up and down in the water column to maximize photosynthesis in the surface layers and nutrient uptake in the deep layers (Ganf and Oliver 1982).

The gas vacuoles are filled with an array of gas vesicles, which have a very low density compared with cytoplasm; as a result, the vacuole has a density approximately one-fourth that of water. The vesicles have a rigid wall and are freely permeable to gases but not liquids. Gas vesicles have an internal pressure related to the atmospheric pressure but are subject to hydrostatic pressure, which increases with depth, as well as the turgor pressure of the cell (Oliver and Ganf 2000). The regulation of buoyancy operates largely through the density of the cell constituents, particularly the carbohydrate and protein content of the cell. Therefore the availability of light, carbon dioxide, nitrogen, and phosphorus will affect cell growth and cell density. Under conditions of abundant nutrients, the buoyancy will be determined by the balance between growth rate and illumination (Oliver and Ganf 2000). If light is limiting, cell growth will deplete carbohydrate stores and reduce cell density, thus increasing buoyancy and helping the cell to rise in the water column into higher light availability. Conversely, if nutrients such as nitrogen and phosphorus are limiting but light availability is high, then the cells will accumulate carbohydrate stores but be less able to grow because of lack of nutrients. They will then become more dense and sink into the nutrient-enriched lower layers of the lake.

A number of toxic species of cyanobacteria will form floating scums under calm, warm weather conditions. One of the most frequent scum-forming species is *M. aeruginosa*, which contains gas vacuoles and can achieve colony floating rates up to 250 m/day (Oliver and Ganf 2000). Under calm conditions, this rapid upward movement will outweigh small wind-driven mixing at the surface and provide the organism with increased light, hence accelerating potential growth. These scums drift downwind and accumulate on shorelines and dam walls. They become a significant problem in recreational water use and in drinking water intakes, which are discussed in Chapter 8 and Chapter 11.

4.4 NUTRIENTS

4.4.1 PHOSPHORUS

The availability of phosphorus to cyanobacteria has a very strong influence on growth rate. If soluble phosphorus concentrations in water drop much below 10 μg/L, the population growth of cyanobacterial cells is likely to be nutrient limited (Cooke, Welch et al. 1993). At the low end of the range of soluble phosphorus concentrations in lakes, below 10 to 20 μg/L in deep lakes and below 50 to 100 μg/L in shallow lakes, the cyanobacterial cell numbers relate linearly to phosphorus concentration (Sas 1989). However, the affinity of cyanobacteria for phosphorus is higher than that of photosynthetic green algae, so that cyanobacteria can outcompete green algae under conditions of phosphorus limitation (Mur, Skulberg et al. 1999). Cyanobacteria have two other advantages in phosphorus metabolism: they can store polyphosphate sufficient for two to four cell divisions in phosphate deficient water and can use their ability to migrate vertically to descend to a depth where phosphate availability is higher. They can even sink down to the sediment surface, where, under the anoxic conditions commonly encountered in eutrophic lakes, otherwise insoluble phosphorus is mobilized into soluble, bioavailable compounds that diffuse into the hypolimnion. Thus the cyanobacteria can recharge their phosphate stores and then ascend into higher light intensities suitable for growth and division. *Microcystis* is one of the cyanobacteria that can well utilize these advantages, with a large capacity for phosphorus storage and high and variable buoyancy (Ganf and Oliver 1982; Kromkamp, Van Den Heuvel et al. 1989).

Much research has been carried out on the relationship between nutrients, phytoplankton growth, and relative abundance of cyanobacteria (Oliver and Ganf 2000). One aspect of this is the influence of the ratio between inorganic nitrogen (ammonia, nitrite, nitrate) and available phosphorus in the water on the relative dominance of diatoms, green algae, and cyanobacteria. In general, the majority of studies indicate that the higher the ratio toward nitrogen excess, the more likely diatoms or green algae will dominate; and the lower the ratio the more likely that cyanobacteria will dominate.

This is discussed at more length in Chapter 11, on mitigation of cyanobacterial growth in lakes and reservoirs.

4.4.2 NITROGEN

Many species of cyanobacteria can "fix" dissolved nitrogen gas into ammonium ions, making them independent of dissolved inorganic nitrogen. The majority of nitrogen-fixing species have specialized thick walled cells, the heterocysts, which have a larger diameter than the normal photosynthetic cell. While these cyanobacterial species will also preferentially utilize ammonium ions, nitrite, or nitrate in water, they are in direct nutrient competition with diatoms and green algae, which will outgrow the cyanobacteria. Under favorable nutrient conditions and adequate light, green algae will grow at about double the rate of cyanobacteria. As the cell density rises, light is shaded, favoring cyanobacteria (Mur, Skulberg et al. 1999); when the available inorganic nitrogen is depleted, the cyanobacteria that fix nitrogen will have a substantial ecological advantage (Tandeau de Marsac and Houmard 1993). Thus cyanobacterial proliferation can occur in highly nitrogen-depleted waters of the surface layers, where light is abundant. The energetics of nitrogen fixation in heterocysts — in which the enzyme system nitrogenase converts dinitrogen into two ammonium groups — is well understood (Berman-Frank, Lundgren et al. 2003). These are immediately incorporated into glutamine, which transfers amino groups into a wide range of reactions of intermediary metabolism. The process is highly energy-demanding, requires photosynthetic electron transport, and operates in competition with carbon fixation (Tandeau de Marsac and Houmard 1993).

4.5 DISTRIBUTION OF *CYLINDROSPERMOPSIS RACIBORSKII* (NOSTOCALES)

This organism is the most widely distributed source of the toxin cylindrospermopsin in drinking and recreational waters worldwide; hence an understanding of its distribution and ecology is of particular relevance to human health.

Padisak (1997) has provided a through review of the worldwide distribution of *C. raciborskii* and a discussion of the likely origin and migration of populations of the organism. While this cyanobacterium does not normally form surface scums of the type seen in *Microcystis* or *Anabaena* blooms, which draw attention to the presence of the cyanobacteria, substantial populations of the organism occur widely in both lakes and rivers. The earlier accounts of the species were from tropical and subtropical regions, particularly Indonesia (Geitler and Ruttner 1936), Philippines, Malaysia, Burma, India, and Pakistan, and largely in ponds, lakes, and reservoirs.

C. raciborskii has been identified across central Asia and Europe, including the Caspian Sea and southern Russia (Padisak 1997). In particular, *C. raciborskii* has been studied in Lake Balaton in Hungary, where it formed heavy blooms in summer from the 1980s to the mid-1990s (Padisak 1992). It has recently been described in a number of lakes in Germany, where it is also a summer occurrence in shallow eutrophic lakes (Weidner and Nixdorf 1997). Since the phytoplankton in lakes in this area have been studied for a considerable time, it appears that *C. raciborskii* is a new invasive species rather than simply a species that has hitherto been unrecorded. One anomaly that has arisen with *C. raciborskii* in Germany is that the toxin cylindrospermopsin has been identified in lakes containing this cyanobacterial species, but cultured strains show no ability to produce the toxin (Fastner, Heinze et al. 2003).

In the central African Great Lakes, *C. raciborskii* has frequently been recorded, both in shallow lakes such as Lake George in Uganda (Ganf 1974) and in very deep, stratifying lakes such as Lake Victoria, Uganda (Komarek and Kling 1991).

In South America, most identifications of *C. raciborskii* have been in Brazil, which has cyanobacterial monitoring in drinking water supply reservoirs as a result of persistent problems from eutrophication. The organism has been found from the far south, in Lagoa dos Patos, a very large natural coastal lake, to the main water supply reservoir serving Brasilia, the capital city. This reservoir has substantial continuous blooms of *C. raciborskii,* which proved difficult to identify, as the characteristic cone-shaped heterocysts were absent. This was attributed to the availability of ammonia in the water body, especially at depth (Branco and Senna 1994). *C. raciborskii* has been reported in reservoirs in Venezuela, Nicaragua, and Cuba (see Padisak 1997 for details).

In the U.S., *C. raciborskii* was first identified in several small lakes in Minnesota in 1966 to 1969 (Hill 1970), at a similar latitude (40° north) to the report of this species in Greece (Hindak and Moustaka 1988). It has also been recorded in Ohio, Michigan, Illinois, Indiana, Texas, and Mexico (see www.in.gov/dnr/fishwild/fish/cylind.htm and Padisak 1997 for details). Not surprisingly, the most abundant location in the U.S. for *C. raciborskii* is the eutrophic lakes and rivers of the subtropical Florida region, which support very large populations of the organism. Some of these lakes are used for the existing drinking water supply and other locations are in rivers, which are likely to be needed for future drinking water supply (Chapman and Schelske 1997; Williams, Burns et al. 2001). Adjacent regions of the U.S. have not yet, as far as published information can indicate, been investigated for this species. Since it occurs in Cuba, Florida, Mexico, and subtropical regions elsewhere in the world, it can be expected to exist throughout the higher rainfall areas of the southern U.S.

4.5.1 In Australia

The distribution of *C. raciborskii* has been extensively investigated in the Eastern States of Australia, in which the great majority of the human population are located. This organism was first brought to prominence in Australia by a substantial human poisoning episode through a bloom of the organism in a drinking water supply reservoir (Byth 1980; Bourke, Hawes et al. 1983; Hawkins, Runnegar et al. 1985).

The organism occurs commonly in drinking water supply sources in the subtropical and tropical regions of Australia, where it can frequently reach bloom proportions. In a study of a series of 47 reservoirs and weir pools on rivers (which are widely used as drinking water sources), 70% contained *C. raciborskii* (McGregor and Fabbro 2000). One of these reservoirs showed year-round dominance, with a median biomass of 8496 mm^3 L. Cell counts of 600,000 cells per milliliter averaged over the top 5-m depth were reported in the summer of 1998, when a rare yellow-green surface slick of the organisms occurred at the shoreline. More commonly the water blooms are below the surface and do not form slicks or scums (Figure 4.4) (Fabbro 1999). In the majority of water sources the blooms were seasonal, peaking in late summer and early autumn. The morphology of the filaments included straight,

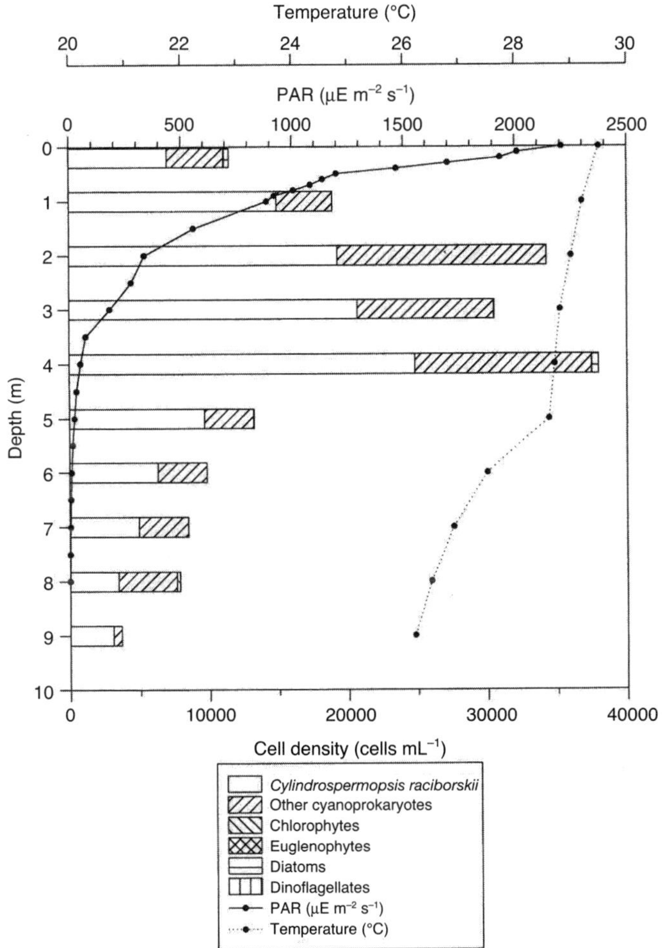

FIGURE 4.4 Distribution of *Cylindrospermopsis raciborskii* and other cyanobacteria down the depth profile of a tropical water storage. Peak concentrations of cyanobacteria are 2 to 4 m below the surface and extend down to the zone of minimal photosynthetically effective radiation (PAR). (From Fabbro 1999. With permission.)

sigmoid, and spiral forms (Figure 4.5). Populations occurred as mixtures of all three forms, or in transition from one form to another throughout the year. The control of the variations in morphology is not clear (McGregor and Fabbro 2000). The ecology of this organism with respect to forming high cell densities is discussed in more detail later in this chapter.

In the more southern rivers and lakes in Australia, the organism is endemic but normally at low cell densities. It has been reported at 72 sampling stations in the Murray Darling river system, including the main river channel, swamps, and lakes (Baker, Humpage et al. 1993). It has been described in a shallow off-river water

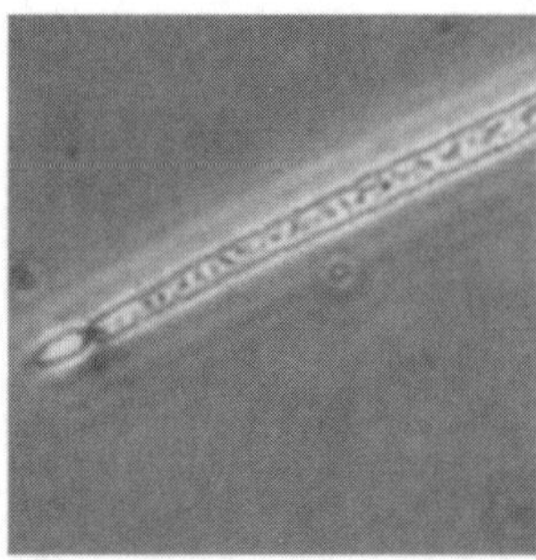
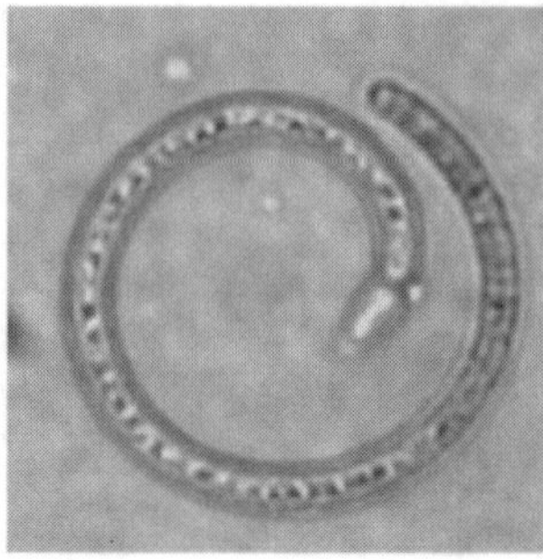

FIGURE 4.5 (See color insert following page 146.) Straight and coiled forms of *Cylindrospermopsis raciborskii*. (From Fabbro and Andersen 2003. With permission.)

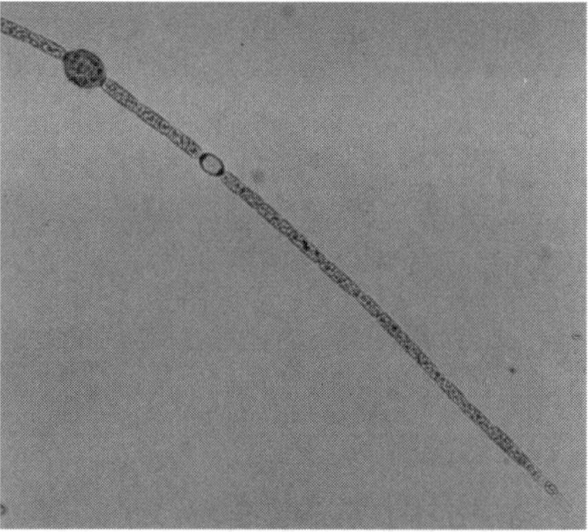

FIGURE 4.6 (See color insert.) *Aphanizomenon ovalisporum*. (From Peter Baker, Australian Centre for Water Quality Research. With permission.)

storage site, where it formed part of a succession of cyanobacterial species all in bloom proportions from early summer to late autumn (Bowling 1994).

The phylogeography of the cyanobacterium shows an interesting pattern, with the strains isolated from Brazil and the U.S. forming one group; the strains from Germany, Hungary, and Portugal forming another; and the strains from Australia the third (Neilan, Saker et al. 2003). This does not directly tie into the differences in toxins present, as the U.S. and Australian strains appear to have cylindrospermopsin alone, the Brazilian strains appear to have saxitoxins and/or cylindrospermopsin, and the European strains have an unknown neurotoxin and possibly cylindrospermopsin.

From this review of the global distribution of the species, it is apparent that it is distributed across the temperate, subtropical, and tropical world. The extent to which the organism forms water blooms that cause it to become apparent or even to pose a hazard can be assessed from a study of the ecology of the organism in locations where it recurs.

4.6 ECOLOGY OF *CYLINDROSPERMOPSIS RACIBORSKII*

Because of the intensity of research on *C. raciborskii* in Australia over the last decade, a considerable body of data is available on the factors affecting the growth and persistence of the species in culture and in natural waters. One of the reservoirs from which a considerable sequence of data has been published is North Pine Dam, which is a major drinking water source for the city of Brisbane, Queensland, Australia (Harris and Baxter 1996).

The location of the dam is 30 km north of the city of Brisbane, in the subtropics (lat 27°16S, long 152° 6E). The volume of the lake is 203×10^6 m^3, with a maximum depth of 35.1 m. Mean summer temperature is 25°C, winter temperature 15°C. The rainfall is extremely erratic, with major flood events interspersed with prolonged droughts. The inflow from the catchment is therefore very variable and has range from 5 to 400×10^6 m^3 per annum. As a consequence, the residence time of water in the dam on an annual basis over the period 1978 to 1994 ranged from 0.58 to 10.82 years (Harris and Baxter 1996).

The lake is highly stratified from spring through summer to autumn, with severely anoxic conditions in the hypolimnion to the extent that 40% of the lake volume is anoxic in midsummer. Storms in summer can wash high nutrient loads into the lake, with observations of 150 to 900 µg/L of phosphate in inflowing streams. Over 16 years of observation, irregular peaks of *C. raciborskii* were seen in summer, with maxima in times of low rainfall, declining dam level, and thermal stability. *Microcystis*, by contrast, was maximal before and after *C. raciborskii*, when the water stratification was less stable.

Cyanobacteria were dominant in the lake in seven major water blooms, with *C. raciborskii* concentrations up to hundreds of thousands of cells per milliliter.

Further north, in central Queensland, at latitude 20°S, the Fitzroy River has been intensively studied. The river is impounded by a barrage in order to supply drinking water to the city of Rockhampton. This impoundment resulted in a lake 50 km long, with a maximum depth of 12 m and a volume of 60×10^6 m^3 (Bormans, Ford et al. 2000). During December (summer) 1992, an increasing *C. raciborskii* bloom of approximately 25,000 cells per milliliter at the drinking water intake was recorded. A depth profile of the organism showed steadily rising cell density between the surface and 4 m, with a rapid fall in cell concentration at 5-m depth in water with a euphotic depth of 4.9 m. The orthophosphate concentration at this time reached a maximum at 6-m depth with none detectable above 3 m; oxidized nitrogen was maximum at 7 m, with none detectable above 6 m; and ammonia was maximal at 8 m, with none detectable above this level. Below 8 m, oxygen saturation was less than 10% (Fabbro and Duivenvoorden 1996). Prior to the peak of the bloom in early January, there was a strong thermocline present at 6 to 7 m, which deepened as a result of wind mixing to 8 to 9 m during the bloom. The surface water temperature was close to 29°C and the bottom temperature 24°C at the time of the maximum thermocline; pH ranged from 8.1 to 9.0.

From these data it is apparent that the stable conditions had enabled the cyanobacterium to progressively grow within the euphotic zone, using nutrients —

particularly phosphate — from the metalimnion. Once the wind conditions caused deep mixing, enhanced nutrients in the euphotic zone accelerated growth, rising to a water bloom with a peak cell concentration at 1-m depth of 40,000 cells per milliliter and falling to 20,000 cells per milliliter at 5-m depth. This was followed by a flood, which washed the cyanobacteria out of the impoundment (Fabbro and Duivenvoorden 1996).

C. raciborskii is a cyanobacterial species that forms akinetes, a large vegetative resting stage, which can carry the organism over the winter (see http//:www.unc.edu/~moisander/image3.htm for a picture). This is of particular importance in temperate regions with low winter temperatures, but it also occurs at the end of high-density blooms in more tropical waters. In the Fitzroy River, during the maximum growth phase of the cells, no akinetes were formed. During the time of peak cell concentration, there were approximately 10 akinetes per trichome, with an indication of even higher numbers on trichomes that were sedimenting down the depth profile (Fabbro and Duivenvoorden 1996).

A second detailed examination of this river system and cyanobacterial growth is given by Bormans (Bormans, Ford et al. 2000), who identified *Aphanizomenon*, *Aphanocapsa*, *Cylindrospermopsis*, *Limnothrix*, and *Planktolyngbya* in mixed cyanobacterial blooms adjacent to the drinking water intake. *C. raciborskii* was again recorded at peak cell densities of 30,000 cells per milliliter in depth-integrated water samples.

In fully tropical lakes, such as Paranoa Lake in Brazil at latitude 15°S, *C. raciborskii* will form high cell concentrations throughout the year in the permanently warm water (Branco and Senna 1994).

In shallow mixed temperate lakes, the growth of *C. raciborskii* is very dependent on warm summer temperatures. In general, the organism will grow to bloom proportions only in water above 25°C (Padisak 1997). In Lake Balaton, Hungary, at latitude 46°N, *C. raciborskii* builds up to high cell densities only in exceptionally warm summers. Studies of akinete germination have demonstrated that temperatures between 22 and 23.5°C are required for laboratory germination of akinetes collected from sediment at Lake Balaton (Gorzo 1987), which is the same temperature as the sediment/water interface during bloom recruitment in the Fitzroy River. The strain of the species isolated from Lake Balaton grew fastest at 30°C in culture (Shafic, Voros et al. 1997), which also corresponds with the optimum temperature range for Australian strains of 27 to 30°C in lakes (Fabbro and Duivenvoorden 1996; Saker and Griffiths 2000).

There is, however, evidence of cold adaptation occurring in the species, which implies that it may become progressively more abundant in cool, temperate locations. In the shallow urban Lake Alte Donau, in Austria, at latitude 48°N, maximum cell density occurred at 15 to 18°C (Dokulil and Mayer 1996). The organism has also occurred in Germany at latitude 52°N, though not in bloom proportions, in a number of shallow and deep lakes that freeze over in winter (Weidner and Nixdorf 1997).

Studies of nutrient requirements for *C. raciborskii* have shown that the species is highly adaptable. In waters containing ammonia, this is the preferred nitrogen source, and the trichomes generally lack heterocysts (Padisak and Istvanovics 1997). However, under conditions of nitrogen limitation (low ratios of nitrogen to phosphorus), the number of heterocysts increases and the organism has a resulting growth advantage over non-nitrogen-fixing species. Phosphorus requirements of the organism, as in the case of all cyanobacteria, can limit growth of cells in the epilimnion. However *C. raciborskii* has a marked ability to absorb phosphorus at low concentrations (Istvanovics, Somlyody et al. 2002) and can migrate to deeper and more nutrient-enriched layers through buoyancy changes (Branco and Senna 1994). *C. raciborskii* appears to be benefited by relatively low nutrient concentrations (Briand, Robillot et al. 2002). Padisak (1997) reports that phosphorus storage in the cells can reach 12 to 24 times higher than the base concentration, allowing up to 5 consecutive cell divisions using stored phosphorus. This phosphorus storage ability, together with variable buoyancy, supports doubling rates in natural ecosystems from 2.9 to 7 days (Padisak and Istvanovics 1997). In culture, growth under optimum conditions was appreciably faster, at 0.9 to 1.2 divisions per day, peaking at 30ºC (Saker and Griffiths 2000).

4.7 CYLINDROSPERMOPSIN PRODUCTION BY *CYLINDROSPERMOPSIS RACIBORSKII*

As a consequence of the human poisoning episode that took place at Palm Island, Queensland, Australia, in 1979 (Byth 1980), the appearance of cylindrospermopsin, the *C. raciborskii* toxin, in Australian drinking water supplies has been increasingly monitored. The highest consistent cylindrospermopsin concentration reported was in a water storage at Hervey Bay, Queensland, latitude 25ºS. In the midsummer period of early November to early January 1997, the peak toxin concentration in the reservoir reached 92 μg/L at a peak cell concentration of 2×10^6 cells per milliliter of *C. raciborskii*. For the whole period from November 7 to December 23, the lowest toxin concentration was 36 μg/L, with a range of cell concentrations from 8×10^4 upward to the peak of 2×10^6. One feature of cylindrospermopsin distribution during water blooms of *C. raciborskii* is that a substantial proportion of the toxin is extracellular in the free water phase. In this reservoir at the time of peak toxin concentration, 68% was free in the water and only 32% in the cells. Prior to this peak, as cell numbers increased, and later in the last 3 weeks of high toxin concentrations, as cell numbers dropped, more than 90% of toxin was free in the water (Chiswell, Shaw et al. 1999). This is in marked contrast to microcystin, the toxin from *Microcystis*, which is almost entirely intracellular until the cells die (Welker, Steinberg et al. 2001).

In a wide survey of *C. raciborskii* and cylindrospermopsin concentrations in 47 water storage sites in Queensland, McGregor and Fabbro (2000) identified the cyanobacterium in 70% of the storage sites investigated. Cell concentrations regularly exceeded 6×10^5 cells per milliliter in a number of reservoirs in midsummer. The reservoir containing the highest average cell biomass of *C. raciborskii* over a 20-month period of monitoring was Cania Dam, where the mean toxin concentration over the period was 18.9 μg/L, with a peak value of approximately 46 μg/L. The surface water temperature reached 28°C in summer with stable stratification, with a mean pH of 7.68. The water retention time was 10.8 years, as this 40-m deep reservoir is located in a low-rainfall zone.

The mean cylindrospermopsin concentration for the 14 reservoirs in which toxin was analyzed was 3.4 μg/mL. In the storage sites that contained above 15,000 cells per milliliter of *C. raciborskii,* the toxin was widely detectable. Examination of the overall distribution of cell and toxin concentrations indicated that toxin concentrations of 1 μg/L could be expected at approximately 20,000 cyanobacterial cells per milliliter of lake water in Queensland, although the relationship between cell density and toxin concentration was widely variable (McGregor and Fabbro 2000). This toxin concentration of 1 μg/L has been recommended as the "Guideline Value" for cylindrospermopsin in drinking water, as is discussed in Chapter 6 and Chapter 8 (Humpage and Falconer 2003).

C. raciborskii producing cylindrospermopsin has been identified in drinking water reservoirs and in finished drinking water in Florida in the U.S. The organism is widely distributed in Florida (Chapman and Schelske 1997), and toxin concentrations of up to 90 μg/L in drinking water have been reported (USEPA 2001).

Studies of cylindrospermopsin production in cultured *C. raciborskii* strains have further clarified the requirements for toxin biosynthesis. Saker and Griffiths (2000) examined seven strains of *C. raciborskii* collected from four drinking water reservoirs, two aquaculture ponds and one farm dam located between 18 and 21°S in Queensland. Cultures grown at temperatures between 20 and 35°C showed a general increase in growth rate to 30°C, but an inverse relationship between temperature and toxin content, expressed as toxin percent of freeze-dried weight of cells, with effectively no toxin produced at 35°C and the highest cell contents at 20°C. They also noted that the strains producing the highest levels of toxin showed progressive toxin leakage into the culture medium as the cultures approached the stationary phase of growth. One of the strains, isolated from a farm dam after cattle poisoning, had 0.15% toxin in the freeze-dried cells when isolated but essentially none in culture 2 years later (Saker and Griffiths 2000).

Examination of the nitrogen requirement for cylindrospermopsin biosynthesis in a highly toxic strain showed that cells supplied with nitrate grew roughly three times as fast as cell cultures without nitrate, but the cylindrospermopsin production was matched to cell division rate (Hawkins, Putt et al. 2001). The maximum yield of cylindrospermopsin (cell and medium content) was obtained from cultures grown to stationary phase at 55 days, which exceeded 2500 μg/L. As cell density increased, the proportion of extracellular toxin rose from approximately 10 to 50% at 2×10^7 cells per milliliter.

Supply of ammonia in the growth medium resulted in the highest growth rates in a group of similar *C. raciborskii* strains in culture, as compared to culture with nitrate or no added nitrogen. The ammonia-supplemented cells were longer, and no terminal heterocysts formed on the trichomes (Saker and Neilan 2001). In this study, cylindrospermopsin concentration in the cells was highest in cultures that had no nitrogen supplementation, and these had the slowest growth rate.

4.8 CYLINDROSPERMOPSIN PRODUCTION BY OTHER CYANOBACTERIAL SPECIES

An intensive cyanobacterial monitoring program on Lake Kinneret in Israel in 1994 identified the organism responsible for an exceptional autumn water bloom peaking at 4000 trichomes per milliliter (approximately 4×10^5 cells per milliliter), evenly distributed through the epilimnion. The organism was *Aphanizomenon ovalisporum* (Nostocaceae) (Figure 4.6), and toxicity testing in mice showed considerable toxicity. The genus *Aphanizomenon* had been earlier identified as neurotoxic in the U.S. (Mahmood and Carmichael 1986), but the symptoms indicated toxicity to the mouse liver, kidneys, lungs, and small intestine. Chemical isolation and identification demonstrated that cylindrospermopsin was the toxin present (Banker, Carmeli et al. 1997). Lake Kinneret is the largest freshwater source in Israel and provides 30% of the country's water requirements, so the appearance of a new toxic species was of concern for the management policy at the lake.

Since that time, toxic blooms of *A. ovalisporum* have been identified in several newly constructed shallow waterways in coastal Queensland housing developments (Shaw, Sukenik et al. 1999). The organism formed brownish scums up to 2 cm thick on the surface of one lake in two consecutive summers. The landowner dosed the lake in February with copper sulfate, completely eliminating the cyanobacteria for the remainder of that year. Two small ponds contained *A. ovalisporum* at 4700 trichomes per milliliter (approximately 5×10^5 cells per milliliter), with a cylindrospermopsin concentration of up to 16 μg/L in the surface 10 cm (Shaw, Sukenik et al. 1999). As this toxic organism has now been recorded in both the Northern and Southern Hemispheres in locations with hot summers, it can be expected to occur in other regions with Mediterranean climates and also in subtropical locations.

Three other cyanobacterial species have also now been shown to produce cylindrospermopsin. An entirely new species, named *Umezakia natans*, was identified in Lake Mikata, Fukui, Japan. This organism is a member of the Stigonemataceae, with branching trichomes, and is able to form numerous heterocysts and akinetes (Watanabe 1987). Examination of a crude extract of the organism demonstrated hepatotoxicity, and cylindrospermopsin was identified as the toxin present (Harada, Ohtani et al. 1994).

Two other members of the Nostocaceae have also been identified as cylindrospermopsin producers: *Raphidiopsis curvata* and *Anabaena bergii.* The first of these was isolated from a fish pond in China and demonstrated to contain both cylindrospermopsin and deoxycylindrospermopsin (Li, Carmichael et al. 2001). The second was identified from the possession of the gene sequences for cylindrospermopsin production as well as the presence of the toxin from cultured cells originally obtained from the Murray-Darling river system in Australia (Schembri, Neilan et al. 2001). When the techniques for measurement of this toxin are more widely used, it is very likely that other species of cyanobacteria will be identified that produce cylindrospermopsin. An illustration of this point is the recent finding of cylindrospermopsin in two German lakes, from which strains of *C. raciborskii* were isolated that do not produce this toxin (Fastner, Heinze et al. 2003).

4.9 PRODUCTION OF OTHER TOXINS BY *CYLINDROSPERMOPSIS RACIBORSKII*

One of the features of toxin-producing cyanobacteria is that some species produce different toxins in different geographical locations. *C. raciborskii* is no exception, with samples of the organism collected in Brazil containing saxitoxin derivatives — the paralytic shellfish poisons — and samples collected in Europe containing as yet unidentified toxins with hepatotoxicity (Lagos, Onodera et al. 1999; Bernard, Harvey et al. 2003; Saker, Nogueira et al. 2003). Examination of the genetic relatedness of the Australian, Hungarian, German, Portuguese, Floridian, and South American strains of *C. raciborskii* has shown that 99.1% similarity exists in 16S rRNA nucleotide sequences (Neilan, Saker et al. 2003). This excludes any misidentification of species, leaving the question of why different isolates produce different toxins. Use of a more discriminatory DNA fingerprinting technique showed how the different isolates related to each other, with each of the strains from a single country closely related. The North and South American strains genetically grouped away from the European and Australian strains, which were related to each other (Neilan, Saker et al. 2003). The toxins still remain puzzling, with the two major genetic groupings both showing genetic subgroups with different toxins.

4.10 DISTRIBUTION OF *MICROCYSTIS AERUGINOSA*

M. aeruginosa is the cyanobacterial species that has received the most worldwide research attention up to the present time. The reason for this is twofold: it ranges from cold, temperate climates to fully tropical environments and it forms highly toxic scums that have killed thousands of farm livestock.

M. aeruginosa does not occur as a significant species in the polar regions, where filamentous forms predominate (Vincent 2000). The most northerly occurrences of this species are in Scandinavia, at latitudes around 60°N, where it occurs with other *Microcystis* species and a range of filamentous species (Skulberg 1996). In Canada, at latitude 50 to 55°N, the organism occurs commonly in lakes and in the water storage sites of small farms (Kotak, Kenefick et al. 1993). At these latitudes, the lakes and reservoirs are frozen in winter, and the organism survives on the sediments until light and temperature increase in early summer. In the Southern Hemisphere, *M. aeruginosa* occurs in New Zealand at latitude 40 to 45°S (Walsby and McAllister 1987). Between these northern and southern limits, *M. aeruginosa* occurs throughout the temperate, subtropical, and tropical regions of the world.

Extensive water blooms have been reported from South Africa; these have been particularly studied in Hartbeespoort Dam, which is a large warm-water hypertrophic lake near Pretoria, latitude 25°S. The organism is present for 10 months of the year, with peak densities in summer (Scott 1983). Massive rafts of *M. aeruginosa* cells form in calm weather, covering 1 to 2 Ha in area with cell densities up to 1.76×10^9 cells per milliliter at 10-cm depth. Within the scum, the conditions are anaerobic, with high ammonia concentrations and negligible light penetration. The cells within the scum remain viable for 11 weeks, indicating survival in the absence of light or dissolved oxygen (Zohary 1985).

Lake George, a shallow central African lake that lies exactly on the equator, has a permanently high *M. aeruginosa* population, which is assisted by the daily stratification of the lake from solar heating (Oliver and Ganf 2000). Even in arid and semidesert regions of North Africa, the organism occurs in rivers, water storage reservoirs (Mohamed 1998; Mohamed, Carmichael et al. 2003), and aquaculture ponds (Mohamed and Carmichael 2000).

In Brazil, *M. aeruginosa* is commonly present in natural and artificial lakes, forming a permanently high population in eutrophic lakes, which are often used to provide drinking water (Azevedo, Evans et al. 1994; Branco and Senna 1994). It has similarly been recorded in Argentina in a major drinking water supply reservoir (Ame, Diaz et al. 2003) and in the La Plata River in Uruguay (De Leon and Yunes 2001).

In Europe, the most systematic monitoring of cyanobacteria in reservoirs, lakes, and rivers has been carried out in Germany (Chorus 2001). *M. aeruginosa* is widely present, largely associated with *Planktothrix* sp., as a summer bloom. Under calm conditions, freshwater bathing beaches are frequently contaminated, posing a health hazard to users of recreational water.

In the U.S., the water industry has recently surveyed cyanobacterial toxins in drinking water supplies across the whole nation, thereby obtaining an overall assessment of the situation. While *Microcystis* is not the sole source of microcystins in U.S. lakes and water supply reservoirs, it is the predominant source. In this survey, carried out from June 1996 to January 1998, a total of 677 water samples were collected by water utilities and 80% were positive for microcystin when tested. This demonstrated the very wide distribution of toxic cyanobacteria, particularly toxic *Microcystis*, in drinking water sources in the U.S. (Carmichael 2001). In China, *Microcystis* was found in 80% of samples tested, with 95% of these showing toxicity (Carmichael, Yu et al. 1988).

Japanese water bodies also frequently contain *Microcystis* blooms in summer, with a range of species including *M. aeruginosa*, *M. viridis*, *M. wesenbergii*, *M. ichthyoblabe*, and *M. novacekii* (Watanabe 1996). In addition to *M. aeruginosa*, *M. viridis* and *M. wesenbergii* appear to produce microcystins (Watanabe, Watanabe et al. 1986; Park and Watanabe 1996; Watanabe 1996).

M. aeruginosa is a major problem in water supply reservoirs and irrigation water storage sites in eastern Australia. Along the southern portion of the Great Dividing range, from latitude 25°S to almost 45°S in Tasmania, *M. aeruginosa* occurs as a summer-to-autumn water bloom in deep-water storage sites that stratify strongly in hot weather. It is a major cyanobacterial species throughout the Murray Darling Basin (Steffensen, Baker et al. 1993), also occurring in shallow off-river storage sites in summer, to the extent that there have been considerable poisonings of livestock (Carbis, Simons et al. 1994; Carbis, Waldron et al. 1995). The species also occurs in small farm dams, resulting in livestock poisoning. The greatly increased intensity of water blooms of this species in Australia is attributed to the general process of eutrophication caused by incoming nutrients from agriculture and urbanization (May 1981).

4.11 DISTRIBUTION OF OTHER MICROCYSTIN-PRODUCING SPECIES OF CYANOBACTERIA

Planktothrix agardhii and *P. rubescens* are significant microcystin producers in cool, temperate climates in the Northern Hemisphere, where they can occur throughout the year. Both species can grow under low light conditions — even under ice in winter. In summer, both species can form substantial water blooms deep in clear-water lakes, at the metalimnion, where nutrients from the anoxic hypolimnion are available (see Figure 4.1). In winter they are found throughout the water column (Sivonen and Jones 1999). The species have been most extensively studied in Germany (Fastner, Neumann et al. 1999) and Scandinavia (Sivonen, Niemela et al. 1990; Willen and Mattsson 1997). Other species of this genus have been described in Australia, but they do not produce toxins (Steffensen, Baker et al. 1993).

Some Scandinavian and North American strains of *Anabaena flos-aquae* can produce microcystins (Krishnamurthy, Carmichael et al. 1986; Harada, Ogawa et al. 1991), as can *Nostoc* sp. from Europe (Sivonen, Carmichael et al. 1990). The cold-climate species *Snowella (Gomphosphaeria) lacustris* is a microcystin producer and is reported from Norway (Berg, Skulberg et al. 1986; Skulberg 1996) and Terra del Fuego (unpublished data). From the evidence of microcystin production by members of the families Chroococcales (*Microcystis* sp., *Snowella* sp.), Oscillatoriales (*Planktothrix* sp.), and Nostocales (*Anabaena* sp., *Nostoc* sp.), it is likely that many more genera of cyanobacteria will be identified as microcystin producers in the future.

4.12 ECOLOGY OF *MICROCYSTIS AERUGINOSA*

M. aeruginosa has been extensively studied, especially through the work of Reynolds on the ecology (and particularly the control of buoyancy) of *M. aeruginosa* colonies in English lakes (Reynolds 1975; Reynolds and Rogers 1976; Reynolds, Jaworski et al. 1981; Reynolds 1984; Reynolds, Oliver et al. 1987; Reynolds 1991; Reynolds 1992; Reynolds 1997). This species does not form akinetes, hence the vegetative cells carry the capacity to survive over winter under cold, dark, and partially anoxic conditions. The roughly spherical colonies of *M. aeruginosa* sediment downward in autumn, as light and temperature decrease. They remain passive during winter on the bottom sediments of temperate lakes. From these overwintering colonies, new cells and small colonies arise in early summer and move upward into the water. The trigger for release from sediments may be increased light or temperature. In some shallow temperate lakes, small populations of *Microcystis* can exist in suspension throughout the winter (Reynolds and Walsby 1975); but in deeper lakes, the sediment population provides the inoculum for the following year's growth. The increase in cell population in the water column occurs with the thermal stratification of the lake and the onset of anoxic conditions in the hypolimnion and on the sediments (Reynolds 1973). Anoxia results in the release of ammonia from sediments and the solubilization of phosphates, both of which provide nutrients for cell growth. It is unclear whether *Microcystis* can utilize energy through heterotrophic nutrition while in the sediments — that is, metabolizing organic molecules available within the sediments for survival and growth while in darkness.

Alternatively, the cells may simply reduce respiration greatly during winter, enabling the conservation of sufficient energy to allow growth and gas vacuole production as the water stratifies and becomes anoxic in early summer. Growth of the organism is most rapid above 17 to 18°C, which requires movement upward into the epilimnion in temperate lakes (Reynolds 1975). As *Microcystis* does not have heterocysts, it is unlikely to be able to fix gaseous nitrogen under aerobic conditions. However, under anaerobic conditions, the availability of ammonia increases, and it is the most effective nitrogen source for cyanobacteria (Tandeau de Marsac and Houmard 1993). It is therefore unlikely that nitrogen fixation occurs under aerobic or anaerobic conditions. No experimental demonstration of nitrogen fixation by this species has been reported.

Nitrogen can be stored very effectively by cyanobacteria in the form of two proteins, cyanophycin and phycocyanin. The former is a specialized storage protein, a polymer of aspartate and arginine. This accumulates under conditions of nitrogen availability together with growth limitation by low temperature, low light intensity, and low phosphorus or sulfur (Oliver and Ganf 2000). Under nitrogen-limiting conditions, which commonly occur during summer, when maximum growth of all photosynthetic plankton is occurring, the *Microcystis* nitrogen stores can be mobilized. In addition to the utilization of cyanophycin for cell growth, the photosynthetic pigment phycocyanin can also be broken down as a nitrogen source. Metabolism of these proteins will not only provide amino groups for biosynthesis of nitrogen-containing molecules, proteins, and nucleic acids, for example, but also energy for biosynthesis; hence it will assist growth under conditions of low light.

As mentioned earlier in this chapter, *Microcystis* has considerable capacity to store phosphorus, so that the species can maximize the enhanced nutrient availability at the sediment surface or at depth in the hypolimnion in stratified lakes in summer.

Because of the presence of gas vacuoles and of large colony size, *M. aeruginosa* has effective buoyancy control, which provides the species with a competitive advantage. In early summer, prior development of green algae and diatoms will deplete the available nutrients in the epilimnion, inhibiting their growth. A cyanobacterial genus such as *Microcystis* that can rapidly move up and down a stationary water column, gathers light energy close to the surface, synthesizes carbohydrates in particular, becomes denser, and sinks down to reach nutrients deep in the water. The cells then grow and divide, use up the carbohydrate stores, gain nitrogen and phosphorus, and float up toward the surface to gather light energy (Oliver and Ganf 2000).

Under very calm, warm conditions *Microcystis* cells can rise to the surface and form scums, which progressively gain density and thickness. Moderate winds will mix the cells back into the water column, but gentle breezes move these scums downwind, so that they accumulate in bays, shallow shorelines, and adjacent to dam walls. The higher the cell population in the epilimnion, the more substantial the scum that can form, so that, in highly eutrophic lakes, scums of 10 cm or more in thickness are observed. The cell population within a scum can reach concentrations 1,000 to 1,000,000 times the average epilimnion concentration of cells (Falconer, Bartram et al. 1999).

It is likely that scum formation is due to excess buoyancy, generated by the cells spending time at low light intensity due to deep mixing or shading or being limited by lack of CO_2 (see discussion in Oliver and Ganf 2000). Scums are also a common phenomenon at night in late summer, as colonies move upward into the warm surface layers. During the day, cells at the immediate surface become damaged by high light intensity and dehydration, which prevents buoyancy regulation and effectively locks the cell into the surface layer (Oliver and Ganf 2000). Cells in the top layer then senesce, liberating pigments that add blue-green and red streaks to the scum. The anaerobic conditions resulting from cell decay in the scum layer cause offensive odors. Unfortunately many water supply reservoirs and natural lakes worldwide have become eutrophic to the extent that *Microcystis* scums occur every hot summer, causing problems in the drinking water supply and the recreational use of lakes (Falconer, Bartram et al. 1999).

In shallow tropical lakes, the population of *Microcystis* remains high all year but is sensitive to weather conditions. In stable climatic conditions, with little change in water volume or flooding, *Microcystis* can form permanent dominant populations. Lake George in Uganda is shallow, with an average depth of 2.4 m, and has low light penetration due to the high turbidity generated by the density of the cyanobacterial population. The depth of the euphotic zone is usually less than 0.8 m (Oliver and Ganf 2000); as a result, the majority of the cyanobacterial population is shaded most of the time. Located on the equator, the lake stratifies each day. At dawn, the temperature is similar from the top to the bottom of the lake at 25°C; but by late afternoon, the temperature gradient may reach 10°C. The surface of the lake becomes alkaline, with a pH above 9, and the water supersaturates with oxygen due to massive cyanobacterial photosynthetic oxygen formation. With nocturnal cooling, the lake mixes, the pH reverts to 7, and oxygen saturation returns to 100% (Oliver and Ganf 2000). The rapid vertical migration of large *Microcystis* colonies enables the species to maximize growth under these conditions, moving upward in the daytime for enhanced light in the dense cyanobacterial population and downward to the nutrient-rich sediments when carbohydrate reserves reduce buoyancy.

4.13 ECOLOGY OF *PLANKTOTHRIX* SPECIES AND *ANABAENA FLOS-AQUAE*

In cold temperate lakes, other toxic cyanobacterial species predominate, which have growth capability at lower temperatures and also effective photosynthesis at lower light intensities. A detailed study of a shallow, turbid, hypereutrophic lake near Berlin, Germany (Weidner 1999), offers an interesting contrast to the analogous tropical Lake George. The Langer See has a maximum depth of 3.8 m and a mean depth of 2.1 m. Water temperature varied widely with season, with a maximum of 20 to 24°C in August in the two seasons measured. In winter the lake froze over for 3 to 4 months, with temperature beneath the ice of less than 5°C. The euphotic depth in winter under the ice was 1.5 m or less; in spring, it reached a maximum of about 3.0 m; while at the peak of cyanobacterial dominance, it reduced to 1.5 m.

In this shallow lake, the cyanobacterial population almost disappeared under the ice in winter, but it rapidly increased in midsummer to a biovolume of 45 mm^3/L, with *Planktothrix agardhii* the predominant organism. Nutrient concentration was maximal under the ice, with dissolved inorganic phosphorus (70 μg/L), ammonia (1000 μg/L) and nitrate (500 μg/L) rising sharply in early winter. During the summer proliferation of cyanobacteria, dissolved inorganic phosphorus dropped to below 10 μg/L, nitrate was not measurable, and ammonia decreased to 50 to 150 μg/L. This filamentous cyanobacterial species is capable of substantial microcystin production. In the two seasons measured, during summer, the toxin content of the cells reached over 2000 μg/g dry weight (Weidner 1999). In a survey of cyanobacterial toxins in German lakes, those lakes with summer blooms of *P. agardhii* were found to contain up to 454 μg of microcystin per liter, with a median of 21 μg/L (Fastner, Neumann et al. 1999; Fastner, Erhard et al. 2001).

In deep, clear, cold mesotrophic lakes, *Planktothrix* species can use their buoyancy regulation to grow in the metalimnion, gaining access to increased nutrients and using their exceptional light-gathering ability to grow in the low-illumination and green light conditions at depth (see Figure 4.1) (Mur, Skulberg et al. 1999). One of these species is *Planktothrix rubescens*, which is red-brown in color through an abundance of phycoerythrin in the light-trapping phycobilisomes and is a major microcystin producer in cold, temperate climates. The organism has the highest concentration of microcystin on a dry-weight basis of any species of cyanobacterium, with a maximum measured in German lakes of over 5000 μg/g dry weight of cells and a mean of approximately 2000 μg/g (Fastner, Erhard et al. 2001). Because it forms dense layers well below the water surface during the period of maximum growth, the organism presents a problem in water supplies drawn off at similar depths. This is discussed further in Chapter 12 with respect to water treatment. At the end of the autumn, as the water temperature drops, the filaments of *P. rubescens* may become buoyant and rise to the surface, forming red scums, as illustrated in Mur, Skulberg et al. (1999, Figure 2.2D). During winter, under mixed conditions with no thermal stratification, *P. rubescens* can also form a high, vertically distributed cell density, even under ice (Sivonen and Jones 1999).

The other cyanobacterium producing microcystins in cool or cold temperate lakes is *A. flos-aquae.* In the German survey of 642 samples of lake and river water, 10 of the samples had *Anabaena* dominant, compared to 39 with *P. agardhii* and 36 with *P. rubescens.* The most common dominance with a single toxic species was *Microcystis,* with 59 samples. In 44 samples, mixtures of the 3 toxic species occurred. Overall, 66% of the samples contained microcystins (Fastner, Erhard et al. 2001).

The genus *Anabaena* presents another of the puzzles of toxic cyanobacterial research; that is, why is it that the same genus or species can produce two quite different toxins in different locations? Within the Northern Hemisphere, *A. flos-aquae* has been demonstrated to produce both microcystins and anatoxin-a. The former is a hepatotoxic cyclic peptide, and the latter is a neurotoxic alkaloid. In the Langer See in Germany, described earlier, an abundance of *A. flos-aquae* is associated with anatoxin-a in the water (Chorus and Mur 1999). In Norway and Finland, *A. flos-aquae* strains were isolated that produced microcystins (Sivonen, Namikoshi et al. 1992). *A. flos-aquae* is a nitrogen-fixing species with heterocysts and the ability to remain in the water column all year. In summer, the species can form surface scums, which can be associated with neurotoxicity, hepatotoxicity, or both (Willen and Mattsson 1997). Laboratory culture of *A. flos-aquae* has shown that the availability of phosphorus is the main nutritional factor increasing microcystin production (Rapala, Sivonen et al. 1997; Rapala and Sivonen 1998).

4.14 ECOLOGY OF MICROCYSTIN PRODUCTION

Analysis of the microcystin content of lake and river water during cyanobacterial blooms of toxic species has shown a wide variation across sampling sites. Several factors contribute to this, all with the potential to have a substantial impact. The most evident is the variation between cell concentrations at different locations and depths, which may vary between very low concentrations suspended evenly through the depth of the water column to highly concentrated surface scums or bands of high concentration at depth in the water. These variations in cell concentration have the potential to give very misleading data on the hazard associated with use of the water for drinking or recreational purposes. The key locations for assessing the cyanobacterial cell concentration or toxicity of raw drinking water will be adjacent to the intake and at the same depth as the water is drawn off. Analogously, for recreational areas, the bathing beaches are the key sampling points, which often have a much higher concentration of cyanobacterial cells than the main water body, because of wind drift of surface scums to the shoreline.

One of the earlier studies of cell density, cyanobacterial species, and toxicity was reported in 1981 by Carmichael and Gorham (1981). This study reported on a hypertrophic lake of a maximum depth of 8 m and a length of about 6 km in Alberta, Canada. Over 3 years of summer and autumn sampling, the predominant species were *M. aeruginosa* and *Anabaena flos-aquae*, with *Aphanizomenon flos-aquae* less abundant. *M. aeruginosa* accounted for the toxicity, which varied on a single day in different samples by more than 600-fold per unit volume of water. The major contributor to this difference was cell concentration in the sample; but at the same cell concentration, samples from different locations in the lake in a single day showed a threefold difference in toxicity (Carmichael and Gorham 1981). The authors concluded that "toxicity, as well as [cell] density and species composition, tend to vary greatly from [sampling] station to station on a single day, on different days throughout a growing season and from one season to the next." Studies on strains of toxic *M. aeruginosa* have shown different toxin-production capabilities, ranging from an inability to biosynthesize toxin, to high sustained toxin production per cell; these characteristics are probably genetically determined (Bolch, Blackburn et al. 1997).

A study of the relationship of colony size, the microcystin content, and the proportion of microcystin-producing genotypes in a Berlin lake (Wannsee) in summer showed that the largest colonies had the highest microcystin content per cell and the highest proportion of toxin-producing genotypes. The authors concluded that the net production of microcystin in the lake was mainly influenced by the abundance of large colonies (Kurmayer, Christiansen et al. 2003). It is interesting to speculate on why the more toxic colonies were also larger, possibly reflecting decreased predation.

Environmental conditions also affect the toxin content of cells. These include temperature, light, nutrient availability, and mixing, which determine both growth rate and toxin biosynthesis. Studies of *M. aeruginosa* in culture have helped to clarify these factors. Microcystin content of batch cultures grown between 10 and 35°C in different investigations indicated that highest toxicity per unit dry weight of cells occurred between 18 and 25°C, with lower and higher temperatures resulting in lower toxicities. The range of microcystin concentration ranged up to fivefold at different temperatures (Sivonen and Jones 1999). *Anabaena* producing microcystins showed a temperature optimum at 25°C, with up to a 10-fold difference in toxin concentration with temperature (Rapala and Sivonen 1998).

Light intensity and toxin production are best studied in continuous culture, as batch cultures increase in turbidity with cell numbers; hence the intensity of the light reaching the cells falls throughout the experiment. Investigation of the relationship between light intensity and toxin production in *M. aeruginosa* growing as isolated cells in continuous culture, with light intensities from 20 to 75 μmol photons m^{-2} s^{-1}, gave peak toxin production per unit dry weight at 40 μmol m^{-2} s^{-1} (Utkilen and Gjolme 1992). Continuous culture of freshly isolated strains that remained in colonies showed increased growth up to 130 μmol m^{-2} s^{-1}, presumably because of self-shading within the colony. The relationship between light intensity and the toxin content of cells was not clear, with the most toxic strain showing no major effects of light intensity, whereas other strains showed evidence of an increase in toxin with light (Hesse and Kohl 2001). Recent research with a continuous culture of isolated *M. aeruginosa* cells producing microcystin showed an increase in growth up to 80 μmol photons m^{-2} s^{-1}, with no inhibition of growth up to the highest intensity tested, of 400 μmol m^{-2} s^{-1}. Toxin production was measured per cell, per biovolume, and per weight of protein over the range of light intensities. All these parameters showed the same response, toxin content increasing with light intensity until the growth peak was reached at 80 μmol m^{-2} s^{-1}, with a linear decrease in toxin as light intensity increased thereafter (Weidner, Visser et al. 2003).

The euphotic zone, which is regarded as the depth over which effective photosynthesis can take place, is defined by the depth at which light intensity is reduced to 1% of incident radiation intensity at the water surface. Surface radiation intensity in temperate regions is in the region of 1000 μmol m^{-2} s^{-1}. Therefore the deep margin of the euphotic zone will have a light intensity of about 10 μmol m^{-2} s^{-1}. *M. aeruginosa* will thus have optimum light for toxin production over approximately 30 to 70% of the euphotic depth, which itself varies greatly due to cyanobacterial cell concentration and other sources of turbidity. Comparison with continuous cultures of *Anabaena* producing microcystins, with an optimum light intensity for toxin production of 25 μmol m^{-2} s^{-1}, shows that *M. aeruginosa* has a higher light requirement for toxin production than *Anabaena* (Rapala and Sivonen 1998). From the depth at which highly toxic *Planktothrix* blooms have been found in stratified lakes, it seems likely that they require even less light than *Anabaena* for optimum toxin production. This suggestion is supported by toxin production of *P. agardhii* in culture, which showed higher microcystin content per unit dry weight at light intensities of 12 and 24 μmol photons m^{-2} s^{-1} than at 50 and 95 μmol m^{-2} s^{-1}

(Sivonen 1990). From these data it is apparent that different species of toxic cyanobacteria have different optimal light regimes for toxin production and growth. These differences can account for some of the changes in dominant species under varying hydrological conditions in natural water bodies; other changes are more related to the availability of nutrients.

Phosphorus availability is a major determinant of growth rate for cyanobacteria, as discussed earlier, and also has a substantial effect on toxin production. In cultures of several cyanobacterial species producing microcystins, phosphorus concentrations of less than 100 µg/L appear to reduce toxin concentration in the cells (Sivonen and Jones 1999). The increase in cell microcystin concentration between phosphorus-limited *Anabaena* cells growing in medium containing 50 µg of phosphate per liter and those growing at 5.5 mg/L was fivefold (Rapala, Sivonen et al. 1997). Both nitrogen-fixing (*Planktothrix*) and non-nitrogen-fixing (*Microcystis*) species of cyanobacteria produce higher toxin concentrations when nitrate is readily available in the culture medium. A threefold increase in microcystin concentration was demonstrated when *Microcystis* was cultured with 1 mg of nitrogen per liter, compared with a culture with 50 µg of nitrogen per liter (Sivonen 1990; Utkilen and Gjolme 1995). An extended review of European research in this field can be found in Chorus (2001).

The influence of mixing or destratification of lakes on toxin production relates to changes in nutrients, light, and temperature. As a consequence, lakes behave differently depending on trophic level, depth, and temperature. Under summer conditions, mixing by freshwater inflow, carrying nutrients, or mixing by wind, bringing nutrient-enriched deeper water toward the surface, may result in major toxic bloom formation when the lake restratifies. By contrast, mixing through cooling in autumn may severely deplete the cyanobacterial population by carrying cells down below the euphotic depth. For a detailed discussion on buoyancy and mixing, see Oliver and Ganf (2000) on freshwater blooms.

4.15 *NODULARIA SPUMIGENA* AND NODULARIN PRODUCTION

N. spumigena is a filamentous, nitrogen-fixing cyanobacterium which occurs worldwide in temperate regions. The organism occupies brackish (semisaline) water environments in coastal lakes and estuaries and is particularly abundant in the Baltic Sea. It produces the hepatotoxic pentapeptide nodularin, and the toxin content of cells varies with the immediate environment. Maximum toxicity occurs when the organism is growing at salinities of 12 to 15 parts per 1000 of sodium chloride. In culture, increased temperature and available phosphorus increased toxin content per cell (Blackburn, McCausland et al. 1996; Lehtimaki, Moisander et al. 1997).

REFERENCES

Ame, V. M., M. d. P. Diaz, et al. (2003). Occurrence of toxic cyanobacterial blooms in San Roque reservoir (Cordoba, Argentina): A field and chemometric study. *Environmental Toxicology* 18: 192–201.

Azevedo, S. M. F. O., W. R. Evans, et al. (1994). First report of microcystins from a Brazilian isolate of the cyanobacterium *Microcystis aeruginosa. Journal of Applied Phycology* 6: 261–265.

Baker, P. D., A. R. Humpage, et al. (1993). *Cyanobacterial Blooms in the Murray Darling Basin: Their Taxonomy and Toxicity.* Adelaide, Australia, Australian Centre for Water Quality and Research: 159.

Banker, P., S. Carmeli, et al. (1997). Identification of cylindrospermopsin in *Aphanizomenon ovalisporum* (Cyanophyceae) isolated from Lake Kinneret, Israel. *Journal of Applied Phycology* 33: 613–616.

Berg, K., O. M. Skulberg, et al. (1986). Observations of toxic blue-green algae (cyanobacteria) in some Scandinavian lakes. *Acta Veterinaria Scandinavica* 27: 440–452.

Berman-Frank, I., P. Lundgren, et al. (2003). Nitrogen fixation and oxygen evolution in cyanobacteria. *Research in Microbiology* 154(3): 157–164.

Bernard, C., M. Harvey, et al. (2003). Toxicological comparison of diverse *Cylindrospermopsis raciborskii* strains: Evidence of liver damage caused by French *C. raciborskii* strain. *Environmental Toxicology* 18(3): 176–186.

Blackburn, S., M. A. McCausland, et al. (1996). Effect of salinity on growth and toxin production in cultures of the bloom-forming cyanobacterium *Nodularia spumigena* from Australian waters. *Phycologia* 35: 511–522.

Bolch, C. J., S. I. Blackburn, et al. (1997). Plasmid content and distribution in the toxic cyanobacterial genus *Microcystis* Kutzing ex Lemmermann (cyanobacteria: Chroococcales). *Phycologia* 36: 6–11.

Bormans, M., P. W. Ford, et al. (2000). Temporal changes in nutrients and cyanobacterial populations in a dammed, stratified tropical river. *Verhandlungen der Internationalen Vereinigung für Theoretische und Angewandte Limnologie* 27: 3239–3242.

Bourke, A. T. C., R. B. Hawes, et al. (1983). An outbreak of hepato-enteritis (the Palm Island mystery disease) possibly caused by algal intoxication. *Toxicon* 21 (suppl 3): 45–48.

Bowling, L. (1994). Occurrence and possible causes of a severe cyanobacterial bloom in Lake Cargelligo, New South Wales. *Australian Journal of Marine and Freshwater Research* 45: 737–745.

Branco, C. W. C. and P. A. C. Senna (1994). Factors affecting the development of *Cylindrospermopsis raciborskii* and *Microcystis aeruginosa* in Paranoa Reservoir, Brasilia, Brazil. *Algological Studies* 75: 85–96.

Briand, J. F., C. Robillot, et al. (2002). Environmental context of *Cylindrospermopsis raciborskii* (cyanobacteria) blooms in a shallow pond in France. *Water Research* 36(13): 3183–3192.

Byth, S. (1980). Palm Island mystery disease. *Medical Journal of Australia* 2: 40–42.

Carbis, C. R., J. A. Simons, et al. (1994). A biochemical profile for predicting the chronic exposure of sheep to *Microcystis aeruginosa*, an hepatotoxic species of blue-green alga. *Research in Veterinary Science* 57: 310–316.

Carbis, C. R., D. L. Waldron, et al. (1995). Recovery of hepatic function and latent mortalities in sheep exposed to the blue-green alga *Microcystis aeruginosa. Veterinary Record* 137: 12–15.

Carmichael, W. W. (2001). *Assessment of Blue-Green Algal Toxins in Raw and Finished Drinking Water.* Denver, AWWA Research Foundation and American Water Works Association.

Carmichael, W. W. and P. R. Gorham (1981). The mosaic nature of toxic blooms of cyanobacteria. *The Water Environment. Algal Toxins and Health.* W. W. Carmichael, ed. New York, Plenum Press: 161–172.

Carmichael, W. W., M. Yu, et al. (1988). Occurrence of the toxic cyanobacterium (blue-green alga) *Microcystis aeruginosa* in Central China. *Archiv für Hydrobiologie* 114, 1: 21–30.

Chapman, A. D. and C. L. Schelske (1997). Recent appearance of *Cylindrospermopsis* (cyanobacteria) in five hypertrophic Florida lakes. *Journal of Phycology* 33: 191–195.

Chiswell, R. K., G. R. Shaw, et al. (1999). Stability of cylindrospermopsin, the toxin from the cyanobacterium, *Cylindrospermopsis raciborskii*: Effect of pH, temperature, and sunlight on decomposition. *Environmental Toxicology* 14(1): 155–161.

Chorus, I., Ed. (2001). *Cyanotoxins: Occurrence, Causes, Consequences*. Berlin, Springer-Verlag.

Chorus, I. and L. Mur (1999). Preventative Measures. *Toxic Cyanobacteria in Water: A Guide to Their Public Health Consequences, Monitoring and Management*. I. Chorus and J. Bartram, eds. London, E & FN Spon (on behalf of WHO): 236–273.

Cooke, G. D., E. B. Welch, et al. (1993). *Restoration and Management of Lakes and Reservoirs*. Boca Raton, FL, Lewis Publishers.

De Leon, L. and J. S. Yunes (2001). First report of a microcystin-containing bloom of the cyanobacteria *Microcystis aeruginosa* in the La Plata River, South America. *Environmental Toxicology* 16(1): 139–141.

Dokulil, M. T. and J. Mayer (1996). Population dynamics and photosynthetic rates of a *Cylindrospermposis-Limnothrix* association in a highly eutrophic urban lake, Alte Donau, Vienna, Austria. *Algological Studies* 83: 179–195.

Fabbro, L. D. (1999). Phytoplankton Ecology in the Fitzroy River at Rockhampton, Central Queensland, Australia. Ph.D. thesis, Rockhampton, Queensland, School of Biological and Environmental Sciences, Central Queensland University.

Fabbro, L. D. and L. E. Andersen (2003). *Baseline Assessment of Water Column Stratification and the Distribution of Blue-Green Algae in Lake Awoonga*. Gladstone, Queensland, Area Water Board: 22.

Fabbro, L. D. and L. J. Duivenvoorden (1996). Profile of a bloom of the cyanobacterium *Cylindrospermopsis raciborskii* (Woloszynska) seenaya and Subba Raju in the Fitzroy River in tropical central Queensland. *Marine and Freshwater Research* 47(5): 685–694.

Falconer, I. R., J. Bartram, et al. (1999). Safe levels and safe practices. *Toxic Cyanobacteria in Water: A Guide to Their Public Health Consequences, Monitoring and Management*. I. Chorus and J. Bartram, eds. London, E & FN Spon (on behalf of WHO): 155–178.

Falkowski, P. G. and J. LaRoche (1991). Acclimation to spectral irradiance in algae. *Journal of Phycology* 27: 8–14.

Fastner, J., M. Erhard, et al. (2001). Microcystin variants in *Microcystis* and *Plantothrix* dominated field samples. *Cyanotoxins: Occurrence, Causes, Consequences*. I. Chorus, ed. Berlin, Springer-Verlag: 148–152.

Fastner, J., R. Heinze, et al. (2003). Cylindrospermopsin occurrence in two German lakes and preliminary assessment of toxicity and toxin production of *Cylindrospermopsis raciborskii* (cyanobacteria) isolates. *Toxicon* 42(3): 313–321.

Fastner, J., U. Neumann, et al. (1999). Microcystins (hepatotoxic heptapeptides) in German fresh water bodies. *Environmental Toxicology* 14(1): 13–22.

Gallon, J. R., D. A. Jones, et al. (1996). *Trichodesmium*, the paradoxical diazotroph. *Archives of Hydrobiology, Supplement, Algological Studies* 83: 215–243.

Ganf, G. G. (1974). Phytoplankton biomass and distribution in a shallow eutrophic lake (Lake George, Uganda). *Oecologia (Berlin)* 16: 9–29.

Ganf, G. G. and R. L. Oliver (1982). Vertical separation of light and available nutrients as a factor causing replacement of green algae by blue-green algae in the plankton of a stratified lake. *Journal of Ecology* 70: 829–844.

Geitler, L. and F. Ruttner (1936). Die Cyanophyceen der Deutschen Limnologischen Sunda-Expedition, ihre Morphologie, Systematic und Okologie. *Archiv für Hydrobiologia Supplement* 14: 308–483, 553–715.

Gorzo, G. Y. (1987). The influence of physical and chemical factors on the germination of spores of heterocystic cyanobacteria in Lake Balaton. (In Hungarian with English summary). *Hidrologiai Kozlony* 67: 127–133.

Harada, K. I., K. Ogawa, et al. (1991). Microcystins from *Anabaena flos-aquae* NRC 525-17. *Chemical Research in Toxicology* 4: 535–540.

Harada, K., I. Ohtani, et al. (1994). Isolation of cylindrospermopsin from a cyanobacterium *Umezakia natans* and its screening method. *Toxicon* 32: 73–84.

Harris, G. P. and G. Baxter (1996). Interannual variability in phytoplankton biomass and species composition in a subtropical reservoir. *Freshwater Biology* 35(3): 545–560.

Hawkins, P. R., E. Putt, et al. (2001). Phenotypical variation in a toxic strain of the phytoplankter, *Cylindrospermopsis raciborskii* (Nostocales, Cyanophyceae) during batch culture. *Environmental Toxicology* 16(6, special issue): 460–467.

Hawkins, P. R., M. T. C. Runnegar, et al. (1985). Severe hepatotoxicity caused by the tropical cyanobacterium (blue-green alga) *Cylindrospermopsis raciborskii* (Woloszynska) Seenaya and Subba Raju isolated from a domestic supply reservoir. *Applied and Environmental Microbiology* 50(5): 1292–1295.

Hesse, K. and J. G. Kohl (2001). Effect of light and nutrient supply on growth and microcystin content of different strains of *Microcystis aeruginosa*. *Cyanotoxins: Occurrence, Causes, Consequences*. I. Chorus, ed. Berlin, Springer-Verlag: 104–115.

Hill, H. (1970). *Anabaenopsis raciborskii* Woloszynska in Minnesota lakes. *Journal of the Minnesota Academy of Science* 36: 80–82.

Hindak, F. and M. Moustaka (1988). Planktic cyanophytes of Lake Volvi, Greece. *Archives of Hydrobiology Supplement* 80: 497–528.

Huber, A. L. (1986). Nitrogen fixation by *Nodularia spumigena* Mertens (cyanobacteria) 1. Field studies on the contribution of blooms to the nitrogen budget of the Peel-Harvey Estuary, Western Australia. *Hydrobiologia* 131: 193–203.

Humpage, A. R. and I. R. Falconer (2003). Oral toxicity of the cyanobacterial toxin cylindrospermopsin in male Swiss albino mice: Determination of no observed adverse effect level for deriving a drinking water guideline value. *Environmental Toxicology* 18(2): 94–103.

Istvanovics, V., L. Somlyody, et al. (2002). Cyanobacteria-mediated internal eutrophication in shallow Lake Balaton after load reduction. *Water Research* 36(13): 3314–3322.

Kohler, J. (1992). Influence of turbulent mixing on growth and primary production of *Microcystis aeruginosa* in the hypertrophic Bautzen reservoir. *Archiv für Hydrobiologie* 123: 413–429.

Komarek, J. and H. Kling (1991). Variation in six planktonic cyanophyte genera in Lake Victoria (East Africa). *Algological Studies* 61: 21–45.

Kotak, B. G., S. L. Kenefick, et al. (1993). Occurrence and toxicological evaluation of cyanobacterial toxins in Alberta lakes and farm dugouts. *Water Research* 27: 495–506.

Krishnamurthy, T., W. W. Carmichael, et al. (1986). Toxic peptides from freshwater cyanobacteria (blue-green algae). I. Isolation, purification and characterization of peptides from *Microcystis aeruginosa* and *Anabaena flos-aquae*. *Toxicon* 24(9): 865–873.

Kromkamp, J., A. Van Den Heuvel, et al. (1989). Phosphorus uptake and photosyntghesis by phosphate-limited cultures of the cyanobacterium *Microcystis aeruginosa*. *British Phycological Journal* 24: 347–355.

Kurmayer, R., G. Christiansen, et al. (2003). The abundance of microcystin-producing genotypes correlates positively with colony size in *Microcystis* sp and determines its microcystin net production in Lake Wannsee. *Applied and Environmental Microbiology* 69(2): 787–795.

Lagos, N., H. Onodera, et al. (1999). The first evidence of paralytic shellfish toxins in the freshwater cyanobacterium *Cylindrospermopsis raciborskii*, isolated from Brazil. *Toxicon* 37: 1359–1373.

Lehtimaki, J., P. Moisander, et al. (1997). Growth, nitrogen fixation, and nodularin production by two Baltic Sea cyanobacteria. *Applied and Environmental Microbiology* 63: 1647–1656.

Li, R. H., W. W. Carmichael, et al. (2001). The first report of the cyanotoxins cylindrospermopsin and deoxycylindrospermopsin from *Raphidiopsis curvata* (cyanobacteria). *Journal of Phycology* 37(6): 1121–1126.

Logan, B. W. (1961). Crytpozoon and associated stromatolites from the recent of Shark Bay, Western Australia. *Journal of Geology* 69: 517–533.

Lukatelich, R. J. and A. J. McComb (1986). Nutrient levels and the development of diatom and blue-green algal blooms in a shallow Australian estuary. *Journal of Plankton Research* 8(4): 597–618.

Mahmood, N. A. and W. W. Carmichael (1986). Paralytic shellfish poisons produced by the freshwater cyanobacterium *Aphanizomenon flos-aquae* NH-5. *Toxicon* 24(2): 175–186.

May, V. (1981). The occurrence of toxic cyanophyte blooms in Australia. *The Water Environment: Algal Toxins and Health*. W. W. Carmichael, ed. New York, Plenum Press: 127–142.

McGregor, G. B. and L. D. Fabbro (2000). Dominance of *Cylindrospermopsis raciborskii* (Nostocales, Cyanoprokaryota) in Queensland tropical and subtropical reservoirs: Implications for monitoring and management. *Lakes and Reservoirs: Research and Management* 5: 195–205.

Mohamed, Z. A. (1998). Studies on Egyptian Toxic Freshwater Cyanobacteria (Blue-Green Algae): Identification, Production and Removal of Microcystins. Ph.D. thesis, Egypt, South Valley University, in collaboration with Wright State University, Dayton, OH.

Mohamed, Z. A. and W. W. Carmichael (2000). Seasonal variation in microcystin levels of river Nile water at Sohag City, Egypt. *Annales de Limnologie — International Journal of Limnology* 36(4): 227–234.

Mohamed, Z. A., W. W. Carmichael, et al. (2003). Estimation of microcystins in the freshwater fish *Oreochromis niloticus* in an Egyptian fish farm containing a *Microcystis* bloom. *Environmental Toxicology* 18(2): 137–141.

Mur, L. R., O. M. Skulberg, et al. (1999). Cyanobacteria in the environment. *Toxic Cyanobacteria in Water: A Guide to Their Public Health Consequences, Monitoring and Management*. I. Chorus and J. Bartram, eds. London, E & FN Spon (on behalf of WHO): 15–40.

Neilan, B. A., M. L. Saker, et al. (2003). Phylogeography of the invasive cyanobacterium *Cylindrospermopsis raciborskii*. *Molecular Ecology* 12: 133–140.

Oliver, R. L. and G. G. Ganf (2000). Freshwater blooms. *The Ecology of Cyanobacteria: Their Diversity in Time and Space*. B. A. Whitton and M. Potts, eds. Dordrecht, Kluwer Academic Publishers: 150–194.

Padisak, J. (1992). Seasonal succession of phytoplankton in a large shallow lake (Balaton, Hungary): A dynamic approach to ecological memory, its possible role and mechanisms. *Journal of Ecology* 80: 217–130.

Padisak, J. (1997). *Cylindrospermopsis raciborskii* (Woloszynska) Seenayya et Subba Raju, an expanding, highly adaptive cyanobacterium: worldwide distribution and review of its ecology. *Archiv für Hydrobiologie* 107 (suppl): 563–593.

Padisak, J. and V. Istvanovics (1997). Differential response of blue-green algal groups to phosphorus load reduction in a large shallow lake: Balaton, Hungary. *Verhandlungen der Internationalen Vereinigung für Theoretische und Angewandte Limnologie* 26: 574–580.

Paerl, H. W. (2000). Marine plankton. *The Ecology of Cyanobacteria: Their Diversity in Time and Space*. B. A. Whitton and M. Potts., eds. Dordrecht, Kluwer Academic Publishers: 121–148.

Park, H.-D. and M. F. Watanabe (1996). Toxic *Microcystis* in eutrophic lakes. *Toxic Microcystis*. M. F. Watanabe, K.-I. Harada, W. W. Carmichael, and H. Fujiki, eds. Boca Raton, FL, CRC Press: 57–77.

Rapala, J. and K. Sivonen (1998). Assessment of environmental conditions that favour hepatotoxic and neurotoxic *Anabaena* spp. strains cultured under light limitation at different temperatures. *Microbial Ecology* 36(2): 181–192.

Rapala, J., K. Sivonen, et al. (1997). Variation of microcystins, cyanobacterial hepatotoxins, in *Anabaena* spp. as a function of growth stimuli. *Applied and Environmental Microbiology* 63: 1–7.

Reed, R. H., J. A. Chudek, et al. (1984). Osmotic adjustment in cyanobacteria. *Archives of Microbiology* 138: 333–337.

Reynolds, C. S. (1973). Growth and buoyancy of *Microcystis aeruginosa* Kutz. Emend. Elenkin in a shallow eutrophic lake. *Proceedings of the Royal Society of London* 184: 29–50.

Reynolds, C. S. (1975). Interrelations of photosynthetic behaviour and buoyancy regulation in a natural population of a blue-green alga. *Freshwater Biology* 5: 323–338.

Reynolds, C. S. (1984). Phytoplankton periodicity: The interactions of form, function and environmental variability. *Freshwater Biology* 14: 111–142.

Reynolds, C. S. (1991). Ecology and control of cyanobacteria (blue-green algae). *PHLS Microbiology Digest* 8: 87–90.

Reynolds, C. S. (1992). Eutrophication and the management of planktonic algae: what Vollenweider couldn't tell us. *Eutrophication: Research and Application to Water Supply*. J. G. Jones and D. W. Sutcliffe, eds. Ambleside, U.K., Freshwater Biological Association.

Reynolds, C. S. (1997). *Vegetation Processes in the Pelagic: A Model for Ecosystem Theory*. Ohlendorf-Luhe, Germany, Excellence in Ecology, Ecology Institute.

Reynolds, C. S., G. H. M. Jaworski, et al. (1981). On the annual cycle of the blue-green alga *Microcystis aeruginosa* Kutz. Emend. Elenkin. *Philosophical Transactions of the Royal Society, London* Ser B. 293: 419–477.

Reynolds, C. S., R. L. Oliver, et al. (1987). Cyanobacterial dominance: The role of buoyancy regulation in dynamic lake environments. *New Zealand Journal of Marine and Freshwater Research* 21: 379–390.

Reynolds, C. S. and D. A. Rogers (1976). Seasonal variations in the vertical distribution and buoyancy of *Microcystis aeruginosa* Kutz. Emend. Elenkin in Rothserne Mere, England. *Hydrobiologia* 48(1): 17–23.

Reynolds, C. S. and A. E. Walsby (1975). Water blooms. *Biological Reviews of the Cambridge Philosophical Society* 50: 437–481.

Saker, M. L. and D. J. Griffiths (2000). The effect of temperature on growth and cylindrospermopsin content of seven isolates of *Cylindrospermopsis raciborskii* (Nostocales, Cyanophycceae) from water bodies in northern Australia. *Phycologia* 39(4): 349–354.

Saker, M. L. and B. A. Neilan (2001). Varied diazotrophies, morphologies, and toxicities of genetically similar isolates of *Cylindrospermopsin raciborskii* (Nostocales, Cyanophyceae) from northern Australia. *Applied and Environmental Microbiology* 67(4): 1839–1845.

Saker, M. L., I. C. G. Nogueira, et al. (2003). First report and toxicological assessment of the cyanobacterium *Cylindrospermoposis raciborskii* from Portuguese freshwaters. *Ecotoxicology and Environmental Safety* 55: 243–250.

Sas, H. (1989). *Lake Restoration by Reduction of Nutrient Loading: Expectations, Experiences, Extrapolations.* St. Augustin, Germany, Academia Verlag Richarz.

Schembri, M. A., B. A. Neilan, et al. (2001). Identification of genes implicated in toxin production in the cyanobacterium *Cylindrospermopsis raciborskii. Environmental Toxicology* 16(5): 413–421.

Scott, W. E. (1983). *The Presence of Toxic* Microcystis *in Hartbeespoort Dam.* Report by the National Institute for Water Research to the Hartbeesport Dam Technical Sub-Committee. Pretoria, South Africa, National Institute for Water Research.

Shafic, H. M., L. Voros, et al. (1997). Growth of *Cylindrospermopsis raciborskii* in batch and continuous culture (in Hungarian with English summary). *Hidrologiai Kozlony* 77: 17–18.

Shaw, G. R., A. Sukenik, et al. (1999). Blooms of the cylindrospermopsin containing cyanobacterium, *Aphanizomenon ovalisporum* (Forti), in newly constructed lakes, Queensland, Australia. *Envrionmental Toxicology* 14(1): 167–177.

Sivonen, K. (1990). Effects of light, temperature, nitrate, orthophosphate, and bacteria on growth of and hepatotoxin production by *Oscillatoria agardhii* strains. *Applied and Environmental Microbiology* 56(9): 2658–2666.

Sivonen, K., W. W. Carmichael, et al. (1990). Isolation and characterization of hepatotoxic microcystin homologs from the filamentous freshwater cyanobacterium *Nostoc* sp. strain 152. *Applied and Environmental Microbiology* 56(9): 2650–2657.

Sivonen, K. and G. Jones (1999). Cyanobacterial toxins. *Toxic Cyanobacteria In Water: A Guide To Their Public Health Consequences, Monitoring and Management.* I. Chorus and J. Bartram, eds. London, E & FN Spon (on behalf of WHO): 41–111.

Sivonen, K., M. Namikoshi, et al. (1992). Isolation and structures of five microcystins from a Russian *Microcystis aeruginosa* strain calu 972. *Toxicon* 30(11): 1481–1485.

Sivonen, K., S. I. Niemela, et al. (1990). Toxic cyanobacteria (blue-green algae) in Finnish fresh and coastal waters. *Hydrobiologia* 190: 267–275.

Skulberg, O. M. (1996). Toxins produced by cyanophytes in Norwegian inland waters — health and environment. *Chemical Data as a Basis for Geomedical Investigations.* J. Lag, ed. Oslo, The Norwegian Academy of Science and Letters: 197–216.

Steffensen, D. A., P. D. Baker, et al. (1993). *Cyanobacterial Blooms in the Murray Darling Basin: Their Taxonomy and Toxicity.* Murray-Darling Basin Natural Resources Management Strategy Project No. S101. Canberra, Murray-Darling Basin Commission.

Tandeau de Marsac, N. and J. Houmard (1993). Adaption of cyanobacteria to environmental stimuli: New steps towards molecular mechanisms. *FEMS Microbiology Reviews* 104: 119–190.

USEPA (2001). *Creating a Cyanotoxin Target List for the Unregulated Contaminent Monitoring Rule.* Cinncinnati, Environmental Protection Agency, Technical Service Center.

Utkilen, H. and N. Gjolme (1992). Toxin production by *Microcystis aeruginosa* as a function of light in continuous cultures and its ecological significance. *Applied and Environmental Microbiology* 58(4): 1321–1325.

Utkilen, H. and N. Gjolme (1995). Iron-stimulated toxin production in *Microcystis aeruginosa*. *Applied and Environmental Microbiology* 60: 797–800.

Vincent, W. F. (2000). Cyanobacterial dominance in the polar regions. *The Ecology of Cyanobacteria: Their Diversity in Time and Space.* B. A. Whitton and M. Potts, eds. Dordrecht, Kluwer Academic Publishers: 321–340.

Walsby, A. E. and G. K. McAllister (1987). Buoyancy regulation by *Microcystis* in Lake Okaro. *New Zealand Journal of Marine and Freshwater Research* 21: 521–524.

Ward, D. M. and R. W. Castenholz (2000). Cyanobacteria in geothermal habitats. *The Ecology of Cyanobacteria.* B. A. Whitton and M. Potts, eds. Dordrecht, Kluwer Academic Publishers: 37–59.

Watanabe, M. (1987). Studies on the planktonic blue-green algae. 2. *Umezakia natans* gen. et sp.nov. (Stigonemataceae) from the Mikata lakes, Fukui Prefecture. *Bulletin of the National Science Museum of Tokyo Series B* 13: 81–88.

Watanabe, M. (1996). Isolation, cultivation and classification of bloom-forming *Microcystis* in Japan. *Toxic Microcystis.* M. F. Watanabe, K.-I. Harada, W. W. Carmichael, and H. Fujiki, eds. Boca Raton, FL, CRC Press: 13–34.

Watanabe, Y., M. F. Watanabe, et al. (1986). The distribution and relative abundance of bloom forming *Microcystis* species in several eutrophic waters. *Japanese Journal of Limnology (Rikusuigaku Zasshi)* 47: 87–93.

Weidner, C. (1999). *Toxische und Nicht-Toxische Cyanobakterian in Gewassern der Scharmultzelseeregion: Ihr Vorkommen in Gewassern Unterschiedlicher Trophie und Morphometrie und Steuermechanismen Ihrer Dynamik in Polymiktischen Flachseen.* Fakultat fur Umweltwissenschaften und Verfahrenstechnik. Cottbus, Germany, Brandenburgischen Technischen Universitat.

Weidner, C. and B. Nixdorf (1997). Verbreitung und Steuerung der Entwicklung von toxischen und nicht-toxischen Cyanobakterien in ostbrandenburgischen Gewassern unterschiedlicher Trophie und Hydrographie im Zuge der reduzierten Belastung. *Toxische Cyanobakterien in deutschen Gewassern: Verbreitung, Kontrollfaktoren und okologische Bedeutung.* Berlin, WaBoLu. Hefte 4/97: 17–26.

Welker, M., C. Steinberg, et al. (2001). Release and persistence of microcystins in natural waters. *Cyanotoxins: Occurrence, Causes, Consequences.* I. Chorus, ed. Berlin, Springer-Verlag: 83–101.

Weidner, C., P. M. Visser, et al. (2003). Effects of light on the microcystin content of *Microcystis* strain PCC 7806. *Applied and Environmental Microbiology* 69: 1475–1481.

Willen, T. and R. Mattsson (1997). Water-blooming and toxin-producing cyanobacteria in Swedish fresh and brackish waters, 1981–1995. *Hydrobiologia* 353: 181–192.

Williams, C. D., J. Burns, et al. (2001). Assessment of cyanotoxins in Florida's lakes, reservoirs, and rivers. *Cyanobacteria Survey Project.* Harmful Algal Bloom Task Force, St. John's Water Management District, Palatka, FL.

Wyman, M. and P. Fay (1986). Underwater light climate and the growth and pigmentation of planktonic blue-green algae (cyanobacteria) I. The influence of light quantity. *Proceedings of the Royal Society of London B* 227: 367–380.

Wynn-Williams, D. D. (2000). Cyanobacteria in deserts-life at the limit. *The Ecology of Cyanobacteria: Their Diversity in Time and Space.* B. Whitton and M. Potts, eds. Dordrecht, Kluwer Academic Publishers: 314–366.

Zohary, T. (1985). Hyperscums of the cyanobacterium *Microcystis aeruginosa* in a hypertrophic lake (Hartbeespoort Dam, South Africa). *Journal of Plankton Research* 7: 399–409.

5 Cyanobacterial Poisoning of Livestock and People

Cyanobacterial toxins first came to attention as a cause of poisoning of domestic animals. Poisonous lakes, ponds, and waterholes have long been known, but the first careful investigation of the cause of a series of livestock deaths to be reported in the scientific literature was that of George Francis in 1878 (Francis 1878). Francis was employed by the South Australian government as an analyst and was asked to report on cases of farm animal poisoning occurring on the shoreline of Lake Alexandrina, a large shallow coastal lake close to the mouth of the Murray River in South Australia. This lake connected to the sea through the river entrance and at that time would have contained partially brackish water. Francis noted that the water level that year was very low, with very slight inflow from the river, and that the water was unusually warm at 74ºF (23.3ºC). He described a "conferva" (a slimy mass of freshwater algae) in excessive quantities in the lake, floating on the surface and "wafted onto the lee shores, where it was forming a thick scum like green oil paint, some two to six inches thick, and as thick and pasty as porridge." He also noted the rafts of scum that passed out through the Murray Mouth and accumulated on the beach as beds of "green stuff" up to 12 in. (300 mm) thick. The decomposing scum was said to make a "most horrid stench like putrid urine" and exude a fluorescent blue pigment. The toxic organism was correctly identified by Francis as *Nodularia spumigena*, a common worldwide species found in the eutrophic brackish waters of the Baltic Sea and in coastal lakes in the present day.

Francis reported that drinking the scum resulted in poisoning and rapidly caused death, with the symptoms of "stupor and unconsciousness, falling and remaining quiet, as if asleep, unless touched when convulsions come on, with head and neck drawn back in rigid spasm, which subside before death." He reported the time from drinking to death of sheep, 1 to 6 or 8 h; horses, 8 to 13 h; dogs, 4 to 5 h; and pigs, 3 to 4 h. He described a postmortem examination of a sheep that had received a dose of 30 oz of scum by mouth (fluid ounces, total 840 ml). The sheep died 15 h later, with no scum visible in the stomach and no reported changes in lungs, liver, kidneys, or brain. Francis reported fluid accumulation in the abdominal cavity and around the heart and changes in the color of the blood. Comparison of this description with more recent pathological examinations of sheep killed by cyanobacterial toxins are presented later in this chapter. This thorough examination of an economically important poisoning of domestic animals has provided the basis for many subsequent field and laboratory investigations of cyanobacterial toxicity; these are explored in this chapter.

5.1 LIVESTOCK AND WILDLIFE POISONING BY CYANOBACTERIAL TOXINS

Since the time of Francis, livestock poisoning in Australia by toxic cyanobacteria has occurred regularly. Most examples have been due to stock drinking from small lakes and farm dams, which are highly eutrophic, with summer blooms of *Microcystis aeruginosa* (McBarron and May 1966). This has been regarded largely as a veterinary issue and only recently became a major public concern in Australia following the 1000-km-long cyanobacterial bloom on the Darling River in Australia in 1991. In this instance cyanobacterial scums accumulated along the river, especially in weir pools, where water flow was negligible due to drought and high summer temperatures. Approximately 2000 sheep and cattle deaths were reported, with high neurotoxicity shown in water samples from the river. There was also evidence of neurotoxicity in the reticulated drinking water supplied to one of the towns. This supply was pumped from a highly contaminated weir pool, and chlorination was the only treatment available. The New South Wales state government declared a state of emergency, which enabled the army to be asked to rapidly deploy portable water purification units. These units used flocculation with dissolved air flotation and were capable of removing intact cyanobacteria. After filtration, the water passed through granular activated carbon for adsorption of toxic organic compounds. No toxicity was detected in water produced by these units (see Box 7.1 in Bartram, Vapnek et al. 1999). The emergency ended with heavy rain in the river catchment, which washed the cyanobacteria downstream into water too turbid for regrowth. The cyanobacterium responsible was *Anabaena circinalis*, which was later shown to contain saxitoxin derivatives similar to those causing paralytic shellfish poisoning (Humpage, Rositano et al. 1994).

In the last 60 years there have been many reports of livestock deaths worldwide due to animals drinking from cyanobacterial scums, indicating that this is a widespread phenomenon in Mediterranean, continental, and temperate climates in both hemispheres (Carmichael and Falconer 1993). Most livestock deaths due to toxic cyanobacteria have been reported from South Africa, where toxic *Microcystis* is abundant in eutrophic water storage sites. The first reports were from Steyn in 1943 and 1945, who stated that many thousands of cattle and sheep had been poisoned over the preceding 25 to 30 years (Steyn 1943, 1944a, 1944b, 1945). One major water storage site that was heavily contaminated with *Microcystis*, the Vaal Dam (supplying drinking water to the Johannesburg area), caused very large numbers of livestock deaths. The Vaal Dam water was described as "green pea soup," with horses, mules, donkeys, dogs, hares, poultry, waterbirds, and fish lying dead on the edge of the lake and on the banks. The stock deaths were related to the prevailing winds, which had driven the scum onto the shore (Harding and Paxton 2001). South African livestock continue to be at risk from toxic cyanobacteria, as an entire dairy herd was poisoned in 1995, and numerous other poisonings have been reported, including wild animals (Harding and Paxton 2001).

Concurrent with the poisoning cases in South Africa, livestock deaths due to toxic blue-green algae in the U.S. and Canada were receiving attention. In a communication to the *American Journal of Public Health* in 1959, T. A. Olson reported that animal deaths due to poisonous algae had been recorded for three consecutive years in the 1880s in Waterville, Minnesota; in 1900 in Fergus Falls, Minnesota; in 1914 in Winnipeg Lake, Michigan; from 1918 to 1934, five poisonings again in Minnesota; in 1939 in Colorado; in 1943 in Missouri River in Montana; and in 1947 at Des Lacs Lake, North Dakota. He found in a survey of cyanobacterial blooms in Minnesota that 87% of the blooms collected contained *M. aeruginosa* and that 49 out of 60 tested (82%) were toxic to mice (Olson 1960). A comprehensive review of livestock poisoning in the U.S. was included in the paper entitled "Medical Aspects of Phycology," by Schwimmer and Schwimmer (1968), which also considered human poisoning and experimental tests on samples of toxic cyanobacteria. These authors quoted references to 25 separate livestock poisonings between 1887 and 1958 in the U.S. and 24 in Canada between 1917 and 1961.

Canadian records of livestock and wildlife deaths from "algal poisoning" commenced when deaths of horses, cattle, pigs, and birds at farms on the shores of three lakes in Alberta were described in 1917 (Gillam 1925). Cattle poisoning in Ontario was reported in 1924 (Howard and Berry 1933) and again in 1948, when occurrences of deaths at five adjacent farms on the shores of Sturgeon Lake in Ontario were published (MacKinnon 1950; Stewart, Barnum et al. 1950). The following year the *Canadian Journal of Comparative Medicine* published case histories for poisonings at two lakes in Alberta, in which deaths of cattle, horses, pigs, chickens, turkey, geese, wild birds, and dogs were reported (O'Donoghue and Wilton 1951). More cases were described in Manitoba in 1952, when the cyanobacterium *Aphanizomenon flos-aquae* was responsible for the deaths of horses, pigs, calves, dogs, and a cat (McLeod and Bondar 1952).

In 1960 a series of toxic water blooms during summer on lakes in Saskatchewan were described, with records of deaths of horses, dogs, geese and cattle. Many of the lakes were in use for recreation, and 10 children at a camp on a lakeshore reported to the local physician with diarrhea and vomiting after swimming in "algae-covered lake-water." Another local physician, deciding to swim at a point on a different lake where the cyanobacterial scum was least heavy, fell off a diving board and swallowed an estimated half-pint of water. Three hours later he developed stomach cramps, vomiting, and diarrhea, followed by a temperature of 102ºF (38.9ºC), a "splitting headache and pains in limb muscles and joints." *Microcystis* cells were abundant in his feces. The health authority issued a warning in newspapers and on radio and TV and posted notices not to bathe at any place where a scum had formed on the shoreline, and lifeguards were instructed to enforce it. As the cyanobacterial blooms continued throughout that summer, the warning was continued, with vigorous objections from resort owners (Dillenberg and Dehnel 1960).

The same summer the problem of water blooms became of concern to the water supply operators for the cities of Regina and Moose Jaw, first because their filters were clogging with cyanobacteria from their supply at Buffalo Pound Lake and second because three cows and six dogs died after a scum formed on the lake shore. *Microcystis* was identified in the scum, and mice died within 10 h after intraperitoneal injection of 0.5 mL, showing liver injury. Toxicity tests on other mice were carried out using water collected at the raw water intake; these animals showed "listless" behavior after injection, while comparable tests using the final, treated water left the subject animals unaffected (Dillenberg and Dehnel 1960). This study was the first to link livestock deaths to illness in the human population, in this case due probably to recreational exposure.

5.2 HUMAN POISONING BY CYANOBACTERIAL TOXINS

Human poisoning caused by toxic cyanobacteria in drinking water has been suspected in the U.S. as well as in Brazil, Europe, Africa, and Australia, but few cases are documented. When the abundantly recorded cases of livestock poisoning are compared with the few cases of suspected human poisoning, one may wonder why there have not been more human cases. The answer has a number of components. The first must be that water contaminated with concentrations of live cyanobacterial cells has an offensive smell and taste, reminiscent of moldy potatoes, due to geosmin and methylisoborneol (Kenefick, Hrudey et al. 1992); there is also a putrid, sulfurous smell when cells are dying and decaying. Bad-tasting water will be avoided unless there is no other choice. This also applies to livestock, which will avoid drinking from cyanobacterial scums by wading into deeper water or moving along the shoreline to clean water. Where cases of livestock death have been investigated in detail, it is often found that fence lines have prevented the animals from moving to areas of clear water, forcing them to drink the scum. Similar considerations apply to birds and wild animals, which will avoid the worst areas of contamination.

The second consideration is that the consequences of consuming a sufficient quantity of cyanobacterial toxins to cause poisoning are most often vomiting, diarrhea, a tender abdomen, and headache. These are also the symptoms of common gastrointestinal illnesses of viral, bacterial, or protozoal origin, which may not even be reported for medical diagnosis or treatment. Most families deal with these illnesses themselves and consult medical practitioners or hospitals only if the symptoms are prolonged. It is assumed in public health that only 20% of gastrointestinal illness is reported (Ministério da Saúde 1986). Only when a substantial number of reported cases in a specific population occur, an initial investigation has been completed, and no infectious disease agents have been found will the possibility of poisoning be addressed. The investigation is then likely to focus on heavy metals and industrial or agricultural chemicals rather than on cyanobacterial toxins. Thus the outcome of medical investigation of actual cases of cyanobacterial poisoning is most likely to be that no causation has been identified, so the illness is reported as gastroenteritis or hepatoenteritis of unknown origin.

5.3 WATERBORNE POISONING IN BRAZIL

An example of suspected cyanobacterial poisoning of a large number of children in several adjacent towns in Brazil illustrates the general points made earlier (Teixera, Costa et al. 1993). In 1988 a new hydroelectric dam was completed, the Itaparica Dam in Bahia, Brazil, resulting in the flooding of towns, villages, plantations, and forests along the river valley. The town and region of Paulo Alfonso, with a population of 213,000 people, drew their drinking water supply from the river upstream of the dam wall. The water was then processed by a conventional filtration and chlorination plant. Flooding of the valley behind the new dam began in mid-February 1988, and in mid-March an epidemic of gastroenteritis and diarrhea began. The health authority immediately implemented a massive campaign involving the issuance of free oral rehydration salts to the community as well as warnings to filter and boil water before use, the latter being broadcast over the radio, by meetings and talks, and by the distribution of educational material. Adjacent towns drawing drinking water from the Itaparica Dam were also affected, but towns that did not receive water from the dam did not have an outbreak.

The local health clinics and the area hospital monitored the outbreak and carried out a detailed epidemiological study of 76 individuals with the disease. The pattern of the outbreak seen by health clinics reached a peak of 76 cases in a single day in mid-April, with monthly totals of 191 cases in February, 436 in March, 1370 in April, and 395 in May. Hospital admissions for severe diarrheal illnesses in the preceding 6 months averaged 41 per month, with 44 in March, which rose to 131 in April and dropped back to 72 in May. Deaths among these patients rose from an average of 5.7 per month over the previous 6 months to 31 in April and 33 in May. The death rate in May from gastroenteritis increased to 45.1% of admissions. Overall, from approximately 2000 cases, 88 deaths resulted. The age distribution of all cases showed 70.6% in children under 5 years of age.

The study group of ill patients was selected from hospital outpatients in the same proportion of age groups to that in the overall distribution of cases. No *Salmonella*, *Shigella*, rotavirus, or adenovirus pathogens were found. Laboratory blood tests for cholinesterase inhibition, used to detect organophosphate poisoning, and measurements for heavy metal contamination of patients, showed results within normal limits. Water samples were collected at a range of locations and tested; the treated, chlorinated water from the distribution system and at domestic taps did not contain significant numbers of fecal coliforms, though the raw water prior to treatment did have high levels of coliforms. The outstanding data from the raw water samples at the treatment plant were that the water contained both *Anabaena* and *Microcystis* colonies at 1104 to 9755 “standard cyanobacterial units” per milliliter. The authors quoted this as 3.7 to 32.5 times the 300 cyanobacterial units per milliliter stated as the maximum acceptable level for drinking water prior to conventional treatment (Pan-American Health Organization, 1984). The term *units* referred to a filament or colony, so that the cell number would be considerably higher. The preferred method now is to carry out a cell count on representative colonies as well as a colony count so as to provide a cell concentration or to disaggregate the cells prior to counting. If an arbitrary conversion from colony count to cell number of 100 cells per colony (Kuiper-Goodman, Falconer et al. 1999) is applied to the upper number of cyanobacterial units reported, then the raw water contained some 10^6 cells per milliliter, which is a very high concentration of organisms and likely to be highly toxic.

The authors found that the study group included gastroenteritis patients who had filtered and boiled their water before use and had no detectable pathogens in their feces. It was therefore concluded that the symptoms of the illness, which resembled a severe upper gastrointestinal irritation, did not indicate any of the common causes of diarrhea endemic in the area, nor were any pathogens found. The lack of evidence for pesticides or heavy metal poisoning led the authors to conclude that cyanobacterial toxins were responsible, since high concentrations of toxic species were in the raw water. As a response, the water supply was dosed during the first week of May with copper sulfate to kill the organisms and again in the third week of May. By this time the epidemic had diminished sharply and the study was terminated (Teixera, Costa et al. 1993).

5.4 GASTROINTESTINAL ILLNESS ASSOCIATED WITH CYANOBACTERIA IN THE U.S.

A sequence of gastrointestinal illnesses in towns in the U.S. along the Ohio River was described in 1931; they appeared to be related to a pulse of water from a heavily contaminated tributary. The preceding year had very low rainfall, and the tributary river, the Kanawha, became anaerobic and carried a heavy cyanobacterial bloom. Rainfall caused this water to flow into the Ohio River and also to enter the drinking water intake for Charleston, West Virginia. Among a population of 60,000 in the town, 4,000 to 7,000 cases of abdominal pain, vomiting, and diarrhea were recorded. Following this, a sequence of brief epidemics of vomiting and diarrhea occurred consecutively along the Ohio River, but only in towns that used the river as the water source. No pathogenic cause was identified, and it was concluded that a chemical irritant in the water was responsible. At the time the illnesses were reported, the downstream drinking water was said to have a musty, decay-like, woody, or moldy odor, which is commonly associated with the presence of cyanobacteria. No measurements of the cyanobacteria were carried out (Tisdale 1931; Veldee 1931).

A more thoroughly investigated gastrointestinal disease outbreak in Sewickley, Pennsylvania, in 1975 was closely tied to cyanobacterial toxicity. Within 2 days of the outbreak, which was reported by a local physician, the State Department of Environmental Resources, the U.S. Environmental Protection Agency (EPA), and the Centers for Disease Control (CDC) in Atlanta were notified. The town water supplied about 8000 inhabitants, and the CDC epidemiological survey found that 62% of the people on that supply system became ill. The symptoms included diarrhea and abdominal cramps, and the illness subsided within 5 days of onset. The apparent incubation period was 2 to 4 days.

Scrutiny of the treatment plant revealed a hole in the groundwater intake structure under the Ohio River, allowing about 40% of the volume of intake to be drawn directly from the river. The turbidity level was five times higher during the outbreak than in samples taken before the hole formed. The treatment followed the sequence of prechlorination, softening/filtration, postchlorination, fluoridation, polyphosphate addition, and pH adjustment with sodium hydroxide. Standard chemical and coliform counts were undertaken daily; no coliforms were detected anywhere in the distribution system. The three distribution reservoirs were concrete and not covered, and the chlorine residual decreased to below 0.1 mg/L in effluent water. In the summer, the reservoirs were treated with copper sulfate 1 mg/L on Mondays, Wednesdays, and Fridays to control "algae." The rapidity with which the outbreak appeared and disappeared indicated that the causative agent was introduced into the distribution system at one or two of the distribution reservoirs and not prior to or at the treatment plant. Inspection of one of the reservoirs showed that a heavy growth of "algae" had occurred recently. Clumps of these organisms were found floating in the water, several piles of them that had been collected from the water were found around the shores, and clumps were growing on the reservoir bottom. A second reservoir had over 100,000 cells of cyanobacteria per milliliter in the open water. The species concerned were the filamentous *Schizothrix calcicola*, *Plectonema*, *Phormidium*, and *Lyngbya*, which normally form mats of filaments on sediments and rocky surfaces underwater. These can become displaced by climatic changes, or, more likely in this case, by dosing with copper sulfate. The investigation concluded that the contamination had entered the distribution system through the open finished-water reservoirs, but it did not point to the cause (Lippy and Erb 1976). Since that time, *Schizothrix*, *Phormidium*, and *Lyngbya* have all been found to be toxic, and the lipophilic toxins debromoaplysiatoxin and lyngbyatoxin have been isolated and characterized (Figure 3.1) (Mynderse, Moore et al. 1977; Baker, Steffensen et al. 2001).

Because these organisms are largely benthic and in most circumstances do not appear in the free water column, they have received little attention from water supply authorities or public health agencies. Only if the filamentous mats become displaced into the bulk water and pass directly into posttreatment drinking water supplies does a public health hazard occur. This was reported in an incident in South Australia in which toxic *Phormidium* was dislodged from the sediments of an open posttreatment holding reservoir due to climatic change and passed directly into the distribution system. As a result of the appearance of discolored lumps and off-flavors in the tap water and consequent complaints from consumers, the supply authority investigated the water source and located clumps of the cyanobacteria dispersed in the reservoir water. Mouse testing for toxicity indicated the presence of toxic compounds in the clumps of filaments. As the cyanobacteria were present in the posttreatment reticulation system, the use of the entire supply for drinking was suspended by the health authority. Free supplies of bottled water were provided through retail shops to the whole population until the system was changed to a noncontaminated source and flushed through with clean water. No unusual cases of gastroenteritis were reported (Baker, Steffensen et al. 2001).

In the cases of both the Sewickley and South Australia incidents, the basic cause was the use of open reservoirs for holding posttreatment water. Even though, in both cases, the water had been chlorinated before discharge into the holding reservoirs, extensive cyanobacterial growth had occurred on the sediments. The practical outcome of these incidents was the covering of the reservoirs by the supply authorities, thus shutting out the light and hence preventing any further cyanobacterial growth.

5.5 GASTROENTERITIS ASSOCIATED WITH CYANOBACTERIA IN AFRICA

Seasonal gastroenteritis in children was investigated by Zilberg (1966) in Harare, Zimbabwe, over the period from 1960 to 1965. He had noted that cases of acute gastroenteritis increased each year in the early winter months. Investigation showed that the gastroenteritis increased in areas with drinking water supplied by one in particular of the several reservoirs providing the city's water supply. Areas supplied by the other reservoirs did not show the increased gastroenteritis. The reservoir concerned had an annual heavy bloom of *M. aeruginosa*, which broke down in the autumn, at a time that coincided with the beginning of the gastroenteritis increase. No pathogens were shown to be responsible for the increased sickness. Zilberg therefore concluded that the disintegrating cyanobacterial cells in the reservoir could be responsible for a proportion of the cases (Zilberg 1966). The natural breakdown of the *Microcystis* cells would be expected to liberate intracellular toxins into solution in the reservoir water, from which they are not removed by normal flocculation and filtration processes in drinking water treatment. This association of the lysis of cyanobacterial cells from a heavy bloom in a drinking water reservoir with injury to the population consuming the water is considered in more detail in the next case.

5.6 LIVER DAMAGE ASSOCIATED WITH *MICROCYSTIS AERUGINOSA* IN AUSTRALIA

The drinking water supply to the city of Armidale, in New South Wales, Australia, was and is drawn from the reservoir formed by Malpas Dam, some 20 km from the city and 150 m higher. The reservoir receives water from an agricultural catchment, largely grazing livestock. When the dam was constructed and for about a decade thereafter, the effluent from the local large abattoir was discharged into a stream supplying the dam. Storm water from part of the town also drained into the reservoir. Superphosphate fertilizer was regularly spread by air onto the pasture adjacent to the reservoir. Regular summer blooms of *M. aeruginosa* occurred from the early 1970s onward and accumulated in the narrow intake area as a result of the prevailing wind, to the extent of forming scums 10 to 15 cm in depth, which surrounded the intake tower.

The resulting complaints of taste and odor in the water caused the supply authority to regularly treat the reservoir with copper sulfate at 1 ppm of copper in the top meter, spread from the air, which effectively terminated the blooms. In 1973, the taste problems were particularly apparent in the drinking water, with visible discoloration and particulate material in water from the tap. Inspection of the open posttreatment water tanks reticulating water to the city showed regrowth of *Microcystis* in the tanks. Inspection of the reservoir showed that a substantial *Microcystis* scum had accumulated around the drinking water intake, which would have put a considerable load on the treatment system, presumably sufficient to allow live cyanobacterial cells to pass through the filters. The water treatment plant for the city of Armidale was a standard construction, with chlorine pretreatment, alum flocculation with pH adjustment, sedimentation, rapid sand filtration, and postchlorination and fluoridation. No facility for activated carbon was provided.

A substantial quantity of *Microcystis* was collected from the reservoir at this time (1973) for investigation. Toxicity was evaluated and the toxin (microcystin) purified and partially characterized as a cyclic peptide of amino acid composition, comprising one molecule of each of L-methionine, L-tyrosine, D-alanine, β-methyl aspartic acid, D-glutamic acid, and a precursor of methylamine (Elleman, Falconer et al. 1978). The novel β-amino acid ADDA was not identified in microcystin until fast-atom bombardment mass spectroscopy was employed in 1985 to completely characterize the molecule as microcystin-YM (Botes, Tuinman et al. 1984; Botes, Wessels et al. 1985). The toxicity (LD_{50}, lethal dose killing 50% of the animals) of the purified material by intraperitoneal injection into mice was 56 μg of toxin per kilogram (Elleman, Falconer et al. 1978).

In 1981 there was a particularly heavy summer bloom of *Microcystis* in Malpas Dam; it was monitored for toxicity during its development and the date of copper sulfate treatment recorded (Falconer, Beresford et al. 1983). A retrospective epidemiological analysis was carried out on the data from all serum enzyme tests carried out by the Regional Pathology Service, covering the population of the whole region. The data for activity of the enzymes used clinically for evaluation of liver function — γ-glutamyl transferase, alanine aminotransferase, aspartate aminotransferase, and alkaline phosphatase — were assessed (Falconer, Beresford et al. 1983). On the basis of the home addresses of the patients whose blood had been tested, the data were sorted into the patients living within the distribution system for Malpas Dam water (Armidale city residents) and those in other towns with independent water supplies from other reservoirs. The data were then grouped for blood taken in the 6-week period prior to the observation of the *Microcystis* bloom, the 6 weeks during the growth of the bloom and its termination by aerial dosing of Malpas Dam reservoir with copper sulfate, and the 6 weeks after that. The data for the serum enzyme changes in the populations are shown in Figure 5.1. A statistically significant increase was shown in γ-glutamyl transferase in the group that received Malpas Dam water during the period of the bloom and its lysis, compared with the same group before and after the bloom, and compared with the patients outside the area supplied by Malpas Dam water. A smaller and not statistically significant increase at the same time and group was shown in alanine aminotransferase, with no clear differences in the other two enzymes. The increased alkaline phosphatase in the group outside the Malpas Dam water supply was due to a higher number of children and repeated enzyme measurements of a renal dialysis patient. Patients with alcoholism were present in all groups in reasonably constant proportions, and there were no major festivals at which high alcohol consumption would be expected. There were no outbreaks of clinical infections, such as hepatitis, or changes in overall health conditions during the periods assessed.

The mean increase in γ-glutamyl transferase activity in the sera of Armidale residents indicated overall minor liver damage; but there were patients within the population who had highly raised enzyme activity, indicating substantial liver damage. Elevation of the activity of γ-glutamyl transferase is an effective monitor for toxic attack on the liver and has been used successfully in experimental studies of *Microcystis* toxicity in pigs as a model for human injury (Falconer, Burch et al. 1994). Epidemiological studies of this type cannot prove a causal relationship, which requires evidence of toxin intake and toxin present in the patients. However, the data are strongly suggestive that the presence and subsequent lysis of the *Microcystis* bloom were responsible for the observed liver damage (Falconer, Beresford et al. 1983).

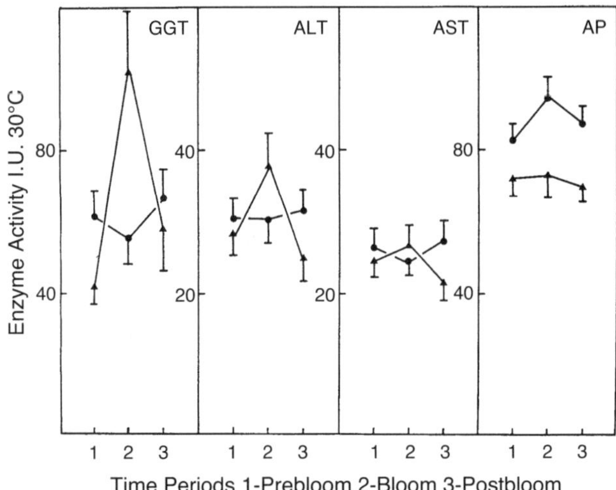

FIGURE 5.1 Serum enzyme activities in patients in the Armidale region of New South Wales before, during, and after a water bloom of *Microcystis aeruginosa* on the water supply reservoir. GGT, γ-glutamyl transferase; ALT, alanine aminotransferase; AST, aspartate aminotransferase; AP, alkaline phosphatase. (From Falconer, Beresford et al. 1983. With permission.)

5.7 RECREATIONAL POISONING IN THE U.K. AND U.S.

A direct case report of poisoning through recreational exposure to a *Microcystis* scum occurred in England in 1989. A group of 20 army recruits were carrying out exercises in Rudyard reservoir, Staffordshire. These involved "Eskimo rolls," in which the canoe is capsized and righted again. They were also swimming while carrying packs. Two recruits were admitted to the medical center after 4 to 5 days of illness with malaise, vomiting, sore throat, blistering around the mouth, dry cough, and pneumonia confirmed by x-ray examination. They also had abdominal tenderness and an elevated temperature. Eight other soldiers who had been canoeing were also examined and found to have sore throat, headache, abdominal pain, dry cough, diarrhea, vomiting, and blistering around the mouth. There was no evidence of pathogens, but the reservoir "contained a mass development of cyanobacteria, dominated by *Microcystis aeruginosa*" (Turner, Gammie et al. 1990). Microcystin-LR was identified in a sample of the bloom. While inhalation toxicity of microcystin has not received much research attention, the evidence available indicates that nasal application in rodents has a toxicity close to that of intraperitoneal injection, or some 10 to 50 times more potent than the oral toxicity (Fitzgeorge, Clark et al. 1994).

A coroner in Wisconsin recently concluded that a teenager who had been wrestling and diving in a golf course pond with a heavy scum of toxic *Anabaena flos-aquae* died through exposure to blue-green algal toxin. The cyanobacterial cells were found in fecal samples, and toxin was measured in scum samples. The teenager and his friend both suffered severe vomiting, diarrhea, and abdominal pain. In the more severely affected individual, this ended in shock and seizure 48 h after exposure (Behm 2003). The toxin implicated, anatoxin-a, is a fast-acting neurotoxin, but there are no data on human exposure. A neurotoxin would not be expected to cause the gastroenteritis, though it could cause death through respiratory inhibition and convulsions. Dog deaths have been reported from anatoxin-a poisoning, due to eating mats of cyanobacteria and from cleaning their coats after swimming in scums containing anatoxin-a (Edwards, Beattie et al. 1992). Allergic reactions to swimming in cyanobacterial blooms are more commonly reported than toxicity, though there is potential for harm by both routes (see discussion in Ressom, 1994). This fatal case in Wisconsin, like many others in which exposure to cyanobacteria has occurred, cannot be regarded as scientifically proved to be due to anatoxin-a. Unless anatoxin-a can be found in the gastrointestinal tract or the tissues in toxic concentrations, the alternatives of an acute allergic reaction or unidentified cyanobacterial toxin must remain.

5.8 THE DIALYSIS TRAGEDY IN BRAZIL

Hemodialysis is widely used to provide clinical support for patients with renal failure. Because of the potential for pathogens and toxins to enter the patients during treatment, great care is needed to ensure that the equipment used in dialysis is fully equipped with effective sterilization and filtration barriers against the entry of unwanted organisms or chemicals. Patients are particularly vulnerable to soluble, low-molecular-weight chemicals in the dialysis water, as some 120 to 150 L are used in the treatment of each patient. An early report of problems in dialysis patients described pyrogenic activity in dialysis water, which was attributed to lipopolysaccharide endotoxins present in water drawn from the Potomac River, U.S., during a cyanobacterial bloom (Hindman, Favero et al. 1975). In Portugal, a more severe case of illness in dialysis patients was described; it was attributed to microcystins from a heavy bloom of *Microcystis* causing liver damage (Araujo, F. personal communication).

In Brazil in 1996, a major disaster occurred at a dialysis clinic in Caruaru, Pernambuco State. Of 131 patients dialyzed over 1 week at one of the two dialysis clinics in the city, 116 were affected by immediate visual disturbances, nausea and vomiting, headache, muscle weakness, and epigastric pain. Of these 116 patients, 100 developed liver failure and 76 died. Of these deaths, 52 could be attributed to liver failure consequent on dialysis (Carmichael, Azevedo et al. 2001). The median age of the patients who died was 47 years, and that of those who survived, 35 years. Serum aspartate aminotransferase activity was raised eightfold and total bilirubin fourfold after dialysis in the patients showing clinical symptoms, clearly demonstrating liver damage (Jochimsen, Carmichael et al. 1998; Pouria, de Andrade et al. 1998).

The water supply was obtained from the Tabocas reservoir, which had a history of cyanobacterial blooms. Phytoplankton samples were not being collected at the time of the dialysis clinic event. Sampling in the following month recorded approximately 20,000 cyanobacterial cells per milliliter, with the predominant organisms *Aphanizomenon manguinii* and *Oscillatoria.* Earlier records show *Microcystis*, *Anabaena*, and *Cylindrospermopsis* present in the reservoir (Azevedo, Carmichael et al. 2002). The exact source of the water used at the dialysis clinic was not clear, as the clinic was not connected to the reticulated water supply. Water was supplied by road tankers, and during the week of the poisoning, the drivers were collecting water before filtration or chlorination at the municipal water treatment plant. The drivers may have put some chlorine into the tanker by hand (Jochimsen, Carmichael et al. 1998). From the road tanker, the water was transferred into a holding tank supplying the dialysis unit at the clinic. The dialysis unit had a water purification sequence of sand filtration, carbon adsorption, cation and anion deionization, and micropore filtration before the water was used for dialysis. The dialysis system had hollow-fiber exchange cartridges, which were replaced after approximately 20 patients. The carbon filter was scheduled to be changed every 6 months and the other filters every 3 months. No filters had been changed in the preceding 3 months, although the unit was receiving visibly turbid water.

No pesticides were detected in the system, though the neurological symptoms were severe in some patients, including deafness and blindness (Pouria, de Andrade et al. 1998). Microcystins were detected in the reservoir water, water in the delivery tanker truck, water in the dialysis unit holding tank, and in the carbon and ion exchange filters. Microcystins were also present in the sera and liver samples from patients both postmortem and while still alive. Histopathology on liver samples showed disruption of liver plates, hepatocyte deformity, extensive hepatocyte necrosis and apoptosis and leucocyte infiltration of the liver (Jochimsen, Carmichael et al. 1998).

Later analysis of material extracted from the filters found up to 2.2 μg of microcystin per gram and 19.7 μg of cylindrospermopsin per gram of filter medium. Analysis of patients' liver samples showed a range from 50 to 471 ng microcystins per gram. Serum content of microcystins was much lower, averaging 2.2 ng/mL. No cylindrospermopsin was found in the limited samples of liver and serum extracts available, which had been prepared for microcystin analysis and may not have been suitable for cylindrospermopsin detection (Carmichael, Azevedo et al. 2001). It is not yet clear what role cylindrospermopsin played in this poisoning event.

It was estimated that 19.5 μg/L of microcystins were present in the water used for dialysis of the affected patients, which compares to the oral intake concentration of the World Health Organizations' Guideline Value of 1 μg/L. Since dialysis is equivalent to an intravenous injection of toxin, it is clear that these patients received a dose of microcystins that would be expected to cause considerable injury (Carmichael, Azevedo et al. 2001).

A parallel study was carried out on patients at a second dialysis clinic at Caruaru, which received its water from the treated supply. Some evidence of symptoms in their dialysis patients was reported, but no serious illness (Jochimsen, Carmichael et al. 1998). Later analysis of the carbon filter at this clinic also showed microcystins (Carmichael, Azevedo et al. 2001).

5.9 PALM ISLAND POISONING BY CYLINDROSPERMOPSIN IN AUSTRALIA

The cyanobacterium *Cylindrospermopsis raciborskii* was responsible for a major poisoning in children at the tropical Palm Island, Queensland, Australia in 1979

(Byth 1980). Two water supplies were available on the island; the larger, Solomon Dam, was used by the majority of the population through a reticulated chlorinated drinking water supply. A smaller distant group used water from an unprotected shallow well. The rainfall had been particularly low, as was the reservoir's water level. For about 2 months before the onset of the poisoning, the odor and taste of the reticulated water had been particularly unpleasant due to the presence of a cyanobacterial bloom, resulting in complaints from the population. Variation in the depth of the water offtake failed to solve the problem, and when the bloom became dense, the supply authority treated the reservoir with 1 part per million of copper sulfate (Bourke, Hawes et al. 1983). Shortly after this action, cases of severe hepatoenteritis commenced in the population on the reticulated water supply. No cases occurred in the distant population without reticulated water.

The clinical picture of the disease was reported in detail (Byth 1980), with 138 children and 10 adults affected. Eighty-five cases were severe enough to be flown to a mainland general hospital and 53 were treated at the island hospital. Sixty-nine percent of affected children required intravenous fluids and electrolytes and 12% required intravenous infusion of plasma protein solution to reverse hypovolemic/acidotic shock (Byth 1980).

The symptoms were unusual, commencing with swollen livers, abdominal pain, constipation, and vomiting. There was little elevation of body temperature. Urinalysis showed evidence of considerable kidney damage, with glycosuria, proteinuria, ketouria and some hematuria. During the next 1 to 3 days, 82% of the children developed acidosis and hypokalemia. Within 2 further days, nearly half the children developed profuse diarrhea containing blood. As a result of considerable clinical care, including relocation into an intensive care unit when needed, no children or adults died, and within 26 days all patients had recovered (Byth 1980).

An epidemiological study was undertaken into the cause of the illness by the Queensland Department of Health, as it was not apparent at the time what had been the cause. No obvious pathogens were found, though the reticulated water supply was clearly the source of the illness and all patients questioned had drunk reticulated water. No secondary infections occurred, though such would have been expected if a pathogen had been responsible, especially as part of the population were using a range of shallow wells for drinking water due to the taste of the reticulated water. A chemical poison was suspected, and copper poisoning was proposed, as the supply reservoir had been dosed with copper sulfate (Prociv 1987).

The epidemiological report concluded that the most likely cause of the outbreak was endotoxins released from the cyanobacteria in the reservoir by copper sulfate dosing, which had been undertaken just prior to the first symptoms of illness (Bourke and Hawes 1983; Bourke, Hawes et al. 1983, 1986). In 1983, slow sand filters and later powdered activated carbon additions were incorporated into the water treatment to minimize risk from cyanobacterial toxicity (D. Griffiths, personal communication).

A 4–year study of the ecology of the Solomon Dam reservoir commencing in 1981 found that the water was subject to periodic cyanobacterial blooms, the predominant species being *C. raciborskii* and *A. circinalis*. Up to this time *C. raciborskii* had not been identified in Australia (Hawkins, Runnegar et al. 1985; Hawkins 1986).

Both species were subsequently cultured for toxicity testing, as they were suspected of causing the previous poisoning of the population on the island. *Anabaena* had earlier been shown to be poisonous (Carmichael, Gorham et al. 1977). The *C. raciborskii* cultured from Solomon Dam proved to be highly toxic, which was a new observation, as toxicity had not previously been reported for this species. Mice given intraperitoneal injections of freeze-dried culture showed severe hepatotoxicity, with damage also to kidneys, small intestine, lungs, and adrenal glands. The authors concluded that the symptoms of the toxicity were quite different from those of microcystin poisoning and that *C. raciborskii* was capable of producing the clinical disease seen at Palm Island (Hawkins, Runnegar et al. 1985). The cultured *Anabaena* strains from Solomon Dam did not show toxicity to mice.

5.10 CONCLUSIONS

There is now overwhelming evidence that cyanobacterial toxins have caused human injury and death. This conclusion is strengthened by the many cases of death of domestic animals from cyanobacterial poisoning, which are distributed worldwide. Since the cases of human injury include those in which the source of the cyanobacterial toxin was the drinking water supply, it is apparent that safe drinking water guideline values are needed. These are under active development and are considered in Chapter 8. Prior to determination of these safe levels, a considerable amount of knowledge is required of the toxicology of the poisons, and this is reviewed for cylindrospermopsin in Chapter 6 and microcystin in Chapter 7.

REFERENCES

Azevedo, S. M. F. O., W. W. Carmichael, et al. (2002). Human intoxication by microcystins during renal dialysis treatment in Caruaru, Brazil. *Toxicology* 181: 441–446.

Baker, P. D., D. A. Steffensen, et al. (2001). Preliminary evidence of toxicity associated with the benthic cyanobacterium Phormidium in South Australia. *Environmental Toxicology* 16(6, special issue): 506–511.

Bartram, J., J. C. Vapnek, et al. (1999). Implementation of management plans. *Toxic Cyanobacteria in Water: A Guide To Their Public Health Consequences, Monitoring and Management*. I. Chorus and J. Bartram, eds. London, E & FN Spon: 211–234.

Behm, D. (2003). Coroner cites algae in teen's death. Milwaukee, WI, *Milwaukee Journal Sentinel*.

Botes, D. P., A. A. Tuinman, et al. (1984). The structure of cyanoginosin-LA, a cyclic heptapeptide toxin from the cyanobacterium *Microcystis aeruginosa*. *Journal of the Chemical Society, Perkin Transactions* 1: 2311–2318.

Botes, D. P., P. L. Wessels, et al. (1985). Structural studies on cyanoginosins-LR, -YR, -YA, and -YM, peptide toxins *Microcystis aeruginosa*. *Journal of the Chemical Society, Perkin Transactions* 1: 2747–2748.

Bourke, A. T. C. and R. B. Hawes (1983). Freshwater cyanobacteria (blue-green algae) and human health. *Medical Journal of Australia* 28: 491–492.

Bourke, A. T. C., R. B. Hawes, et al. (1983). An outbreak of hepato-enteritis (the Palm Island mystery disease) possibly caused by algal intoxication. *Toxicon* 21 (suppl 3): 45–48.

Bourke, A. T. C., R. B. Hawes, et al. (1986). Palm Island mystery disease. *Medical Journal of Australia* 145: 486.

Byth, S. (1980). Palm Island mystery disease. *Medical Journal of Australia* 2: 40–42.

Carmichael, W. W., S. M. F. O. Azevedo, et al. (2001). Human fatalities from cyanobacteria: Chemical and biological evidence for cyanotoxins. *Environmental Health Perspectives* 109(7): 663–668.

Carmichael, W. W. and I. R. Falconer (1993). Diseases related to freshwater blue-green algal toxins, and control measures. *Algal Toxins in Seafood and Drinking Water.* I. R. Falconer, ed. London, Academic Press Limited: 187–209.

Carmichael, W. W., P. R. Gorham, et al. (1977). Two laboratory case studies on the oral toxicity to calves of the freshwater cyanophyte (blue-green alga) *Anabaena flos-aquae* NRC-44-1. *Canadian Veterinary Journal* 18(3): 71–75.

Dillenberg, H. O. and M. K. Dehnel (1960). Toxic waterbloom in Saskatchewan, 1959. *Canadian Medical Association Journal* 83: 1151–1154.

Edwards, C., K. A. Beattie, et al. (1992). Identification of anatoxin-a in benthic cyanobacteria (blue-green algae) and in associated dog poisonings at Loch Insh, Scotland. *Toxicon* 30(10): 1165–1175.

Elleman, T. C., I. R. Falconer, et al. (1978). Isolation, characterization and pathology of the toxin from a *Microcystis aeruginosa* (*Anacystis cyanea*) bloom. *Australian Journal of Biological Science* 31: 209–218.

Falconer, I. R., A. M. Beresford, et al. (1983). Evidence of liver damage by toxin from a bloom of the blue-green alga, *Microcystis aeruginosa*. *Medical Journal of Australia* 1(11): 511–514.

Falconer, I. R., M. D. Burch, et al. (1994). Toxicity of the blue-green alga (cyanobacterium) *Microcystis aeruginosa* in drinking water to growing pigs, as an animal model for human injury and risk assessment. *Environmental Toxicology and Water Quality* 9: 131–139.

Fitzgeorge, R. B., S. A. Clark, et al. (1994). Routes of intoxication. First International Symposium on Detection Methods for Cyanobacterial (Blue-Green Algae) Toxins. Cambridge, U.K., Royal Society of Chemistry.

Francis, G. (1878). Poisonous Australian lake. *Nature* 18(2): 11–12.

Gillam, W. G. (1925). The effect on livestock of water contaminated with freshwater algae. *Journal of the American Veterinary and Medical Association* 67: 780–784.

Harding, W. and B. Paxton (2001). Cyanobacteria in South Africa: A Review. Pretoria, South Africa, Water Research Commission: 165.

Hawkins, P. R. (1986). Some Aspects of the Limnology of a Small Tropical Impoundment and an Assessment of Two Techniques for Managing Water Quality, with Special Reference to the Growth of Cyanobacteria. Ph.D. thesis, Townsville, Australia, James Cook University.

Hawkins, P. R., M. T. C. Runnegar, et al. (1985). Severe hepatotoxicity caused by the tropical cyanobacterium (blue-green alga) *Cylindrospermopsis raciborskii* (Woloszynska) Seenaya and Subba Raju isolated form a domestic supply reservoir. *Applied and Environmental Microbiology* 50(5): 1292–1295.

Hindman, S. H., M. S. Favero, et al. (1975). Pyrogenic reactions during haemodialysis caused by extramural endotoxin. *Lancet* 18: 732–734.

Howard, N. J. and A. E. Berry (1933). Algal nuisances in surface waters. *Canadian Public Health Journal* 24: 377–384.

Humpage, A. R., J. Rositano, et al. (1994). Paralytic shellfish poisons from Australian cyanobacterial blooms. *Australian Journal of Marine and Freshwater Research* 45: 761–771.

Jochimsen, E. M., W. W. Carmichael, et al. (1998). Liver failure and death after exposure to microcystins at a hemodialysis center in Brazil. *New England Journal of Medicine* 338(13): 873–878.

Kenefick, S. I., S. E. Hrudey, et al. (1992). Odorous substances and cyanobacterial toxins in prairie drinking water sources. *Water Science and Technology* 25: 147–154.

Kuiper-Goodman, T., I. Falconer, et al. (1999). Human health aspects. *Toxic Cyanobacteria in Water: A Guide to Their Public Health Consequences, Monitoring and Management.* I. Chorus and J. Bartram, eds. London, E & FN Spon (on behalf of WHO): 113–153.

Lippy, E. C. and J. Erb (1976). Gastrointestinal illness at Sewickley, Pa. *Journal of the American Water Works Association* 68: 606–610.

MacKinnon, A. F. (1950). Report on algae poisoning. *Canadian Journal of Comparative Medicine* 14(6): 208.

McBarron, E. J. and V. May (1966). Poisoning of sheep in New South Wales by the blue-green alga *Anacystis cyanea* (Kuetz.) Dr. and Dail. *Australian Veterinary Journal* 42: 449–453.

McLeod, J. A. and G. F. Bondar (1952). A case of suspected algal poisoning in Manitoba. *Canadian Journal of Public Health* 43: 347–350.

Ministério da Saúde (1986). Programa de assistencia integral à saúde da criança: a doença diarréica aguda. Brasilia, Brasil.

Mynderse, J. S., R. E. Moore, et al. (1977). Antileukemia activity in the Oscillatoriaceae: isolation of debromoaplysiatoxin from *Lyngbya*. *Science* 196: 538–540.

O'Donoghue, J. G. and G. S. Wilton (1951). Algal poisoning in Alberta. *Canadian Journal of Comparative Medicines* 15(8): 193–198.

Olson, T. A. (1960). Water poisoning: A study of poisonous algae blooms in Minnesota. *American Journal of Public Health* 50(6): 883–884.

Pan-American Health Organization (1984). Guidelines for Drinking Water Quality: Vol. 1. Recommendations. Washington, D.C., PAHO.

Pouria, S., A. de Andrade, et al. (1998). Fatal microcystin intoxication in haemodialysis unit in Caruaru, Brazil. *Lancet* 352: 21–26.

Prociv, P. (1987). Palm Island reconsidered. Was it copper poisoning? *Australian and New Zealand Journal of Medicine* 17: 345–349.

Ressom, R., F. S. Soong, et al. (1994). *Health Effects of Toxic Cyanobacteria.* Canberra, Australia, National Health and Medicinal Research Council.

Schwimmer, M. and D. Schwimmer (1968). Medical aspects of phycology. *Algae, Man and the Environment.* D. F. Jackson, ed. Syracuse, NY, Syracuse University Press: 278–358.

Stewart, A. G., D. A. Barnum, et al. (1950). Algal poisoning in Ontario. *Canadian Journal of Comparative Medicine* 14(6): 197–202.

Steyn, D. G. (1943). Poisoning of animals by algae on dams and pans. *Farming in South Africa* 18: 489–510.

Steyn, D. G. (1944a). Vergiftiging deur slyk (algae) op damme en panne. *South African Medical Journal* 18: 378–379.

Steyn, D. G. (1944b). Poisonous and non-poisonous algae (waterbloom, scum) in dams and pans. *Farming in South Africa* 19: 465–472.

Steyn, D. G. (1945). Poisoning of animals and human beings by algae. *South African Journal of Science* 41: 243–244.

Teixera, M. G. L. C., M. C. N. Costa, et al. (1993). Gastroenteritis epidemic in the area of the Itaparica Dam, Bahia, Brazil. *Bulletin of the Pan-American Health Organization* 27(3): 244–253.

Tisdale, E. S. (1931). The 1930–1931 drought and its effect upon public water supply. *American Journal of Public Health* 21: 1203–1218.

Turner, P. C., A. J. Gammie, et al. (1990). Pneumonia associated with contact with cyanobacteria. *British Medical Journal* 300: 1440–1441.

Veldee, M. V. (1931). An epidemiological study of suspected water-borne gastroenteritis. *American Journal of Public Health* 21(9): 1227–1235.

Zilberg, B. (1966). Gastroenteritis in Salisbury European children: A five-year study. *Central African Journal of Medicine* 12(9): 164–168.

6 Cylindrospermopsin Toxicity

This cytotoxic alkaloid, found in several genera of cyanobacteria, is a recent addition to the field of toxin research. Evaluation of the toxicity of extracts of the cyanobacterium *Cylindrospermopsis raciborskii* began when it was identified as the probable cause of the Palm Island mystery disease, discussed in Chapter 5 (Hawkins, Runnegar et al. 1985). Evaluation of the purified toxin started in only 1992, when it was first isolated and the structure identified (Ohtani, Moore et al. 1992). Research activity into the toxicity of this alkaloid is rapidly intensifying as a result of its detection in drinking water sources in the U.S., Europe, Israel, Brazil, Southeast Asia, Japan, and Australia.

6.1 TOXICITY OF CYLINDROSPERMOPSIN: WHOLE-ANIMAL STUDIES

The oral and intraperitoneal toxicity to mice of *C. raciborskii* extracts and of purified cylindrospermopsin have been investigated. The first report on the toxicity of the organism cultured from the Solomon Dam, Palm Island, was published by Hawkins, Runnegar et al. (1985). This examined the effect of intraperitoneal injection of a saline extract of freeze-dried cells on young male mice. The lethal dose killing 50% of the animals (LD_{50}) over 24 h was 64 mg of freeze-dried cells per kilogram of body weight per mouse. Mice surviving beyond 24 h after receiving the LD_{50} died up to 5 days later, indicating a prolonged action of the poison. At a dose of 168 mg/kg, survival time was 6 to 9 h; increasing the dose beyond this by 2 to 10 times increased survival time to 10 to 12 h — an unexpected observation, as the expected effect of increasing the dose of a toxin is to shorten survival.

Postmortem examination of mice that received doses at or below the LD_{50} showed pale livers with white foci and focal hemorrhages in the lungs. Histopathological examination of these livers showed centrilobular necrosis. Higher doses resulted in swollen livers, mottled red, with a lobular pattern of congestion and hemorrhage. The histology showed massive general hepatocyte necrosis, sinusoidal congestion, and debris in the central veins. Embolic material and thrombi were seen postmortem in portal veins, lungs, and kidneys in mice that received a wide range of doses.

At the highest doses tested, liver injury diminished, with only scattered small groups of hepatocytes showing necrosis, and survival time was prolonged.

A number of other tissues demonstrated injury. The kidneys showed epithelial cell necrosis in the proximal tubules, the adrenal glands showed epithelial cell degeneration and necrosis, and the small intestine displayed congestion and edema. Other organs appeared normal (Hawkins, Runnegar et al. 1985).

A second source of highly toxic *C. raciborskii* (strain AWT 205) was cultured from an ornamental pond in Sydney, Australia. The LD_{50} for this material and the postmortem injury to mice were very similar to those shown by the Palm Island strain. The 7-day LD_{50} was 32 mg/kg compared to the 24-h LD_{50} of 52 mg/kg, showing progressive toxic action. However, analysis of the cylindrospermopsin content indicated a considerably greater toxicity than could be ascribed to the known toxin content, which suggested the possible presence of other toxic agents (Hawkins, Chandrasena et al. 1997).

Further examination of the effects on mice of intraperitoneal administration of *C. raciborskii* extracts focused on renal injury, as this was the main life-threatening component of the Palm Island poisoning. Dose-dependent damage was again seen in liver and kidney. Kidney damage was characterized by a reduction in the number of erythrocytes in the glomerulus and an increase in the space around it. The proximal tubules showed epithelial necrosis, increased diameter of the tubular lumina, and proteinaceous contents in the distal tubules. Electron microscopy demonstrated some tubules in which the lining cells were extensively vacuolated with shortened microvilli and others with extensive cellular degradation and cell debris in the tubular lumen at 24 h after an intraperitoneal dose of approximately half of the 24-h LD_{50}. By 72 h, the distal tubules were blocked by proteinaceous material. By 5 days, kidney tissue repair was evident in animals receiving sublethal doses of extract (Falconer, Hardy et al. 1999).

Oral administration of *C. raciborskii* extract resulted in the same damaging consequences to liver and kidney as intraperitoneal dosing but required much larger doses. It was noted that there were marked differences in toxicity between batches of *C. raciborskii* strain AWT 205, though all had been cultured from the same source in Australia. Different batches were seen to be of various colors, from deep greenish-red to grass-green and golden. The filaments were straight and few akinetes were seen (Falconer, unpublished observations). The relative toxicity to kidney and liver differed between batches, implying again that there may be more than one toxin in this species (Falconer, Hardy et al. 1999).

An investigation of the toxicity of oral dosing of *C. raciborskii* used a freeze-dried culture of known cylindrospermopsin content (Seawright, Nolan et al. 1999). The median lethal dose for mice was 4.4 to 6.9 mg/kg of toxin-equivalent, with death occurring between 2 and 6 days after dosing. Postmortem examination showed that the stomach after 4 days still contained culture material together with food, as did the proximal third of the small intestine. The large intestine contained some abnormal fecal pellets. Some mice had blood in the stomach and small intestine. The esophageal portion of the stomach showed ulceration. The liver was swollen and uniformly pale, and the kidneys were swollen and pale. The thymus was atrophic and the spleen shrunken. These results are compatible with the symptoms seen in children on Palm Island, which began with constipation and vomiting, a swollen liver, and kidney damage, only later followed by bloody diarrhea.

The histopathological study carried out on the tissues of dosed mice showed that there was marked fat accumulation in the hepatocytes, with perilobular necrosis and macrophage invasion. The kidneys showed extensive damage to tubular epithelium and debris in the ducts. Lymphoid tissue in the thymus and spleen was necrotic. The authors concluded that tissue anoxia and general systemic responses to intoxication may have been responsible for the extrahepatic effects apart from the stomach ulceration, which reflected lack of motility (Seawright, Nolan et al. 1999).

Coincident with these studies in Australia, the toxicity of *C. raciborskii* has been investigated in strains cultured from Brazil and from European lakes and rivers. One Brazilian strain was found to be poisonous to mice through saxitoxin-type toxins, commonly known as paralytic shellfish poisons (Lagos, Onodera et al. 1999; Molica, Onodera et al. 2002). European isolates from Hungary, Portugal, France, and Germany were also toxic, showing hepatic injury when dosed intraperitoneally to mice, and some also gave evidence of neurotoxicity. These *C. raciborskii* strains were examined for the presence of cylindrospermopsin, but none was found (Hiripi, Nagy et al. 1998; Kiss, Vehovszky et al. 2002; Bernard, Harvey et al. 2003; Fastner, Heinze et al. 2003; Saker, Nogueira et al. 2003). The toxicity of these strains was considerably lower than that of the Australian strains, but the absence of cylindrospermopsin indicates that other, so far uncharacterized toxins also occur in this organism.

6.2 ORAL TOXICITY OF CYLINDROSPERMOPSIN: STUDIES OF THE NO OBSERVED ADVERSE EFFECT LEVEL

To assess the minimum oral dose of cylindrospermopsin giving an adverse effect in mice (the Lowest Observed Adverse Effect Level, or LOAEL) and the maximum dose giving no adverse effect (the No Observed Adverse Effect Level, or NOAEL), subchronic toxicity studies were performed (Humpage and Falconer 2003). These studies were carried out to provide an experimental basis for determining a Guideline Value for safe drinking water and followed the protocol of the Organization for Economic Cooperation and Development (OECD) for toxicological evaluation (OECD 1998). In the first trial, cylindrospermopsin-containing extracts of *C. raciborskii* were given to mice in their drinking water for 10 weeks with a dose range of 0 to 657 μg cylindrospermopsin per kilogram of body weight per day. In the second trial, purified cylindrospermopsin was dosed by gavage for 11 weeks with a dose range of 0 to 240 μg/kg/day. Clinical observation was carried out throughout the trial, urine analysis during the latter part, and detailed postmortem evaluation, serum and whole-blood analysis, and histopathology at the termination of the period.

Toxic changes were apparent in different clinical parameters at different doses. Body-weight changes, which reflect overall toxicity, showed an increase in weight at low doses (30 and 60 μg/kg/day) but a decrease at higher doses (432 and 657 μg/kg/day). Liver and kidney weights increased at 240 and 60 μg/kg/day, respectively. Serum bilirubin, indicating liver damage, significantly increased at 216 μg/kg/day, while serum bile acids, also reflecting liver damage, decreased at the

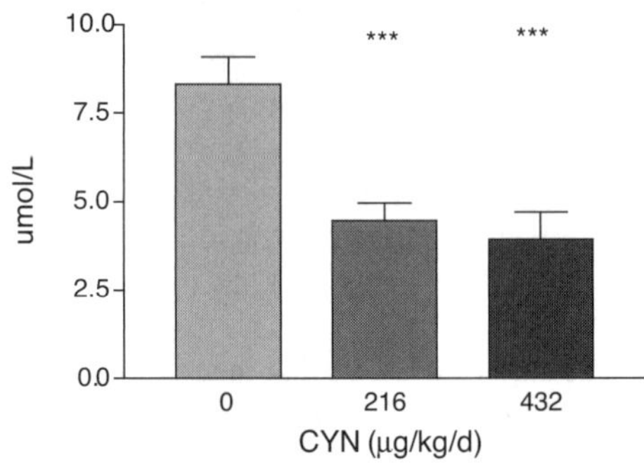

FIGURE 6.1 Serum bile acid concentration (μmol/L) in male mice exposed to cylindrospermopsin at two dose rates in drinking water for 10 weeks. (From Humpage and Falconer 2003. With permission.)

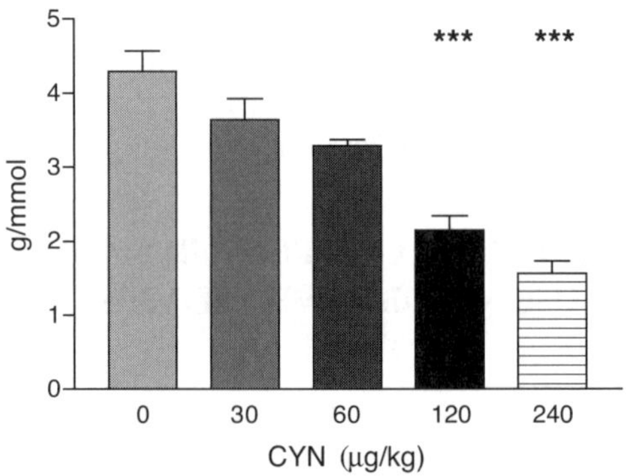

FIGURE 6.2 Urine protein/creatine (g/mmol) in male mice dosed by mouth with cylindrospermopsin at four dose rates over 11 weeks. (From Humpage and Falconer 2003. With permission.)

same dose (Figure 6.1). Urine protein content significantly decreased at 120 μg/kg/day (Figure 6.2). The data indicated that the kidney was more sensitive to cylindrospermopsin toxicity than the liver and that the NOAEL based on the total data was 30 μg/kg/day in male mice. These data are considered further in Chapter 8, which considers risk assessment and drinking water Guideline Values for cylindrospermopsin in safe potable water.

6.3 CYLINDROSPERMOPSIN UPTAKE AND EXCRETION

^{14}C-cylindrospermopsin was used to study the organ uptake and excretion of this toxin (Norris, Seawright et al. 2001). The labeled alkaloid was obtained by growing *C. raciborskii* with ^{14}C-bicarbonate in the culture medium in sealed flasks, followed by extraction and purification. The major excretion route following an intraperitoneal injection of a nonlethal dose of ^{14}C-cylindrospermopsin was through the kidneys, with urine content ranged from 30 to 100% of dose over 24 h, with a mean of 64%. Only 15% was recovered in the feces. The liver content was also highly variable; the mean at 48 h was 13%, with kidney content approximately one-tenth of liver content. Approximately half of the liver content of radioactivity after 48 h was extracted with methanol, which gave two peaks on high-performance liquid chromatography separation. The larger peak, containing about 80% of the radioactivity, appeared earlier off the column, indicating greater hydrophilicity than cylindrospermopsin; the later peak was coincident with cylindrospermopsin. The early peak was also seen in urine samples, which would be expected if it was an inactive conjugate of the parent molecule produced by detoxification processes in the liver. The chemical nature of this compound has not been identified.

The mechanism by which cylindrospermopsin enters cells is not clear. Microcystin has been shown to enter hepatocytes by active transport related to the bile acid transporter, which is discussed in Chapter 7. Incubation of rat hepatocytes with a lethal concentration of cylindrospermopsin (approximately 2 μM) and a range of concentrations of bile acids showed some protection by 100 μM bile acids at 48 h of incubation but none at 24 or 72 h. Similar incubation of epithelial carcinoma (KB) cells, which lack the bile acid transporter system, showed toxicity at higher cylindrospermopsin concentrations, showing that the toxin can enter cells independently of bile acid transport (Chong 2002). This is supported by the whole-animal toxicity studies, which show damage by cylindrospermopsin in a number of tissues that do not have the bile acid transporter. To resolve this issue, measurement of ^{14}C-labeled cylindrospermopsin uptake into incubated cells from different tissues in the presence of a range of transport inhibitors will be required. The difficulty with inferring toxin uptake from changes in toxicity is the possibility of intracellular metabolism of the toxin, which may generate toxic metabolites and thus affect any results in which cell death is the measure of entry.

6.4 MECHANISM OF CYLINDROSPERMOPSIN TOXICITY

Evaluation of the mechanism of toxicity of cylindrospermopsin commenced shortly after the initial isolation and purification of the alkaloid by Ohtani, Moore et al. (1992). Cultures of isolated primary rat hepatocytes were found to be a useful research tool for examination of cylindrospermopsin toxicity, which caused cell death and glutathione depletion, reaching 60% cell death in 18 h of culture at 5 μM concentration. Addition of agents depleting glutathione in the cultures potentiated cell death (Runnegar, Kong et al. 1994). Further evaluation of the role of glutathione in cylindrospermopsin toxicity found that the decrease in cell glutathione content was largely due to inhibition of synthesis rather than a major increase in consumption (Runnegar, Kong et al. 1995). Depleting glutathione in mice by dosing with buthionine sulfoximine and diethylmaleate prior to administration of cylindrospermopsin had no effect on survival rate, so it is unclear whether glutathione depletion plays a significant role in cylindrospermopsin toxicity (Norris, Seawright et al. 2002).

The production of cylindrospermopsin metabolites was implicated in toxicity, as inhibition of cytochrome P450 (a major pathway for oxidative metabolite formation from alkaloids) in cultured rat hepatocytes provided partial protection from cell death caused by cylindrospermopsin toxicity (Runnegar, Kong et al. 1995).

A significant development in the understanding of cylindrospermopsin toxicity resulted from electron microscopic studies of liver from mice dosed with the pure toxin. A dissociation of ribosomes from the rough endoplasmic reticulum of hepatocytes was seen within 16 h of intraperitoneal dosing of the mice, which was followed by a proliferation of the smooth endoplasmic reticulum into concentric whorls of membrane and an increase in autophagic vacuoles. Fatty droplet accumulation followed in the hepatocytes at 72 h after dosing (Terao, Ohmori et al. 1994). A decrease in cytochrome P450 was seen in livers of dosed mice. The action of cylindrospermopsin on protein synthesis, illustrated by ribosomal dissociation in hepatocytes, was confirmed by experiments with *in vitro* protein synthesis systems. The well-characterized methodology of the use of rabbit reticulocyte lysate for synthesis of proteins allows exploration of inhibitors of protein synthesis. Cylindrospermopsin was shown to totally inhibit globin synthesis *in vitro* at a concentration of 50 ng/mL (120 nM) (Terao, Ohmori et al. 1994). The investigations of Runnegar et al. and Terao et al. provide a coherent picture of one aspect of cylindrospermopsin toxicity, that of the inhibition of protein synthesis. This may be expected to result in the decrease of enzyme content in the liver, or isolated hepatocytes, due to the continuous degradation of enzymes in the cells, without replacement. It will also cause injury to other tissues, the extent depending on the entry of the toxin into the cell and the rapidity of protein turnover in the cell. Thus cells with a high synthesis or turnover of protein will be most affected. The liver and the gut lining as well as the immune system are thus particularly vulnerable. The accumulation of fat in the hepatocytes following low doses of cylindrospermopsin can be attributed to two elements. One is the lack of the protein component of lipoprotein transport out of the liver, leading to an accumulation of lipid. The other is a decrease in the ATP requirement due to inhibition of protein synthesis, reducing the use of acetyl groups for energy supply and thus enhancing fat synthesis.

6.5 INHIBITION OF PROTEIN SYNTHESIS

The effect of cylindrospermopsin on the mechanism of protein synthesis has been explored in isolated hepatocytes and in the rabbit reticulocyte system (Froscio, Humpage et al. 2001; Froscio 2002; Froscio, Humpage et al. 2003). The characteristics of the inhibition of protein synthesis in reticulocyte lysates are reproducible and can be used as an assay system for the measurement of cylindrospermopsin. The sensitivity allowed a detection limit of 50 nM in the assay solution, with an inhibition constant (IC_{50}) of 120 nM and a correlation of $r^2 = 0.99$ with the results by tandem mass spectroscopy. The time course of protein synthesis in the cell-free system was used to explore the location of the inhibition in the three main steps of protein synthesis. The first step is initiation, the association between the two subribosomal

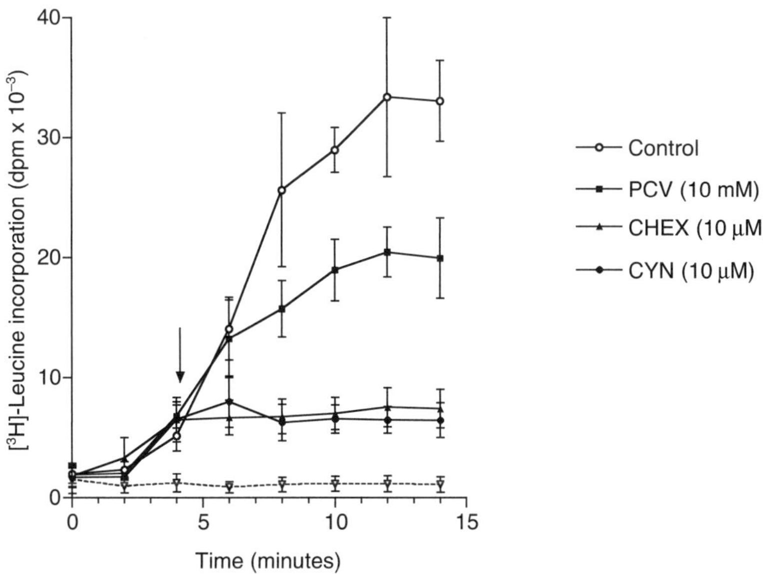

FIGURE 6.3 Inhibition of protein synthesis (measured by ^{3}H-leucine incorporation into protein) in reticulocyte lysates by pyrocatechol violet, cycloheximide, and cylindrospermopsin. Blank incubation (--∇--) contained no added mRNA. (From Froscio 2002. With permission.)

components and messenger RNA (mRNA) with initiation factors to form a ribosome initiation complex. This is followed by elongation of the growing peptide chain by sequential addition of amino acids, which requires activated amino acids attached to transfer RNA (tRNA) binding to the complex, formation of the peptide bond, and movement of the mRNA with respect to the ribosome. Protein elongation factors and energy in the form of guanosine triphosphate are needed. The final step is the termination of the peptide synthesis on the mRNA reaching a stop codon, at which point release factors assist in the dissociation of the peptide from the ribosome, the separation of the ribosome into its subunits, and the release of the mRNA. To identify which step is inhibited, the reticulocyte lysate system was used and the protein synthesis followed for 15 min after the commencement of synthesis by the addition of mRNA. Inhibitors were added after 4 min of incubation. The well-characterized inhibitors pyrocatechol violet, which is an initiation inhibitor, and cycloheximide, which is an elongation inhibitor, were compared with cylindrospermopsin. The results are shown in Figure 6.3, demonstrating the immediate stopping of protein synthesis by cycloheximide and cylindrospermopsin as compared to a time lag of 2 min before the inhibition by pyrocatechol violet commenced (Froscio 2002). This clearly showed that cylindrospermopsin is an elongation inhibitor. Further differentiation of the detailed point at which the toxin acts in peptide chain elongation remains to be explored.

The presence of the hydroxymethyl uracil residue on the cylindrospermopsin molecule provides an opportunity for hydrogen bonding (base pairing) with adenine on the tRNA, mRNA, or ribosomal RNA. Cycloheximide is known to bind to the larger ribosome subunit and prevent translocation, the relative movement of ribosome and mRNA as each amino acid is added to the chain. As the cycloheximide structure has similarities with cylindrospermopsin in having a glutarimide ring (resembling uracil) linked to the molecule through a hydroxyethyl group, this may be a point to commence investigation (Jiminez 1976). The rapidity of the inhibition by cylindrospermopsin in the cell-free system indicates that the toxin does not require metabolism for development of inhibitory activity. The reticulocyte lysate lacks the cytochrome P450 enzyme system used in hepatocytes for oxidative drug metabolism, supporting the concept that the unaltered toxin is responsible for protein synthesis inhibition.

Isolated primary mouse hepatocytes in monolayer culture also demonstrated inhibition of protein synthesis, with a maximum inhibition at 500 nM cylindrospermopsin observed from 4 h after toxin addition. The IC_{50} for cylindrospermopsin in hepatocyte cultures was 200 nM, approximately double the concentration required in the cell-free reticulocyte lysate system. The sigmoid shape of the inhibition curves was almost identical (Froscio 2002). This indicates that hepatocytes are freely permeable to cylindrospermopsin but that no active concentration mechanism was involved to transfer the toxin into the cells.

Examination of the reversibility of inhibition was carried out by incubating hepatocytes for 60 min in 5 μM toxin, then washing them and continuing incubation. The reversible inhibitor cycloheximide caused initial inhibition of protein synthesis with recovery to control rate of synthesis by 120 min after washing. Both emetine, a known irreversible inhibitor of protein synthesis, and cylindrospermopsin caused a total inhibition of protein synthesis with no recovery after washing (Froscio, Humpage et al. 2003). This lack of reversibility may be due to a covalent binding of cylindrospermopsin to the ribosome, secondary to uracil-mediated hydrogen bonding.

6.6 CYTOCHROME P450 IN CYLINDROSPERMOPSIN TOXICITY

Hepatocytes treated with cylindrospermopsin in culture showed dose- and time-dependent changes in cell morphology. These changes included rounding of the formerly flattened and adhering cells of the monolayers and irregularity of outline. There was a loss of cell viability as assessed by the leakage of lactate dehydrogenase from the cells, indicating loss of membrane integrity in dying cells. The effect is clearly concentration-dependent, with a significant enzyme leakage at 1 μM toxin, increasing up to 5 μM. The time course of leakage shows minor changes up to 10 h and significant change at 18 h. This compares with significant inhibition of protein synthesis at 0.5 μM at 4 h after toxin addition (Froscio, Humpage et al. 2003). Taken together, these differences indicate a dissociation between protein synthesis inhibition and cell death, which in cultured hepatocytes requires higher concentrations and longer time.

To explore the possibility that cylindrospermopsin metabolites are responsible for cell death whereas the unaltered cylindrospermopsin molecule causes protein synthesis inhibition but not death, a series of experiments were undertaken. Runnegar, Kong et al. (1995) showed that addition of cytochrome P450 inhibitors (α-naphoflavone and cimetidine) to rat hepatocyte cultures provided protection from cell death caused by cylindrospermopsin. This approach was extended with the use of two other cytochrome P450 inhibitors (ketoconazole and proadifen), both of which substantially protected cultured mouse hepatocytes from cylindrospermopsin toxicity, confirming that cylindrospermopsin toxicity in these systems requires oxidative metabolism of the original toxin to a reactive metabolite (Froscio, Humpage et al. 2003). Proadifen had no effect on protein synthesis in cultured hepatocytes and also provided no protection from protein synthesis inhibition by cylindrospermopsin, demonstrating again that unaltered cylindrospermopsin and not a metabolite was responsible for protein synthesis inhibition (Froscio, Humpage et al. 2003).

Studies with experimental animals treated with a cytochrome P450 inhibitor (piperonyl butoxide) have given ambiguous results, with apparent protection of one strain of mice from cylindrospermopsin toxicity but not of a second strain (Norris, Seawright et al. 2002). Induction of cytochrome P450 activity in the mouse liver had no effect on cylindrospermopsin toxicity, so that this enzyme system does not appear to be limiting if toxic metabolites are important in death from cylindrospermopsin poisoning (Norris, Seawright et al. 2002). There also appear to be differences in sensitivity of cells from different origins to cylindrospermopsin toxicity, as shown by the substantial time difference required to exert toxicity to hepatocytes as compared to the slower onset of toxicity in epithelial carcinoma cells (Chong 2002). This is likely to reflect the differences in activity of cytochrome P450 in cells of different organs, and since the liver is the main site of drug metabolism in the body, hepatocytes are likely to be the most active. The enzymology of cytochrome P450 metabolism of xenobiotics is complex because of the multiplicity of isoforms of the system, responding to different substrates and inhibited by different inhibitors (Meyer 1996). It is also possible that the cytochrome P450 system, which metabolizes cylindrospermopsin into reactive toxic products in mouse liver, has different characteristics than that in human liver. To clarify this point, it will be necessary to repeat the studies using human liver cells in culture and human liver microsome preparations containing cytochrome P450.

The markedly lower LD_{50} (increased toxicity) in mice at 7 days after dosing compared to 24 h after dosing may relate to initial toxic metabolites causing mortality at 24 h, whereas protein synthesis inhibition results in mortality at 7 days (Hawkins, Chandrasena et al. 1997). This is supported by the evidence that a toxic metabolite of cylindrospermopsin generated by cytochrome P450 oxidation causes death in cultured hepatocytes over 8 to 24 h in culture, whereas the simultaneous inhibition of protein synthesis does not (Froscio, Humpage et al. 2003). This phenomenon is not limited to mammals, since it was shown even more clearly in the brine shrimp, with an LD_{50} for exposure over 24 h of 8.1 μg cylindrospermopsin per milliliter as compared to 0.7 μg/mL after 72 h exposure (Metcalf et al. 2002). The apparently increased toxicity seen over several days may again reflect death through inhibition of protein synthesis as compared to more rapid death at higher doses through metabolite toxicity.

6.7 DNA DAMAGE, CHROMOSOME DAMAGE, AND CARCINOGENICITY

There is preliminary evidence of covalent adducts of cylindrospermopsin in livers of mice dosed with ^{14}C-labeled toxin, as approximately half of the radioactivity 6 h after dosing was in a washed methanol precipitate from livers (Norris, Seawright et al. 2001). This is supported by preliminary evidence of a DNA adduct in liver from cylindrospermopsin-dosed mice, shown by DNA extraction, hydrolysis, ^{32}P-labeling, and two-dimensional electrophoresis (Shaw, Seawright et al. 2000). Both protein and DNA adducts may occur by reaction of cylindrospermopsin itself with groups on the macromolecules, since the toxin molecule has several potential sites for reactivity. Cytochrome P450–generated reactive products from cylindrospermopsin will also be likely to form adducts with protein and DNA, as well as forming low-molecular-weight conjugates that are on the excretion pathway.

Since the cylindrospermopsin molecule contains a nucleic acid base, uracil, it is apparent that the possibility for interaction with adenine groups in RNA and DNA exists. Construction of models of cylindrospermopsin and DNA using matching dimensions has shown the possibility of intercalation of cylindrospermopsin between nucleotides of the DNA double helix (personal observation). Such intercalation may predispose the molecules to the formation of adducts to DNA and suggests the potential for DNA strand cleavage or mutation during DNA replication. It therefore can be hypothesized that cylindrospermopsin is likely to be mutagenic, clastogenic (causing DNA strand breakage), and even carcinogenic.

6.8 MICRONUCLEUS FORMATION IN THE PRESENCE OF CYLINDROSPERMOPSIN

To investigate these possibilities, *in vitro* and *in vivo* experiments have been carried out. The formation of micronuclei during cell division has been explored as a model for potential carcinogenicity. These micronuclei are chromosome fragments or whole chromosomes enclosed in a membrane outside the cell nucleus but within the cytoplasm in cells during division. The cell has thus undergone substantial genetic damage with loss of DNA (Fenech 1993).

A human lymphoid cell line in culture was used to explore the effect of cylindrospermopsin on micronucleus formation. The method employed a cytokinesis block (blocking the separation of daughter cells) by incubating dividing cells with cytochalasin-B. This allowed the identification of newly divided nuclei formed by chromosome duplication during the experimental incubation, as the subsequent prevention of separation of daughter cells resulted in accumulation of binucleate cells. In this way, newly divided nuclei could be identified from those in cells dividing prior to addition of the blocking agent. Incubation with or without cylindrospermopsin was carried out over 24 h, and the frequency of micronuclei in the binucleate cells was counted under a microscope. A significant and dose-dependent eightfold increase in micronuclei in binucleate cells was seen after incubation with cylindrospermopsin, demonstrating that the toxin caused chromosome damage and DNA loss from the nucleus. Examination of the frequency of centromeres (the chromosome component which attaches to the spindle fibers in nuclear division) in micronuclei also showed a dose-related threefold increase in the presence of cylindrospermopsin. Centromeres in a micronucleus demonstrate the loss of whole chromosomes, also induced by cylindrospermopsin. The increase in centromeres in micronuclei did not account for the greater increase in numbers of micronuclei, showing that an increased loss of DNA by chromosome breakage at higher concentrations of cylindrospermopsin occurred as well as whole-chromosome loss (Humpage, Fenech et al. 2000). These data demonstrated that cylindrospermopsin caused cytogenetic damage in human cells in culture by two mechanisms: by inducing DNA strand breaks with chromosome fragment loss into micronuclei and by affecting kinetochore/spindle function inducing loss of whole chromosomes. DNA strand breakage may be a consequence of adduct formation with cylindrospermopsin. Both processes will cause mutation or lethality in affected cells and are potentially carcinogenic.

A direct examination of DNA strand breakage in mouse liver following cylindrospermopsin dosing of male and female mice showed an approximately 50% decrease in median molecular length of extracted DNA at 6 h. This decrease reached a maximum at 24 h after dosing, when DNA from dosed mice showed about 25% of median molecular length of DNA from control livers. After this time DNA length appeared to be recovering at 72 h (Shen, Lam et al. 2002). This study, together with that of Humpage et al. (2000), demonstrates considerable cytogenetic damage resulting from cylindrospermopsin exposure.

6.9 WHOLE-ANIMAL CARCINOGENICITY

Whole-animal studies of carcinogenicity can be carried out using a range of experimental protocols, the most exhaustive being lifetime low-dose trials with two species of animal. These studies are very expensive to undertake, costing more than $1 million each. National agencies, for example the U.S. National Toxicology Program, carry out these trials after extensive evaluation of the risks posed by the potential carcinogen.

To carry out a preliminary exploration of carcinogenicity, a different protocol was employed (Falconer and Humpage 2001). Mice were orally dosed once, twice, or three times, with the doses separated by 2 weeks, with an extract of *C. raciborskii* containing a known concentration of cylindrospermopsin. Fourteen mice received one dose of cell extract of 1500 mg/kg body weight, 17 mice two doses of 1500 mg/kg, and 34 mice three doses of 500 mg/kg. Twenty-seven mice received no cylindrospermopsin. A proportion of each group was given a tumor promoter, O-tetradecanoylphorbol acetate, in the diet for the duration of the experiment. The study continued for 30 weeks, when the mice were euthanized and subjected to postmortem examination and histopathological investigation. Twelve of the mice that received two doses of 1500 mg/kg of extract were euthanized shortly after the second dose due to signs of toxicity.

Five tumors were found in the 53 cylindrospermopsin-treated mice after 30 weeks, while none were found in the controls. No effect of the tumor promoter was observed on treated or control mice. The number of mice used in this experiment was too low to provide statistical significance, though the calculated relative risk of cancer from cylindrospermopsin dosing (6.6) indicates a potential public health significance (Falconer and Humpage 2001).

6.10 ASSESSMENT OF CARCINOGENICITY

An overall assessment of the potential for carcinogenicity of cylindrospermopsin points to considerable risk from this toxin. The molecular structure implies the ability to base-pair to adenine nucleotides, and the shape indicates the possibility of intercalation into DNA. The inhibition of protein synthesis by cylindrospermopsin points to interaction with RNA. Preliminary evidence for DNA adducts in the mouse liver after cylindrospermopsin dosing and the considerable covalent binding of radiolabeled cylindrospermopsin to macromolecules in the livers of dosed mice show reactivity in tissues. Micronucleus formation in human lymphoid cells incubated with cylindrospermopsin showed DNA strand breakage as well as whole-chromosome loss, both of which are mutagenic and potentially carcinogenic. Finally, oral dosing of mice with cylindrospermopsin resulted in tumor growth in 10% of the animals, with no tumors seen in any control animal.

To provide statistical proof of cylindrospermopsin's carcinogenicity in rodents requires substantial additional experimentation. In particular, a lifetime study with group sizes of 200 or more mice or rats of both sexes and with three dose levels of toxin administered orally by gavage should provide statistical significance if a relative risk of 5 is to be demonstrated with 95% confidence. This type of study is essential to clarify the carcinogenic potential of cylindrospermopsin.

6.11 TERATOGENICITY, IMMUNOTOXICITY, AND REPRODUCTIVE INJURY

Other important areas of cylindrospermopsin toxicity that have yet to be investigated are teratogenic, immunological, and reproductive effects. The developing embryo is very vulnerable to toxins that can readily cross the placental barrier. Whether cylindrospermopsin can move into the embryo and fetus remains to be elucidated, requiring a focused study of teratogenic effects.

As this toxin inhibits protein synthesis, there is the potential for any cells actively secreting protein or undergoing hypertrophy to be particularly affected. Immune cells and the cells responsive to endocrine stimulation, such as the endometrial lining, are such possible targets. Investigation of immunological and reproductive injury by this toxin is urgently needed.

REFERENCES

Bernard, C., M. Harvey, et al. (2003). Toxicological comparison of diverse *Cylindrospermopsis raciborskii* strains: Evidence of liver damage caused by a French *C. raciborskii* strain. *Environmental Toxicology* 18: 176–186.

Chong, M. W. K. (2002). Toxicity and uptake mechanism of cylindrospermopsin and lophyrotomin in primary rat hepatocytes. *Toxicon* 40: 205–211.

Falconer, I. R., S. J. Hardy, et al. (1999). Hepatic and renal toxicity of the blue-green alga (cyanobacterium) *Cylindrospermopsis raciborskii* in male Swiss Albino mice. *Environmental Toxicology* 14(1): 143–150.

Falconer, I. R. and A. R. Humpage (2001). Preliminary evidence for in-vivo tumour initiation by oral administration of extracts of the blue-green alga *Cylindrospermopsis raciborskii* containing the toxin cylindrospermopsin. *Environmental Toxicology* 16(2): 192–195.

Fastner, J., R. Heinze, et al. (2003). Cylindrospermopsin occurrence in two German lakes and preliminary assessment of toxicity and toxin production of *Cylindrospermopsis raciborskii* (cyanobacteria) isolates. *Toxicon* 42(3): 313–321.

Fenech, M. (1993). The cytokinesis-block micronucleus technique: A detailed description of the method and its application to genotoxicity studies in human populations. *Mutation Research* 285: 35–44.

Froscio, S. M. (2002). Investigation of the Mechanisms Involved in Cylindrospermopsin Toxicity: Hepatocyte Culture and Reticulocyte Lysate Studies. Ph.D. thesis, Adelaide, Australia, University of Adelaide.

Froscio, S. M., A. R. Humpage, et al. (2001). Cell-free protein synthesis inhibition assay for the cyanobacterial toxin cylindrospermopsin. *Environmental Toxicology* 16(5): 408–412.

Froscio, S. M., A. R. Humpage, et al. (2003). Cylindrospermopsin-induced protein synthesis inhibition and its dissociation from acute toxicity in mouse hepatocytes. *Environmental Toxicology* 18(4): 243–251.

Hawkins, P. R., N. R. Chandrasena, et al. (1997). Isolation and toxicity of *Cylindrospermopsis raciborskii* from an ornamental lake. *Toxicon* 35(3): 341–346.

Hawkins, P. R., M. T. C. Runnegar, et al. (1985). Severe hepatotoxicity caused by the tropical cyanobacterium (blue-green alga) *Cylindrospermopsis raciborskii* (Woloszynska) Seenaya and Subba Raju isolated from a domestic supply reservoir. *Applied and Environmental Microbiology* 50(5): 1292–1295.

Hiripi, L., L. Nagy, et al. (1998). Insect (*Locusta migratoria migratorioides*) test monitoring the toxicity of cyanobacteria. *Neurotoxicology* 19(4–5): 605–608.

Humpage, A. R. and I. R. Falconer (2003). Oral toxicity of the cyanobacterial toxin cylindrospermopsin in male Swiss albino mice: Determination of no observed adverse effect level for deriving a drinking water guideline value. *Environmental Toxicology* 18: 94–103.

Humpage, A. R., M. Fenech, et al. (2000). Micronucleus induction and chromosome loss in WIL2-NS cells exposed to the cyanobacterial toxin, cylindrospermopsin. *Mutation Research* 472: 155–161.

Jiminez, A. (1976). Inhibitors of translation. *Trends in Biochemical Sciences* 1: 28–29.

Kiss, T., A. Vehovszky, et al. (2002). Membrane effects of toxins isolated from a cyanobacterium, *Cylindrospermopsis raciborskii,* on identified molluscan neurones. *Comparative Biochemistry and Physiology Part C* 131(2): 167–176.

Lagos, N., H. Onodera, et al. (1999). The first evidence of paralytic shellfish toxins in the freshwater cyanobacterium *Cylindrospermopsis raciborskii*, isolated from Brazil. *Toxicon* 37: 1359–1373.

Metcalf, J. S., K. A. Beattie, et al. (2002). Effects of organic solvents on the high performance liquid chromatographic analysis of the cyanobacterial toxin cylindrospermopsin and its recovery from environmental eutrophic waters by solid phase extraction. *FEMS Microbiology Letters* 216: 159–164.

Meyer, U. A. (1996). Overview of enzymes of drug metabolism. *Journal of Pharmacokinetics and Biopharmacology* 24: 449–459.

Molica, R., H. Onodera, et al. (2002). Toxins in the freshwater cyanobacterium *Cylindrospermopsis raciborskii* (Cyanophycae) isolated from the Tabocas reservoir in Caruaru, Brazil, including demonstration of a new saxitoxin analogue. *Phycologia* 41(6): 606–611.

Norris, R. G. L., A. A. Seawright, et al. (2001). Distribution of 14C cylindrospermopsin *in vivo* in the mouse. *Environmental Toxicology* 16: 498.

Norris, R. L. G., A. A. Seawright, et al. (2002). Hepatic xenobiotic metabolism of cylindrospermopsin *in vivo* in the mouse. *Toxicon* 40: 471–476.

OECD (1998). OECD Guideline for the Testing of Chemicals. Paris, Organization for European Cooperation and Development: 10.

Ohtani, I., R. E. Moore, et al. (1992). Cylindrospermopsin: A potent hepatotoxin from the blue-green alga *Cylindrospermopsis raciborskii*. *Journal of the American Chemical Society* 114: 7941–7942.

Runnegar, M. T., S. M. Kong, et al. (1994). The role of glutathione in the toxicity of a novel cyanobacterial alkaloid cylindrospermopsin in cultured rat hepatocytes. *Biochemical and Biophysical Research Communications* 201: 235–241.

Runnegar, M. T., S. M. Kong, et al. (1995). Inhibition of reduced glutathione synthesis by cyanobacterial alkaloid cylindrospermopsin in cultured rat hepatocytes. *Biochemical Pharmacology* 49: 219–225.

Saker, M. L., I. C. G. Nogueira, et al. (2003). First report and toxicological assessment of the cyanobacterium *Cylindrospermopsis raciborskii* from Portuguese freshwaters. *Ecotoxicology and Environmental Safety* 55: 243–250.

Seawright, A. A., C. C. Nolan, et al. (1999). The oral toxicity for mice of the tropical cyanobacterium *Cylindrospermopsis raciborskii* (Woloszynska). *Environmental Toxicology* 14(1): 135–142.

Shaw, G. R., A. A. Seawright, et al. (2000). Cylindrospermopsin, a cyanobacterial alkaloid: Evaluation of its toxicologic activity. *Therapeutic Drug Monitoring* 22(1): 89–92.

Shen, X., P. K. S. Lam, et al. (2002). Genotoxicity investigation of a cyanobacterial toxin, cylindrospermopsin. *Toxicon* 40(10): 1499–1501.

Terao, K., S. Ohmori, et al. (1994). Electron microscopic studies on experimental poisoning in mice induced by cylindrospermopsin isolated from blue-green alga *Umezakia natans*. *Toxicon* 32: 833–843.

7 Microcystin Toxicity

The microcystin family of peptides has received more research attention than have any other cyanobacterial toxins. This is a consequence of the many livestock deaths caused by consumption of toxic *Microcystis aeruginosa* (earlier known as *Anacystis cyanea* and *Microcystis toxica*) and the potential risk to human health from microcystin-contaminated drinking water. The recent inclusion of microcystin-LR as a toxic chemical in the World Health Organization (WHO) drinking water guidelines has further accelerated investigation of the adverse effects of these peptides on mammals.

7.1 ACUTE TOXICITY OF MICROCYSTIN TO RODENTS

Careful research into this toxicity commenced in the U.S. with a histopathological study of poisoning by *M. aeruginosa* extracts containing microcystin, which was published in 1946 in the *American Journal of Pathology.* In these experiments, a toxic extract of *M. aeruginosa* was administered by intraperitoneal injection into rats; detailed examination was then carried out on these animals from 15 min to 1 month after injection. The findings of this study were that changes in the liver could be discerned within 15 min, with increased blood content, which by 3 h had increased liver weight by 25%. The hepatic parenchyma was engorged with blood, with partial liquefaction of the cells. In animals surviving to 2 to 3 days, the liver was shrunken, yellow, and mottled; by 5 days, it was recovering; and after 1 month, it was apparently normal. Histology of the liver showed centrilobular necrosis by 4 h, with breakup of the liver architecture and sinusoids filled with red cells. Cell cytoplasm was seen in the central vein area of lobules, and the sinusoidal endothelium was destroyed in some areas. By 3 to 5 days, the liver architecture was reformed and mitotic figures were commonly seen, indicating regeneration of the hepatocyte population. Kidney injury was also seen, with necrosis of epithelial cells in the convoluted tubules. Heart and lungs showed small hemorrhages; other organs were unchanged. The authors concluded that death after short periods of exposure was probably due to shock; after longer exposure, death was likely due to hepatic insufficiency (Ashworth and Mason 1946). A later study of the cause of acute death to mice by intraperitoneal injection of purified microcystin also concluded that shock induced by intrahepatic hemorrhage was responsible (Elleman, Falconer et al. 1978).

Indicators of liver damage used very widely in human and veterinary clinical practice are measurements of serum enzymes. Two of these enzymes characteristic of liver injury were measured after intraperitoneal injection of purified microcystin-YM into mice. Aspartate aminotransferase increased 50-fold and lactate dehydrogenase

20-fold, with changes seen at 15 min after injection and increasing enzyme activity to 45 min. Ultrastructural changes in the liver demonstrated disruption of the endothelial lining of the sinusoids and hemorrhage into the hepatocyte architecture within 15 min of toxin exposure. By 60 min, hepatocyte nuclei were disrupted, the cytoplasmic structure was damaged, plasma membranes had disintegrated in an increasing proportion of cells, and cytoplasm could be seen adjacent to red cells (Falconer, Jackson et al. 1981). The observed enzyme changes were due to the leakage of hepatocyte contents into the venous drainage of the liver. Secondary lung damage can occur in this circumstance as capillaries become blocked by cell debris or small emboli. Similar studies in rats and mice in which microcystin-LR was dosed intraperitoneally showed a similar pattern of damage, with a rapid increase in alanine aminotransferase in serum and later increases in alkaline phosphatase, total bilirubin, blood urea nitrogen, and creatinine (Hooser, Beasley et al. 1989).

An ultrastructural study of microcystin-LR damage to rats showed a time sequence of changes in hepatocytes. The earliest changes were seen 10 min after intraperitoneal dosing: centrilobular cells showed a widening of intracellular spaces. At 20 min, the microvilli were diminishing on the sinusoidal surfaces, the cells were separated, the plasma membrane invaginated, and intracytoplasmic vacuoles formed. By 60 min, the centrilobular region of the liver showed cellular organelles in among red cells and platelets with necrotic hepatocytes. Cellular debris was seen in lung vasculature, and smaller amounts in the capillaries of the kidneys (Hooser, Beasley et al. 1990). The authors concluded that the changes seen were consistent with microcystin-induced alterations in the hepatocyte cytoskeleton. These changes and their mechanism are discussed in detail later.

7.2 SUBCHRONIC AND CHRONIC TOXICITY

When a drinking water reservoir has a *Microcystis* bloom, a category of poisoning to animals or human consumers that may occur is subchronic exposure to toxin. This type of injury by microcystin was initially assessed by daily dosing of mice by intraperitoneal injection of a range of doses of purified microcystin (LD_{50} 56 μg/kg body weight) for 6 weeks. At lower doses, this resulted in progressive enlargement of the liver, with hepatocyte degeneration and necrosis, fibrosis, and mononuclear leukocyte infiltration. At the higher doses, there was additionally weight loss and jaundice, with progressive liver failure (Elleman, Falconer et al. 1978).

A trial of chronic toxicity caused by the consumption of drinking water containing microcystins was later carried out; in it, mice were exposed for over 1 year. The doses selected ranged from one that resulted in 50% mortality within 5 weeks to one with no observed adverse effect. In general, cumulative mortality increased with dose, though the animals largely died of infection (pneumonia due to *Mycoplasma*), not liver failure. Histological examination of livers showed chronic active liver injury, which increased with age or duration of exposure. Male mice showed much greater mortality and also sensitivity to microcystin toxicity compared to female mice (Falconer, Smith et al. 1988).

7.3 DETERMINATION OF THE NO OBSERVED ADVERSE EFFECT LEVEL FOR MICROCYSTIN IN MICE

In order to provide data for determination of a No Observed Adverse Effect Level (NOAEL) for microcystin, a toxicity trial following Organization for Economic Cooperation and Development (OECD) guidelines was carried out in mice (Fawell, James et al. 1994). Groups of 15 male and female mice were administered by gavage doses of 0, 40, 200, or 1000 μg of microcystin-LR per kilogram of body weight daily for 13 weeks. An extensive set of serum enzymes, other blood biochemistry parameters and hematology parameters were assessed. All tissues from control and high-dose animals were assessed postmortem and microscopically, together with liver, lungs and kidneys from all animals. Male mice showed more sensitivity to poisoning, with raised serum enzymes (alkaline phosphatase, alanine aminotransferase, and aspartate aminotransferase) in the 200- and 1000-μg/kg groups and reduced serum protein and albumin. Histopathology showed hepatocyte degeneration and chronic inflammation in the liver only in the highest dose group in both sexes. No other organs showed any dose-related changes. It was concluded that the clear NOAEL in these experiments was 40 μg microcystin-LR per kilogram of body weight (Fawell, James et al. 1994).

7.4 LARGE-ANIMAL TOXICITY

The clinical picture in poisonings of large domestic animals is not identical to that of rodents. A study of the oral toxicity of microcystins to sheep was carried out by intraruminal infusion of *M. aeruginosa* extract over a range of doses. The consequences to the sheep were followed by serum analysis, postmortem pathology, histopathology, and electron microscopy of tissues. The earliest death was 18 h after administration, with the other high-dose animals surviving only up to 24 h. These animals showed a dramatic increase in aspartate aminotransferase, lactate dehydrogenase, glutamate dehydrogenase, and alkaline phosphatase in blood commencing 12 h after dosing.

The mid-dose group showed enzyme changes reaching a peak at 48 to 72 h, with return to normal by 150 h except for lactate dehydrogenase, which was still high at day 6. Serum bilirubin showed changes similar to those of the enzymes, with recovery by 150 h. The animals that died or were euthanized within 24 h showed a striking and uniform accumulation of yellow fluid in the body cavities, with a volume of 150 mL in the lungs and 300 mL in the abdomen, which rapidly clotted on exposure to air. The livers were increased in weight up to 50% above normal, pale, and with a lobular pattern. The interior of the body cavity and organs were patterned with small hemorrhages. Petechial hemorrhages were seen subcutaneously, intermuscularly, on the liver, around the heart, in the thymus, and in the mucosa of the abomasum, small intestine, and large intestine, with blood leaking into the contents of the large intestine. Histology of the liver showed centrilobular hepatocyte necrosis, containing cells with pyknotic nuclei and neutrophil and mononuclear leukocyte infiltration. The epithelium of the bile ductules was hyperplastic. Transmission electron microscopy of liver showed hepatocytes with large aggregations of smooth endoplasmic reticulum (Jackson, McInnes et al. 1984).

Two groups of sheep were accidentally poisoned by drinking scum on a lake shore heavily contaminated with *M. aeruginosa*. Over a period of 26 to 178 days, a total of 12 animals died adjacent to the lake and another 12 died after removal from the lake. The sheep surviving the acute poisoning developed photosensitivity, with facial swelling and dermatitis. The ongoing mortality shown in this case indicates that the initial liver injury had persisting consequences, causing death up to 6 months later. This lasting damage was not detected by serum enzyme analysis for aminotransferases, which had returned to normal after 3 weeks, though serum bile acids were raised for almost 3 months (Carbis, Waldron et al. 1995).

7.5 DETERMINATION OF THE NO OBSERVED ADVERSE EFFECT LEVEL OF MICROCYSTIN IN PIGS

Toxicity of microcystins to pigs has been used to explore the toxicokinetics and distribution of toxin and also to provide an animal model for human toxicity.

Pigs have digestive systems that are similar to those of humans; they also eat similar foods, have similar adult body weights, and have a similar relative lack of thermal insulation. Their kidney function is also similar. Their physiology and metabolism are therefore comparable to those of humans, so that they provide a closer model than a rodent or ruminant animal would offer. However, in terms of handling, feeding, and housing, as well as quantity of toxin needed for the equivalent dose, pigs are much more demanding than rodents as experimental animals. All of these factors make pig research costly; hence pigs are used only for critical studies.

To provide data for the development of a WHO Guideline Value for microcystin, two studies were carried out — one in mice and the other in pigs. The mouse study by Fawell et al. (1994) was described above. The pig study used four groups of five male pigs during growth from a body weight of 25 to 60 kg, which was a period of 7 weeks. The dose rate of microcystins averaged 280 to 1312 μg of microcystins per kilogram per day, consumed in drinking water. The source of the microcystins was an extract from a naturally occurring bloom that had earlier been associated with sheep deaths (Carbis, Simons et al. 1994). An 8000-L tanker truckload of *Microcystis* scum was collected, mixed evenly in an above-ground swimming pool, and divided into 5-L aliquots for blast-freezing. The frozen, lysed cells were thawed, diluted as required, and supplied as drinking water to the pigs. The microcystins present were a mixture of at least nine variants of the microcystin molecule, the major component being microcystin-YR (L-tyrosine, L-arginine). No microcystin-LR or -RR was present. As microcystin variants have a range of toxicities (Table 3.2; Sivonen and Jones 1999), with microcystin-LR among the most toxic at an LD_{50} of 50 μg/kg body weight for mice, an assumed LD_{50} of 100 μg/kg was used to calculate dose rates of the mixture.

Blood samples were collected at regular intervals over 8 weeks and analyzed for liver function indicators. Microcystin exposure lasted for 6 weeks. After euthanasia, all animals were subjected to detailed postmortem examination and organs were removed for histopathological examination. No external clinical symptoms were observed during dosing apart from a brief reduction in water and food intake in the highest-dose-rate pigs on introduction of the cyanobacterial extract. This was attributed to the taste and odor of the drinking water and resulted in a brief reduction in weight gain. No gastrointestinal effects, such as diarrhea, were seen. Only the pigs' livers showed any indication of injury, with raised γ-glutamyl transferase, alkaline phosphatase, and bilirubin in the plasma of the two higher-dosed groups and a reduction in albumin (Figure 7.1). There were no clear changes in aspartate aminotransferase, alanine aminotransferase, or lactate dehydrogenase. Histopathological examination showed moderate hepatic cord disruption, centrilobular degeneration, single-cell necrosis, and cytoplasmic degeneration of hepatocytes in livers from most pigs of the two higher-dosed groups but in only one pig of the low-dose group. It was concluded that clear evidence of liver injury was shown in the pigs of the two higher-dosed groups but only in one of the five pigs in the lowest-dosed group. The Lowest Observed Adverse Effect Level (LOAEL) was therefore 280 μg/kg/day. On the basis of only one pig showing changes at this dose, it was likely to be close to the NOAEL (Falconer, Burch et al. 1994).

This dose rate, standardized for toxicity against microcystin-LR, reduced to a microcystin-LR equivalent of approximately 100 μg/kg/day (Kuiper-Goodman, Falconer et al. 1999). This value provided supporting evidence to the microcystin-LR dose of 40 μg/kg/day for the NOAEL in mice, which was used by WHO for the determination of the Guideline Value for safe drinking water, as discussed in Chapter 8.

7.6 TOXICOKINETICS OF MICROCYSTIN

The toxicokinetics and organ distribution of microcystins in animals were first studied by the use of ^{125}I-iodinated microcystin-YM. This radioisotope can be incorporated into peptides containing tyrosine through gentle enzymatic oxidation of iodide 125 to iodine 125, which then reacts with tyrosyl residues to form iodotyrosine 125. The iodinated toxin was shown to have normal toxicity. Radiolabeled microcystin was administered to mice by intraperitoneal injection and was found to locate preferentially in the liver. At 24 h after dosing, it was noted that badly damaged livers had retained a greater proportion of the radioactivity than less damaged livers, implying excretion of toxin from undamaged liver tissue (Runnegar, Falconer et al. 1986).

The toxicokinetics of uptake and excretion of ^{125}I-microcystin-YM were studied after intravenous injection in rats. A biphasic blood disappearance curve gave half-lives of 2.1 and 42 min, the fast phase interpreted as distribution from blood into the extracellular fluid pool and the slow phase as organ and tissue uptake. The liver showed a concentration ratio of 4.7:1 compared to the blood concentration after 120 min, and ratios of 14.6:1 and 21:1 were shown in the fluid contents of the upper and lower duodenum at the same time. As this radioactivity appeared first in the upper duodenum, it was considered to be from bile excretion by the liver. Kidney and urine also showed concentration of radioactivity, but this was likely to be

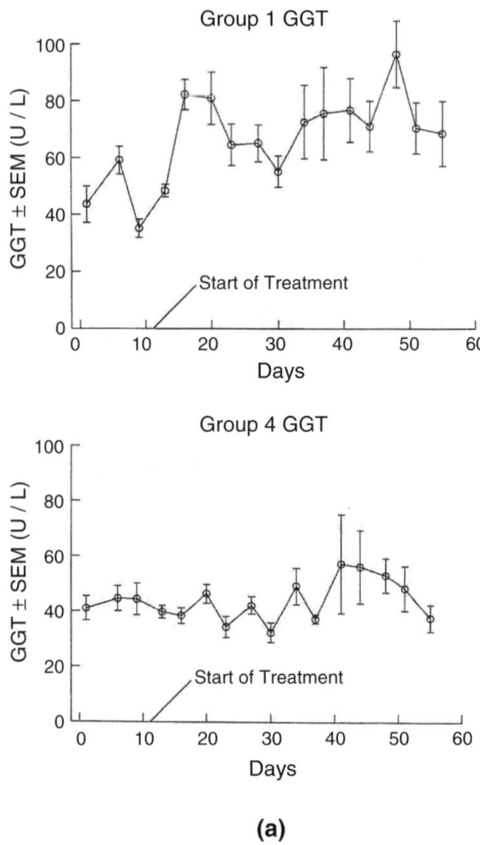

FIGURE 7.1 Serum changes in pigs exposed the microcystins in their drinking water. Group 1, high dose; Group 4, no microcystin: (a) GGT, γ-glutamyl transferase; (b) ALP, alkaline phosphatase; (c) total bilirubin; (d) serum albumin. (From Falconer, Burch et al. 1994. With permission.)

overestimated, as any contaminating iodide 125 would be rapidly excreted through the kidney. Only 1.9% of the total dose accumulated in the urine. No other organs showed significant concentration of dose. The proportion of injected dose found in the liver at 120 min was 19.25%, with 9.4% in the gut contents. From these data it was clear that the liver was the main target for both the accumulation and excretion of labeled microcystin (Falconer, Buckley et al. 1986).

Tritiated microcystin was used to explore the toxicokinetics of microcystin distribution in fasted mice (Robinson, Miura et al. 1989). This was prepared by an exchange of hydrogen between $^3H_20$ and microcystin-LR. The radiolabel was identified in the glutamate and aspartate residues of the toxin and the LD_{50} of tritiated toxin was similar to that of the native toxin. There is the possibility of subsequent reverse exchange, which would result in the formation of tritiated water, with loss

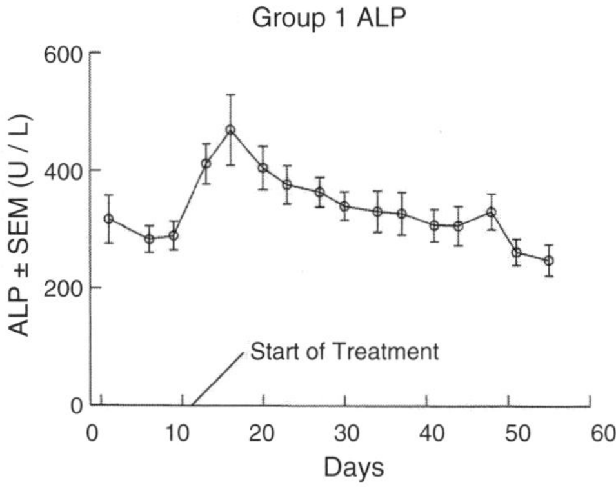

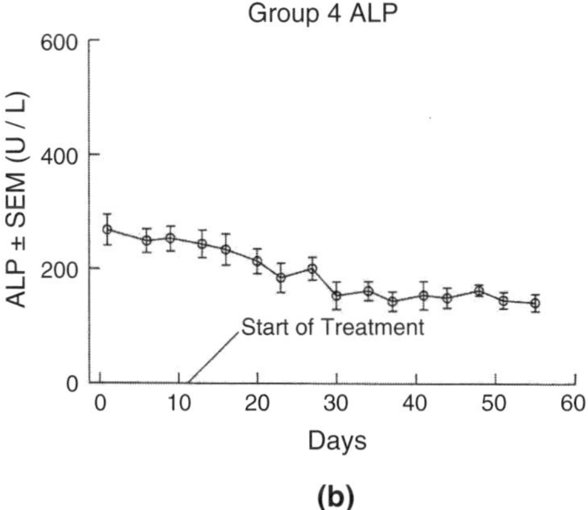

FIGURE 7.1 (CONTINUED)

of label from the radiolabeled microcystin. This would raise the excretion of radiolabel in the urine, leading to an overestimation of urinary excretion.

Six hours after intraperitoneal injection of tritiated toxin in mice, 56% of the radioactivity was in liver, 7% in intestine, and 0.9% in kidneys (Robinson, Miura et al. 1989). After intravenous injection, the half-lives of radioactivity in the blood were 0.8 min for the fast phase and 6.9 min for the slow phase. At 60 min, the liver contained 67% of the dose. Over 6 days, 14% of the radioactivity was excreted in the feces and 9% in urine. The liver essentially retained the radioactivity, which was taken up, initially, for the 6 days of observation. In the liver after 1 day, 83% of the radioactivity in the cytosol of hepatic cells was covalently bound, and there was evidence of the presence of potential detoxification products (Robinson, Pace et al. 1991).

Because of the possibility of racemization of the D-glutamate and D-aspartate into L-amino acids during the tritiation of the α-hydrogen atom, a second approach to study toxin uptake was developed. The unsaturated amino acid *N*-methyldehydroalanine in the peptide ring will reduce to *N*-methylalanine if carried out with

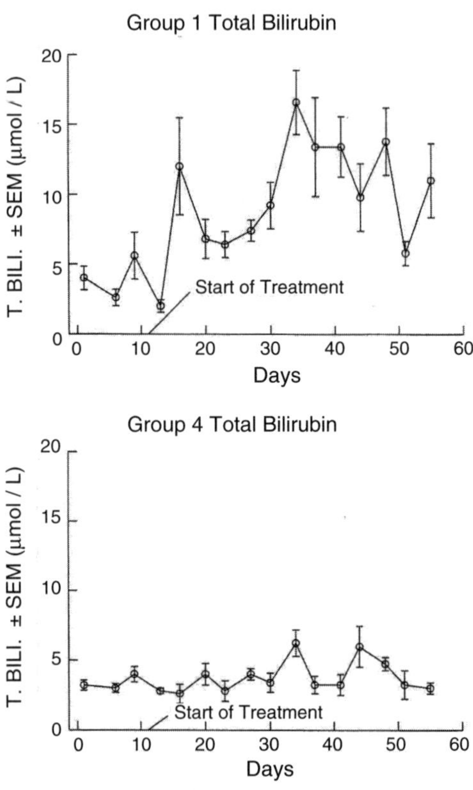

FIGURE 7.1 (CONTINUED)

sodium boro[^{3}H]hydride. The [^{3}H]hydrogen atoms in the alanine will be inherently stable and not readily exchanged. This reaction produced two epimers of [^{3}H]dihydromicrocystin, with a toxicity of the mixture roughly one-half to one-third that of the original toxin and with similar toxic action on the liver. The cellular uptake of one of the epimers was three times higher than that of the other, possibly accounting for the difference in toxicity. Forty-five minutes after intravenous injection of a sublethal dose in male mice, approximately 35% of radioactivity was in the liver, 4% in the intestine and kidneys, and less than 2% in other locations (Meriluoto, Nygard et al. 1990).

Pigs have provided a good model for microcystin uptake, also using [^{3}H]dihydromicrocystin-LR, prepared by reduction with sodium boro [^{3}H]hydride of microcystin-LR (Stotts, Twardock et al. 1997a,b). Two dose rates of intravenous infusion in anesthetized pigs were used, 25 or 75 μg/kg of unlabeled dihydromicrocystin-LR plus ^{3}H-dihydromicrocystin-LR/kg (both epimers), infused over 1 min to pigs of approximately 20 kg. Radioactivity in the blood was monitored for 4 h, after which the animals were euthanized and tissues removed for analysis and histopathology.

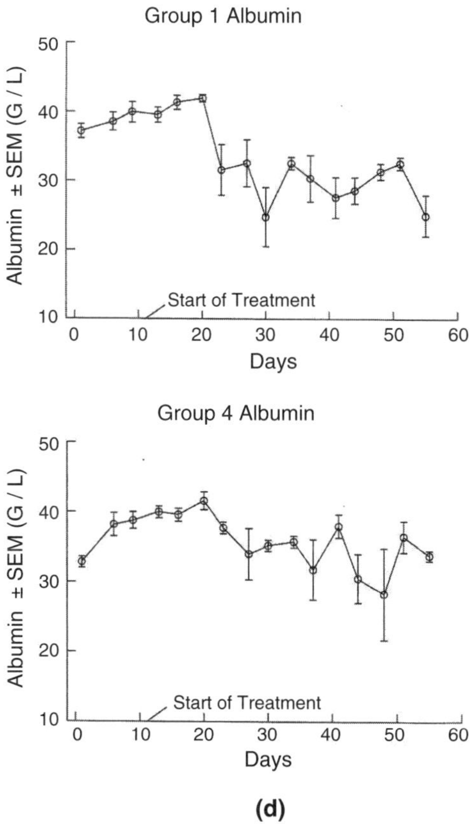

FIGURE 7.1 (CONTINUED)

Histopathological injury to the liver was severe at the higher dose. At the lower dose the half-life of radioactivity in the blood during the fast phase of distribution was 3.7 min, the slower phase 134 min. At the high dose rate, the fast-phase half-life was unchanged but the slow phase was doubled, indicating that the damaged liver was less capable of toxin uptake (Stotts, Twardock et al. 1997b). This relates to the fast phase being redistribution of toxin from blood into the extracellular fluid, while the slow phase is largely uptake into the liver. At 4 h after administration of the lower dose, 65% of the dose was in the liver, with less than 2% of the dose in any other tissue. Radioactivity was present in the bile within 30 min of dosing, showing rapid excretion of the toxin into the duodenum (Stotts, Twardock et al. 1997b).

Infusion of tritiated toxin through an isolated ileal loop in pigs demonstrated transfer into the portal venous blood, with 3.6:1 ratio of portal blood radioactivity to systemic radioactivity 90 min after dosing. After 5 h, half of the total dose infused into the ileal loop was relocated into the liver (Stotts, Twardock et al. 1997a,b).

The evidence from these studies of microcystin uptake and distribution therefore indicates that oral exposure of the animals to microcystin would result in transfer of toxin through the ileal wall into the portal bloodstream, from which it is rapidly taken up into the liver, generating a considerable concentration gradient between liver and blood. This implies an active uptake mechanism in hepatocytes. Almost immediately after the liver is exposed to microcystin, the toxin — or a metabolite or conjugate — begins to be excreted into the bile, from which it passes down the duodenum and ileum. Little excretion of microcystin or metabolites occurred in the urine. Fecal excretion was therefore the main route, though the possibility of an enterohepatic circulation exists, with microcystin passing from the bile duct being taken up again lower in the ileum. Conjugates of microcystin may be bacterially degraded to the native molecule in the intestine and also taken up into the portal vein.

7.7 CONJUGATION AND EXCRETION OF MICROCYSTIN

It was demonstrated that microcystin caused decreases in the glutathione concentration in rat hepatocytes in culture, indicating a role for glutathione in the detoxification of this peptide (Runnegar, Andrews et al. 1987).

After administration of microcystin to a live rat or mouse or perfusion of a rodent liver with microcystin, very little free unaltered toxin can be found in the liver. The majority of toxin present is covalently bound, attached through the methyldehydroalanine residue to a specific cysteine on protein phosphatase 1 or 2A (Mackintosh, Dalby et al. 1995; Runnegar, Berndt et al. 1995b). A small amount of microcystin in the liver occurs as toxin conjugates. The reaction of the methyldehydroalanine residue with cysteine is also an excretion route, since the enzyme glutathione S-transferase in the liver formed a glutathione (GSH) conjugate of microcystin within 3 h of exposure of the animal to toxin (Kondo, Matsumoto et al. 1996). The liver also contained cysteine (CYS) conjugate, and both compounds are likely to be major excretion routes for unbound microcystin from the liver. A third metabolite of microcystin was also identified, in which the ADDA residue had been sulfated and the methyldehydroalanine conjugated with glutathione. It was presumed, for the formation of this conjugate, that the unsaturated chain of the ADDA had been oxidized via an epoxide, which then hydrolyzed and esterified to form a sulfate (Kondo, Matsumoto et al. 1996). All three of these conjugates can be expected to pass rapidly into the bile duct through the action of the ATP-driven conjugate excretion pump (Ito, Takai et al. 2002).

Toxicity measurement of microcystin-GSH and microcystin-CYS conjugates, dosed by intratracheal administration, demonstrated a major reduction in toxicity to 1/12 of the toxicity of the original microcystin-LR. This was unexpected, as *in vitro* inhibition of protein phosphatases 1 and 2A by the conjugates was similar to that of microcystin-LR (Ito, Takai et al. 2002). The interpretation was unclear, though probably related to lowered uptake of conjugates into the liver and more rapid elimination into the bile. Immunostaining showed that the conjugates appeared in the lung, the route of administration; the stomach, small intestine, large intestine, and cecum; and in the kidneys. Less appeared in the liver. The CYS conjugate appeared concentrated in the kidney proximal tubules and medulla, indicating an excretion route through the kidneys for this compound. Microcystin-RR and the -GSH conjugate appeared to be mainly excreted through the feces (Ito, Takai et al. 2002).

Recent investigation of the role of glutathione in microcystin-LR metabolism has shown that in mice, both glutathione-S-transferase activity and total reduced glutathione in the liver increase in response to the toxin and that the transcription of γ-glutamyl transferase (the rate-limiting enzyme in glutathione synthesis) also increases. It is therefore apparent that glutathione plays a major role in microcystin detoxification in the hepatocyte and that the cells respond to the toxin by new enzyme synthesis and raised glutathione production. Glutathione peroxidase is also activated, indicating that the peroxidase is also involved in repair mechanisms following microcystin toxicity (Gehringer, Shephard et al. 2004).

7.8 STUDIES WITH ISOLATED HEPATOCYTES

The uptake of microcystin has been further explored by the use of suspensions of freshly isolated hepatocytes, which provide a valuable *in vitro* model for cell metabolism. Rat hepatocytes were prepared by collagenase perfusion, washed and incubated at 37ºC for up to 2 h. Concentrations of purified microcystin-YM were added to the incubation, and time- and dose-dependent deformation of the cells was observed by light and scanning electron microscopy. At 50 ng/mL (approximately 50 nM) of toxin, changes in cell surface microvilli were visible after 5 min of exposure to the toxin. At 5 min of exposure to 1.0 µg/mL, gross surface blebbing of hepatocytes was seen, which at 20 min had resulted in complete loss of microvilli and extrusion of large blebs from the cell body (Figure 7.2) (Runnegar, Falconer et al. 1981).

These deformations resembled those caused by the mushroom poison phalloidin (which is also a peptide toxin) in isolated hepatocytes. Investigation of the mechanism of phalloidin entry into hepatocytes showed that the bile acid transport

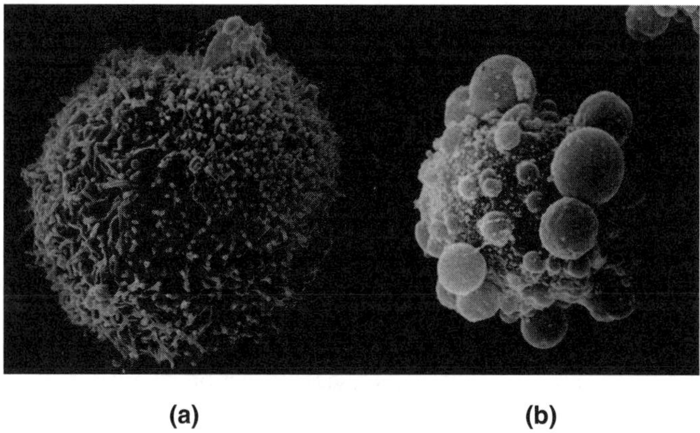

FIGURE 7.2 Isolated hepatocytes incubated for 30 min in the absence or presence of microcystin: (a) control hepatocyte; (b) hepatocyte incubated with microcystin.

mechanism in the plasma membrane was responsible, which carried anionic bile acids and the anionic phalloidin peptide into the cells (Petzinger and Frimmer 1980). This is an energy-requiring active transport, which will concentrate the transported compound into hepatocytes against a concentration gradient. Microcystin is also an anionic peptide at physiological pH due to the dicarboxylic acids glutamate and aspartate in the heptapeptide ring. Transport mechanisms are subject to competition between the different molecules transported, as the carrier has a finite maximum rate of transport. Hence incubation of hepatocytes with microcystin and varying concentrations of bile acids was expected to identify whether or not the bile acid transporter was carrying the microcystin into the cells. Cell deformation was used as the indicator of microcystin entry. At 100 ng/ml (approximately 100 nM) of toxin, a concentration of 20 μM deoxycholate reduced the proportion of deformed hepatocytes to 25% of the number in the absence of the bile acid. As the microcystin concentration increased, the extent of protection by the bile acid decreased (Runnegar, Falconer et al. 1981; Runnegar and Falconer 1982). The conclusion that the bile acid transporter carried microcystin into hepatocytes was supported by the prevention of deformation of hepatocytes by rifampicin and sulfobromophthalein, both of which interfere with bile acid uptake.

A study of the kinetics of microcystin uptake into isolated hepatocytes was carried out using ^{125}I-labeled microcystin-YM. Uptake with time was linear at low microcystin concentrations (17 nM) up to 60 min but rose linearly and then stopped after 10 min at higher concentrations (417 nM). This was associated with cell-surface deformation occurring at this time at the higher concentration. Rate of uptake was dependent on temperature, with an activation energy calculated from initial uptake velocity of 77 kJ/mol. Measurements of the concentration ratio of intracellular microcystin to concentration in the incubation medium after 60 min of incubation showed a greater than 80-fold concentration in the cells. Both the concentration ratio and the activation energy demonstrate an active, carrier-mediated uptake process.

Inhibition of ^{125}I-microcystin uptake into hepatocytes was also demonstrated, with rifampicin and sulfobromophthalein as well as with sodium deoxycholate. That the inhibitor effects were of physiological significance was confirmed by measurement of cell blebbing and of the activity of the enzyme phosphorylase-a in the cells. A rapid effect of microcystin in hepatocytes is a substantial activation of this enzyme, the biochemistry of which is discussed later. The inhibitors of microcystin transport also prevented blebbing and phosphorylase-a activation (Runnegar, Gerdes et al. 1991).

Tritiated [^{3}H]dihydromicrocystin-LR has also been used in uptake studies with hepatocytes and with perfused rat liver (Hooser, Kuhlenschmidt et al. 1991). Cold (0°C) and rifampicin were shown to inhibit uptake. Fractionation of hepatocytes and liver after [^{3}H]dihydromicrocystin exposure demonstrated that 65 to 77% of the label was in the cytosolic fraction and 13 to 18% in the nuclear/membrane fraction. Trichloroacetic acid precipitation of the cytosol fraction showed that 50 to 60% of the label was bound to protein, indicating a covalent addition product of microcystin.

In a separate study, the protein-microcystin complex from liver cytosol was subjected to heat denaturation and pronase digestion. This released 80% of the bound radioactivity, which constituted 22% of the unchanged toxin, together with 52 and

6% of different radioactive metabolites. Both of the major components appeared to be bound to a protein monomer of molecular weight 40,000. Binding studies with microcystin showed the binding capacity present in the cytosol of a variety of tissues as well as their capacity to form the microcystin metabolite. Okadaic acid showed concentration-dependent inhibition of microcystin binding, indicating that the proteins concerned were protein phosphatases, which are discussed in detail in the next section (Robinson, Matson et al. 1991).

The relationship between microcystin uptake into liver and isolated hepatocytes and protein phosphatase inhibition has also been studied using a range of uptake inhibitors. Predosing of mice with cyclosporine, rifamycin, and trypans blue and red provided protection for protein phosphatases in the liver against later toxic doses of microcystin. *In vitro* in hepatocyte suspensions, cholate, taurocholate, and sulfobromophthalein additionally gave protection of phosphatases from inhibition. Of particular interest in these experiments was a protective effect shown by pretreatment with microcystin, which showed that prior inhibition of intracellular phosphatases suppressed the subsequent uptake of microcystin into the cells while not reducing the uptake of taurocholate. This indicates that the microcystin transporter is itself sensitive to phosphatase inhibition and that it is not identical to the bile acid transporter (Runnegar, Berndt et al. 1995a).

7.9 MECHANISMS OF MICROCYSTIN TOXICITY

Toxins with a very high biological potency often act as specific inhibitors of key enzymes in cell metabolism. To be active at very low intracellular concentrations, the affinity between the toxin and the catalytic site of the enzyme must be very high, the number of enzyme molecules in the cell low, and the nonspecific binding of the toxin to other cell constituents low. If the enzyme inhibited by the toxin is at the apex of a cascade of enzymatic activation or deactivation reactions involving other enzymes, the toxic effect is greatly amplified. Microcystin poisoning of liver cells meets these criteria for highly potent toxins through the specific inhibition of particular phosphatase enzymes that are part of an enzyme cascade. These enzymes remove phosphate (dephosphorylation) from either serine or threonine hydroxyl groups on specific proteins, including key enzymes whose activity is regulated by the presence or absence of the phosphate group (Barford 1996).

7.10 MICROCYSTIN AND PHOSPHATASE INHIBITION

The role of protein phosphorylation and dephosphorylation in cells is complex, as it is the interaction between two sets of enzymes acting in opposite directions, and both form cascades of reactions. The enzymes that add phosphate groups from ATP to serine/threonine hydroxyl groups on specific proteins, the serine/threonine protein kinases, are major cell control enzymes. Many of these reactions are under the influence of hormones. Addition of phosphate to the enzyme protein activates some enzymes such as glycogen phosphorylase, which is responsible for the breakdown of glycogen to glucose phosphate. Phosphorylation inhibits other enzymes, such as glycogen synthetase, which polymerizes glucose phosphate into glycogen. These protein kinases include enzymes, for example protein kinase C, that are activated by compounds such as the phorbol esters, which stimulate tumor growth (Suganuma, Fujiki et al. 1988; Leithe, Cruciani et al. 2003).

The protein phosphatases hydrolyze phosphate groups off phosphorylated proteins, thus reversing the effect of the kinases. The balance between the activities of these two groups of enzymes powerfully influences cell structure and function, intermediary metabolism, cell division, and cancer cell growth (Cohen and Cohen 1989).

Evidence that microcystins might act as stimulators of cancer growth (tumor promoters) (Falconer, Smith et al. 1988) led to investigation of the possibility that they may operate in a similar manner to the tumor promoter okadaic acid (Falconer and Buckley 1989; Nishiwaki-Matsushima, Nishiwaki et al. 1991). Okadaic acid is a marine algal toxin, causing diarrheic shellfish poisoning (see Falconer 1993), which is a potent stimulator of tumor tissue growth. Okadaic acid had earlier been shown to interact with specific protein receptors in cells, leading to excess phosphorylation of cell proteins (Suganuma, Fujiki et al. 1988; Cohen, Holmes et al. 1990). Similarly, microcystin-LR, when added to a mouse liver cytosol fraction, caused an apparent increase in protein phosphorylation. Microcystin also blocked [3H]okadaic acid binding to its receptors and *in vitro* inhibited protein dephosphorylation (Yoshizawa, Matsushima et al. 1990).

The okadaic acid receptors in the cell cytosol are protein phosphatases 1 and 2A, enzymes that remove phosphate groups from specific structural cell proteins and from specific phosphorylated enzymes. Phosphatase activity was measured *in vitro* by removal of [^{32}P]phosphate from [^{32}P] phosphorylated histone H1, a nuclear protein binding to DNA. Microcystin was shown to inhibit the activity of this enzyme by 50% at 1.6 nM concentration. A similar concentration inhibited [^{3}H]okadaic acid binding to the phosphatase, showing that both toxins were acting at the same site on the enzyme protein (Yoshizawa, Matsushima et al. 1990).

Further characterization of the microcystin inhibition of protein phosphatase showed very tight binding between the phosphatases and microcystin. The effectively stoichiometric binding resulted in an inhibition constant (Ki) of 0.06 nM for protein phosphatase 1 and below 0.01 nM for protein phosphatase 2A (Mackintosh, Beattie et al. 1990). The substrate used for the phosphatase assays was [^{32}P]glycogen phosphorylase, the enzyme which controls the hydrolysis of cellular glycogen stores. This enzyme had earlier been shown to be activated by microcystin, as one of the early responses of the liver to this toxin was glycogen hydrolysis (Runnegar, Andrews et al. 1987).

Microcystin was shown to have no effect on a range of other protein phosphatases or on protein kinases. However, phosphatases 1 and 2A from plants were inhibited as much as mammalian enzymes, as was protein phosphatase 1 from the protozoan *Paramecium*. Neither bacterial nor cyanobacterial protein phosphatases were inhibited (Mackintosh, Beattie et al. 1990). These results on phosphatase inhibition were independently confirmed by Honkanen et al. and Eriksson et al. (Eriksson, Meriluoto et al. 1990; Honkanen, Zwiller et al. 1990) and also shown in cytosol extracts from skin cells (Matsushima, Yoshizawa et al. 1990). The molecular enzymology of these phosphatase inhibitors is well understood and indicates that microcystins, okadiac acid, and the toxin calyculin all bind to the same catalytic site on protein phosphatase-1 (Holmes, Maynes et al. 2002).

Confirmation that the acute toxicity of microcystin was due to phosphatase inhibition was undertaken by the simultaneous measurement of the enzyme inhibition and the clinical features of microcystin toxicity after intraperitoneal administration of toxin in mice. At 60 min after a lethal dose of microcystin, the activity of hepatic protein phosphatase 1 and 2A activity was reduced to 22% of controls, and the reduction was time- and dose-dependent (Runnegar, Kong et al. 1993).

It was therefore clear that the cyanobacterial toxin microcystin is a specifically eukaryotic cell poison, which acts through very potent inhibition of two key phosphatase enzymes involved in cell regulation. This inhibition leads to a multiplicity of effects, as discussed below. It does not preclude the possibility that microcystin also acts in other ways, but the complexity of outcomes from phosphatase inhibition makes identification of independent toxic effects difficult to achieve.

7.11 CYTOSKELETAL EFFECTS

The rapid deformation of isolated hepatocytes by microcystin was identified as a consequence of hyperphosphorylation of structural proteins in the cell, particularly the cytokeratins (Ohta, Nishiwaki et al. 1992). The assembly of cytokeratins into the intermediate filaments of the cytoskeleton is determined by the extent of phosphorylation of the proteins. As excess phosphate groups accumulate on the cytokeratins through the inhibition of the phosphatases, the filaments progressively disaggregate into collapsed, unstructured fibers in a ring around the nucleus and finally into spherical granules of cytokeratin (Eriksson, Meriluoto et al. 1990; Falconer and Yeung 1992). Since the intermediate filaments play a major role in both cell structure and cell–cell adhesion, their breakdown allows the actin microfilaments to contract into fibrous masses causing cell contraction, blebbling of the cell plasma membrane, and loss of the microvilli. Actin polymerization is not itself affected by microcystin, as the actin microfilaments remain intact but contract away from the plasma membrane into clumps at the base of the largest protrusions of the plasma membrane (Runnegar and Falconer 1986; Wang 1987; Eriksson, Paatero et al. 1989; Falconer and Yeung 1992).

More recent studies using immunostaining and confocal microscopy have provided very clear and informative images of the architectural changes in the cell brought about by microcystin. In particular, the desmosome proteins, desmoplakins, forming the junctions between adjacent cells, have been demonstrated to show the quickest change after microcystin exposure of hepatocytes. Within 12 min of exposure, the cell–cell junctions have largely gone; by 22 min, there was no desmoplakin visible at the cell boundary, while diffuse fluorescence in the cell indicated that the protein had dissociated into solution in the cytoplasm (Toivola, Goldman et al. 1997). This was associated with phosphorylation of the desmoplakins, molecular weight 215 and 250 kDa and the phosphorylation of the two cytokeratins 8 and 18 (45 and 55 kDa). The loss of the desmoplakins from the cell junction was followed by intermediate filament detachment and retraction, possibly caused by microfilament contraction (Toivola, Goldman et al. 1997). Structural damage to the nucleus has also been observed in cultured cells exposed to microcystin-LR at high concentrations. The nuclear spindle showed deformity and the chromosomes were hypercondensed in mitotic cells (Lankoff, Banasik et al. 2003).

These studies of the breakdown of the cell cytoskeleton caused by microcystin exposure provide a mechanistic basis for the observed acute hepatotoxicity. Electron microscopic studies of the liver damage caused by acute microcystin exposure have shown loss of lobular architecture as cells separate and rupture, leading to loss of cell contents into the venous drainage and massive bleeding into the tissue as sinusoids disintegrate (Falconer, Jackson et al. 1981). This bleeding rapidly leads to death from hemorrhagic shock quite independently of loss of liver function, which causes death later.

7.12 NUCLEAR ACTIONS OF MICROCYSTIN

Recent research on the intracellular localization of microcystin in hepatocytes, using immunostaining techniques, has led to the discovery that the toxin is not restricted to the cytoplasm but also occurs in the nucleus. Use of an immunospecific antibody to microcystin-LR demonstrated that centrilobular hepatocytes were both damaged and stained for the presence of toxin after a lethal intraperitoneal injection. Nuclear staining was observed as well as cytoplasmic staining. Apoptotic cells and rounded centrilobular cells stained for microcystin, whereas cells showing coagulative necrosis in the midzonal region did not stain. This lack of staining, indicating absence of toxin, may imply that these cells died from secondary consequences of microcystin toxicity, such as oxygen deprivation. Cell extracts showed phosphatase-microcystin adducts, indicating covalent binding of the enzymes with the toxin, but no measurable free toxin (Yoshida, Makita et al. 1998).

Studies of sublethal microcystin toxicity in mice confirmed that intense immunostaining for microcystin occurred in hepatocyte nuclei at 4 h after dosing, which faded by 12 h. Repeated low doses of toxin resulted in less nuclear staining. Extracts of the nuclei showed 40 kDa proteins carrying microcystin, identified as protein phosphatases 1 and 2A (Guzman and Solter 2002). Measurement of nuclear protein phosphatases in liver after a single lethal dose of microcystin in mice or after incubation of isolated hepatocytes with microcystin showed almost total inhibition. Sublethal doses of microcystin showed only nonsignificant inhibition of nuclear phosphatases; however, more sensitive responses — discussed later — confirmed the importance of nuclear effects (Guzman, Solter et al. 2003).

Similar results for the immunochemical localization of microcystin in hepatocytes were obtained from studies in trout liver (Fischer, Hitzfeld et al. 2000).

7.13 MICROCYSTIN AND APOPTOSIS

The consequences of inhibition of nuclear phosphatases are beginning to be investigated, particularly as they relate to the apoptosis (programmed cell death) of hepatocytes, which is seen in liver sections after microcystin dosing of rodents. In histological sections, apoptotic cells show loss of the normal nuclear structure, with condensed granules of chromatin, reduced cell diameter, and loss of cytoplasmic detail. Apoptosis is an inherent component of tissue turnover and regeneration, in which cells die, are hydrolyzed, and in most tissues replaced by new cells. The processes occurring in apoptosis are genetically regulated and can be triggered by extrinsic factors acting on cell-surface receptors or by intracellular changes including toxin action (Lockshin and Zakeri 2001).

Microcystin has been shown to rapidly stimulate apoptotic cell death in isolated or cultured hepatocytes, as have nodularin and okadaic acid. A range of other cell types are also responsive to microcystin but require microinjection, as the microcystin transport mechanisms are absent. The proteolytic enzymes named caspases, integral components of the processes of cell lysis and DNA fragmentation occurring during apoptosis, are activated by microcystin but do not appear to have a key role in the mechanism of initiation of apoptosis by microcystin (Fladmark and Brustugun et al. 1999).

Since these toxins act by phosphatase inhibition, which results in hyperphosphorylation of regulatory proteins, experimental blocking of the key kinase enzymes that carry out the hyperphosphorylation may be expected to protect cells from apoptosis. Investigation of a wide range of kinase inhibitors for ability to block both hyperphosphorylation and apoptosis demonstrated a major role for the calcium/calmodulin-dependent protein kinase (Fladmark, Brustugun et al. 2002). This enzyme is a major regulatory enzyme in the cell, particularly responding to hormonal stimuli. Inhibitors of this protein kinase were demonstrated to prevent, to the same extent, both the protein phosphorylation and apoptosis that followed microcystin exposure. Blocking of this enzyme by microinjection of an inhibitory peptide or by cell-permeant inhibitors prevented apoptosis, whereas overexpression of the active enzyme induced apoptosis. The activity of the enzyme itself in normal cell homeostasis is controlled by two separate modulators. One is by fluctuating levels of phosphorylation and dephosphorylation of the enzyme protein, with the phosphorylated enzyme protein being the activated form. The other is by intracellular calcium concentration, which had earlier been shown to be increased in liver cells exposed to microcystin; these mobilized intracellular calcium stores probably arose from the mitochondria (Falconer and Runnegar 1987). Which of the enzyme substrates of the calcium/calmodulin dependent protein kinase are responsible for the apoptotic cascade of reactions leading to cell death remains to be clarified.

Mitochondria are also involved in the mechanism of apoptosis, in particular through changes in the permeability of the mitochondrial membrane. This is associated with the generation of reactive oxygen groups, such as hydroxyl radicals and hydrogen peroxide. Exposure of isolated hepatocytes in culture to 1 μM microcystin-LR caused rapid cell changes, including an increase in reactive oxygen groups (species), which could be prevented by incubation with an iron-chelating compound; this resulted in some cell protection. The formation of reactive oxygen groups was followed by a mitochondrial permeability transition in which depolarization of the membrane and marked changes in solute permeability occur (Ding, Shen et al. 2000). The permeability transition was preceded by entry of calcium into the mitochondria, which, when blocked, reduced cell death (Ding, Shen et al. 2001). One of the key events in mitochondrial activation of apoptosis is the release of cytochrome *c* from the mitochondria, which in some cells activates the proteolytic caspase enzymes. In hepatocytes, this does not appear to occur, and an alternative calcium-activated proteolytic enzyme called calpian appears to respond to cytochrome *c*, leading to cell death (Ding, Shen et al. 2002).

Oxidative stress also appears to be involved with apoptosis, with rapid production of peroxides after exposure of cells to microcystin-LR, which has damaging effects on cell components (Ding and Ong 2003; Botha, Gehringer et al. 2004).

One of the elements inducing apoptosis in cells is damage to DNA, which can be caused, for example, by mutagenic chemicals or radiation. In response to DNA damage, the cell produces a potent apoptosis inducer called tumor-suppressor protein p53 (Levine 1997). The activity of this nuclear protein is regulated by the balance of its phosphorylation and dephosphorylation, which involve protein kinases and phosphatases (Gottifredi, Shieh et al. 2000). Hence this protein may also be a key link between microcystin's effects in the nucleus and cell death by apoptosis, in parallel with the apoptotic effects of okadaic acid, also a potent inhibitor of phosphatase 1 and 2A, which acts on p53 phosphorylation (Milczarek, Chen et al. 1999).

Measurement of the extent of phosphorylation of p53 after sublethal dosing of mice with microcystin showed considerable sensitivity to the toxin, with p53 phosphorylation increasing, even in experiments in which bulk inhibition of phosphatases could not be demonstrated (Guzman, Solter et al. 2003).

Because a variety of cell types will respond if the extracellular concentration of microcystin is high enough and the time of incubation sufficiently prolonged, it appears that microcystin will stimulate apoptosis in any cell into which the toxin can enter (Mankiewicz, Tarczynska et al. 2001). As the cells of the intestinal lining have already been shown to be damaged by microcystins, it is likely that a component of the gastroenteritis caused by ingesting microcystins is due to apoptosis of intestinal epithelial cells, including those of the colon (Falconer, Dornbusch et al. 1992; Botha, Gehringer et al. 2004).

While cell death is an important element of the toxicity of microcystin, effects of the toxin at lower doses on the cell cycle and — more critically for human health — on tumor promotion or carcinogenesis are considered in more detail in the next section.

7.14 CELL-CYCLE EFFECTS OF MICROCYSTIN

Studies of microcystin's effects on the cell cycle of hepatocytes have been carried out in cultures of freshly isolated cells. Hepatocytes are not a uniform cell population, either in mice or in humans. The dividing immature cells are smaller mononucleate cells and have not developed the full xenobiotic-metabolizing capacity of older cells. They also have a lower capacity to transport bile acid and microcystin, hence a different sensitivity to toxin. Older and more biochemically differentiated hepatocytes have two or more nuclei, each of which may contain diploid, tetraploid, or higher chromosome numbers (Brodsky and Uryvaeva 1977). In the 1-month-old

C3H mouse, approximately 60% of the hepatocytes were tetraploid and the same proportion were binucleate. Older animals have an increased proportion of polyploid cells. The great majority of hepatocytes at 1 year of age were binucleate cells, each nucleus containing tetraploid chromosomes, which resulted in octaploid cells (Severin, Meier et al. 1984). The consequences of microcystin exposure are therefore likely to depend on the level of differentiation of the individual cell and the concentration of toxin applied.

Monolayers of hepatocytes isolated from immature mice were exposed to very low concentrations of microcystin-LR for 65 h. These cultures, which showed spontaneous hepatocyte proliferation, were exceptionally sensitive to exposure to microcystin in the culture medium. The outcome of continued exposure to a concentration of 10 to 100 pM microcystin was a significant increase in cell numbers compared to control incubations, together with a decrease in DNA content per cell. The explanation of this cell proliferation was through cytokinesis (cell division), in which preexisting binucleate cells and/or tetraploid mononucleate cells divided — without compensating chromosome replication —under stimulation from this very low toxin concentration. Apoptosis was inhibited, allowing the increase in cell numbers. At higher toxin concentrations (1 to 10 nM) cell numbers were close to controls due to a combination of cell proliferation together with increased apoptosis and DNA synthesis rate was raised, indicating chromosome replication in the dividing cells. At 100 nM concentration of microcystin, all cells were killed (Humpage 1997; Humpage and Falconer 1999).

Further clarification of the changes occurring in this cell population was undertaken, using cell-sorting techniques to identify differential variations in cell ploidy caused by microcystin exposure. In these experiments, concentrations of microcystin up to 10 nM resulted in an increase in the number and proportion of binucleate tetraploid cells at the expense of octaploid cells which decreased. Diploid cell numbers remained constant (Humpage 1997). The preferential killing of octaploid cells probably resulted from the higher transportability for microcystin of these more differentiated cells, leading to higher and more toxic concentrations of intracellular microcystin.

Overall, these data can be interpreted as demonstrating that microcystin caused cell proliferation at low concentrations and in immature cells, with a reduction in apoptotic cell death. As concentration increased, selective apoptosis of more differentiated octaploid cells occurred together with replication of less differentiated cells, forming an increased population of tetraploid cells. It is also possible that a decrease in cell contact inhibition contributed to the cell proliferation as the older hepatocytes dissociated from the cell monolayer.

The complex of enzymes involved in cell-cycle regulation in hepatocytes and other cells is substantially affected by phosphorylation/dephosphorylation reactions, which, in turn, are sensitive to inhibition of cytoplasmic and nuclear phosphatases (Lorca, Labbe et al. 1992; Tiwari, Jamal et al. 1996). Due to the complexity of these changes, a great deal of further biochemical investigation will be required to clarify the exact mechanisms involved (Gehringer 2004).

7.15 TUMOR PROMOTION BY MICROCYSTIN

It is likely that both acute cell death from intracellular disruption and apoptotic cell death occur in the livers of animals and people in response to microcystin poisoning. Chronic exposure to microcystin at sublethal doses is likely to cause continual apoptotic cell death, leading to invasion of the liver tissue by scavenging white cells, as is characteristic of active liver injury (Falconer, Smith et al. 1988). The consequence of both processes of cell death is proliferation of surviving hepatocytes, leading to the recovery of liver function.

That microcystin may affect tumor induction or growth was indicated by a study of the effects of exposure of mice to microcystins in their drinking water for periods up to 1 year. Of 71 mice receiving the highest oral concentration of toxin, 4 developed tumors, compared with only 2 mice of 223 receiving lower concentrations (Falconer, Smith et al. 1988). None of these were tumors of the liver.

That microcystin is a tumor promoter was demonstrated by investigations based on the theory of multistage carcinogenesis. This hypothesises rests on the concept that the development of cancers can be separated into a series of sequential stages. The first stage of cancer formation is a mutation event, which may remain dormant, kill the cell, or mark the beginning of a progression into cancer. For growth, the mutated cell will require further stimulus, which can be either natural or from an exogenous tumor-promoting compound, leading to a cluster or focus of proliferating cells, usually of a less differentiated type. These cells, in turn, progress by growth and division into a tumor and may become metastatic and migrate into other locations, causing cancer.

Skin mutations caused by the potent carcinogen dimethylbenzanthracene have been used as model systems to study tumor promotion, as low doses of carcinogen alone, painted onto the backs of mice, will not result in papilloma (a common precancerous skin tumor) formation. However when this treatment is followed by dosing or topical application of a tumor promoting compound such as phorbol ester or okadiac acid, a dose response curve of tumor growth results (Suganuma, Fujiki et al. 1988).

This approach was used to investigate whether microcystin exposure through drinking water would affect the growth of skin papillomas in mice treated with dimethylbenzanthracene. Fifty-two days after treatment with dimethylbenzanthracene, the number and weight of skin papillomas in mice was significantly increased in animals consuming a toxic *Microcystis* extract (Falconer and Buckley 1989). That this effect was unlikely to be due to cyanobacterial components other than microcystin was shown by a lack of response in papilloma growth to an extract of another abundant cyanobacterial species, *Anabaena circinalis* (Falconer 1991). These results may not, however, reflect a direct action of microcystin on the mutated epithelial cells, as skin cells are relatively resistant to microcystin; rather, they may indicate an indirect response — for example, through the immune system (Shen, Zhao et al. 2003). The outcome by either route is clearly of concern for the health risk posed by the consumption of microcystin.

Treatment of rats or mice with liver carcinogens will turn hepatocytes into potential cancer cells. With additional treatment with a tumor-promoting compound, these cells will proliferate into precancerous nodules. Administration of nitrosamines, which are potent liver mutagens, to rats at low doses resulted in preneoplastic foci of proliferating hepatocytes showing the placental form of glutathione S-transferase. This enzyme is a useful immunological marker for potentially cancerous liver cells. With microcystin dosing, these foci were more abundant and grew faster (Nishiwaki-Matsushima, Ohta et al. 1992), demonstrating the tumor-promoting effect of microcystin in the liver. This effect was comparable to that of okadaic acid, also a protein phosphatase inhibitor and earlier shown to be a potent tumor promoter (Katoh, Fitzgerald et al. 1990). More recently, both diethylnitrosamine and aflatoxin B1 were used as cancer initiators in rats; when followed by intraperitoneal injection of microcystin for 6 weeks, both increased the development of preneoplastic foci in the liver (Sekijima, Tsutsumi et al. 1999).

In an experiment not using mutagenic compounds, sublethal intraperitoneal doses of microcystin to mice repeated 100 times over 28 weeks resulted in three different types of hyperplastic nodules in the liver, one of which was a precancerous type (Ito, Kondo et al. 1997). This indicates that microcystin may be a nonmutagenic carcinogen if tissues are exposed to a high enough dose for sufficient time. Another group of mice was treated with microcystin as before but continued for a further 8 weeks without microcystin dosing. These animals had more but smaller nodules and a decrease in the proportion of the precancerous type. It would have been informative to carry these animals through to natural death and observe whether the nodules became actual hepatocellular carcinomas.

These data do not, however, provide a definitive demonstration of carcinogenicity, as the untreated control animals in this experiment also showed liver nodules. Since cell mutation can occur through defective cell division, viral infection, exposure to dietary carcinogens, or radiation, there will always be a population of potentially cancerous cells in an organ. These mutated cells can then be stimulated to replicate by tumor promoters or by further mutation.

Stronger evidence of direct carcinogenesis has been shown for nodularin in an experiment comparing microcystin with nodularin, with and without pretreatment of rats with the mutagen diethylnitrosamine. Preneoplastic foci of cells in the liver were seen after mutagen dosing followed by both microcystin and nodularin; however, nodularin dosing without the mutagen also showed foci, whereas microcystin did not. Several of the gene loci involved in carcinogenesis also showed a direct response to nodularin (Ohta, Sueoka et al. 1994).

Experiments with higher health relevance were the investigations of tumor promotion by microcystin in tumors of the gastrointestinal tract. Gastrointestinal cancers form a major component of human cancer mortality, particularly in western societies, where hepatocellular carcinomas are relatively rare. Two investigations were carried out, both using mice exposed to microcystins in their drinking water. In the first study, the carcinogen *N*-methyl-*N*-nitrosourea was dosed by mouth, giving rise to tumors of the duodenum, liver, and lymphoid system. The experiment continued to 154 days of microcystin exposure in mice, leading to minor clinical outcomes. No demonstrable effect of oral microcystin on the growth of any of these tumors was observed, even at the high dose in which inhibition of growth was apparent (Falconer and Humpage 1996).

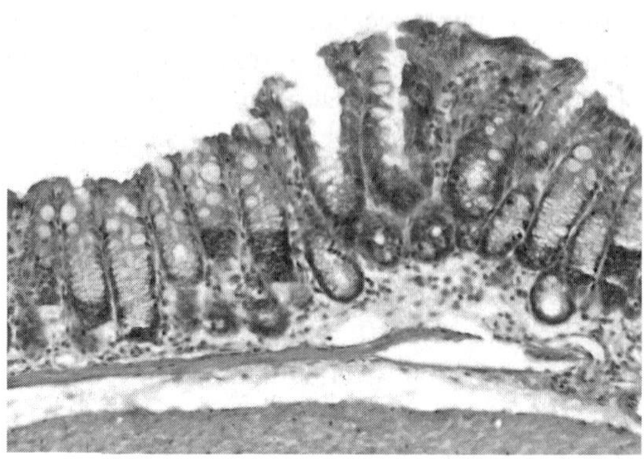

FIGURE 7.3 (See color insert following page 146.) Precancerous colon crypt in mouse dosed with azoxymethane to initiate hypertrophic crypt formation and then exposed to microcystin in drinking water for 212 days. (From Humpage, Hardy et al. 2000. With permission.)

In the second study, cancer of the lower intestinal tract was investigated by dosing mice with the colon carcinogen azoxymethane. This carcinogen gives rise to hypertrophic crypts in the colon lining, which later develop into carcinoma. Their number and size can be readily quantified by microscopy. Three injections of azoxymethane were administered, followed by zero or two concentrations of microcystin in the drinking water for 212 days. Exposure to microcystins in the drinking water produced a significant dose-dependent increase in crypt size due to increased hyperplasia (Humpage, Hardy et al. 2000). Figure 7.3 illustrates these hypertrophic colon crypts in a mouse given azoxymethane followed by microcystin in the drinking water.

The colon lining is likely to be highly exposed to microcystins from drinking water, as unabsorbed toxin will be present in the gut contents as well as toxin excreted from the liver through the bile. Intestinal cells have been shown to be vulnerable to microcystin injury, both in *in vitro* experiments (Falconer, Dornbusch et al. 1992) and in cases of human exposure, in which diarrhea is a notable outcome of swallowing water contaminated by *Microcystis* (Zilberg 1966; Turner, Gammie et al. 1990; Teixera, Costa et al. 1993).

To verify whether microcystins in drinking water act as tumor promoters for colon cancer in human populations, careful epidemiological research is needed. This requires measures of exposure to microcystins in drinking water, for which data are not widely available. Few water utilities measure microcystins in drinking water, and when measurements have been done in the past, the results were confidential to the company. The cancer rates for populations are, however, recorded and are available for research access. These are sufficiently detailed to allow populations to be identified through residential postal codes, which can be used to locate groups supplied with water from known sources. The cancer rates are classified by type of tumor and patient's age and gender, so that the incidence of a range of cancers can be compared between populations and corrected for age and gender. The only published attempt at this in a western nation was initiated by measurements of cyanobacterial toxins in drinking water supplies in Florida (Williams, Burns et al.

2001). The epidemiological results were indicative of an effect on liver cancer rate of consumption of surface water; but in the absence of widespread toxin measurements for these waters, this could not be directly related to cyanobacteria and remains inconclusive (Fleming, Rivero et al. 2002). An unpublished epidemiological assessment of cancers related to *Microcystis* in drinking water supply reservoirs in Australia was also inconclusive. This may have been because the drinking water sources used for comparison were selected by the presence of *Microcystis* in the supply reservoirs, with no measurements of microcystin in the tap water. While microcystin-producing cyanobacteria in a reservoir are the first step toward microcystins in tap water, effective water treatment (as discussed in Chapter 12) will minimize the toxin concentration. Hence it is essential for an epidemiological study to measure actual toxin exposure of the target population.

This has been done in China, which has long-term data on cancer incidence as well as measurements of cyanobacterial toxins in drinking water supplies, as discussed below.

7.16 CARCINOGENESIS, LIVER DAMAGE, AND CANCER IN CHINA

Villages in the less developed countries of the world draw their drinking water from local sources. These include ditches, ponds, canals, lakes, and rivers as well as shallow and deep wells. As a consequence of population density, some of these water sources are highly eutrophic, including water with persistent cyanobacterial blooms. These include *Microcystis* blooms, which are widespread in village water sources in China. Microcystins are stable to boiling, so that local populations will be directly exposed to toxin if it is present in the raw water.

In an epidemiological study of liver cancer incidence in China, it was demonstrated that particular groups of villages showed greatly elevated rates, some above 60 per 100,000 (Yu 1989, 1995). Overall, liver cancer is a major source of cancer deaths in China, in some areas second only to stomach cancer.

Identification of the risk factors correlated with the elevated liver cancer showed that aflatoxin contamination of maize in the diet and prevalence of hepatitis B and C, were the major components. Drinking water drawn from ditches and ponds was also significantly correlated to the rate of hepatocellular carcinoma, with lower incidence in populations drinking well water. To reduce the incidence of liver cancer, the populations were encouraged to eat rice instead of maize, and wells were dug to supply cleaner drinking water. Children were vaccinated against hepatitis B. These measures were effective, as the liver cancer rates dropped close to the nonendemic rates, particularly in villages with new wells for water supply (Yu 1989, 1995).

It was proposed that the reason that the drinking of ditch and well water posed a higher risk of liver cancer was due to microcystin contamination. Surveys of microcystins in these water supplies have shown relatively low concentrations, up to 0.46 μg/L (Ueno, Nagata et al. 1996).

These are appreciably lower than recorded concentrations in lakes and rivers during *Microcystis* blooms elsewhere in the world. It is, however, possible that the people of these villages were particularly vulnerable to microcystin as a consequence of their constant exposure to the powerfully carcinogenic aflatoxin and infection with hepatitis B virus. There is also the potential for the presence of other carcinogens, pesticides, and natural toxins in shallow surface waters. The question of which were the active components in surface water that stimulated liver cancer in these cases may never be answered because of the effective public health measures that were implemented.

A recent epidemiological study of colorectal cancer and its relationship with microcystin in drinking water in China has proved very informative. This was carried out as a retrospective case control study of eight townships in Haining City. The colorectal cancer rate was 8.37 per 100,000 averaged over the accumulated population of nearly 5 million. The relative risk increased sharply with the drinking of river (7.9) and pond (7.7) water compared to well water. Tap water showed a relative risk of 1.9. Analysis of microcystins in the drinking water samples showed negligible concentrations in well and tap water, with average concentrations of 0.14 and 0.11 μg/L in river and pond water, respectively. The maximum concentrations measured were 1 μg/L of microcystins in rivers and 2 μg/L in ponds. The concentration of microcystin in the water supply significantly correlated with colorectal cancer rate (Zhou, Yu et al. 2002).

These epidemiological data for both human liver cancer and colorectal cancer in China are supported by the experimental demonstrations of tumor promotion by microcystin in the livers and colons of rodents. The issue of human risk from these sources of potential harm in drinking water is explored in Chapter 8.

7.17 MICROCYSTIN, TERATOGENESIS, AND REPRODUCTIVE TOXICITY

Early work carried out in the USSR by Kirpenko and colleagues indicated that injecting rats with microcystin-containing extracts of *M. aeruginosa* from the first day of gestation until the nineteenth day caused a mild degree of teratogenesis in the fetuses, expressed as hydrocephaly and internal hemorrhages. no skeletal abnormalities were seen; however, some fetuses from treated groups were small though viable (Kirpenko, Sirenko et al. 1981).

Teratogenic effects in mice were investigated by chronic exposure of male and female mice to *Microcystis* extracts in drinking water throughout the period of postweaning growth and continued until after the female mice had given birth and suckled their young for 5 days. No effect on number, sex ratio, or size of young was seen. There were no skeletal abnormalities in the neonatal young. Investigation of internal organs similarly did not show any changes. However, the brains of 7 of 73 young of microcystin-exposed parents were reduced in size, and histopathological examination of the brain sections revealed a region of extensive cell damage in the outer hippocampus (Falconer, Smith et al. 1988).

Recent investigations of the effects of microcystin-LR on teratogenesis have produced negative results. Careful studies were made in which sublethal intraperitoneal doses of 2 to 128 μg/kg microcystin-LR were administered to female mice over 2 days of pregnancy: on days 7 and 8, 9 and 10, or 11 and 12. No changes in

17-day-old mouse fetuses or in neonates were found (Chernoff, Hunter et al. 2002). Whole mouse embryos were also cultured for 24 h in the presence of microcystin-LR at up to 1 μM concentration without effect. These investigators also exposed developing toad embryos for 10 days to microcystin-LR at concentrations up to 20 mg/L (approx 20 μM) with no effect. They concluded that there was no evidence of teratogenic activity of microcystin-LR in the systems studied (Chernoff, Hunter et al. 2002).

Shortly after the study described above, a group working in China published the results (in Chinese) of teratogenesis studies on the rat using microcystin-LR. They administered toxin at 4, 16, and 62 μg/kg daily to newly pregnant rats for 10 days. Twenty days later, the rats were killed and the fetuses examined. Fetal deformities were found in the offspring, with 0.68% and 1.2% of deformed fetuses in the middle and higher dose ranges, respectively. In this latter group, severe liver degeneration and maldevelopment of the renal glomerulus and medulla were also observed. The investigators concluded that microcystin passed the placental barrier, causing fetal damage (Zhang, Lian et al. 2002).

While there is no clear consensus on microcystin's teratogenicity in mammals, there is increasing evidence of deformity in fish and amphibian embryos exposed to cyanobacterial extracts and pure toxins; this is of ecological concern (Oberemm, Fastner et al. 1997; Oberemm, Becker et al. 1999). There is also evidence of teratogenic effects of both cyanobacterial biomass and purified microcystin-LR on *Xenopus* (African clawed toad) embryos. At high doses, cyanobacterial biomass with or without microcystin caused more than 50% malformations. However, 53% of embryos were malformed at the highest tested concentration of purified microcystin-LR alone, showing that both unidentified cyanobacterial components and microcystin were teratogenic in this assay (Dvorakova, Dvorakova et al. 2002). Microinjection of microcystin-LR into fish embryos resulted in reduced survival, accelerated hatching, and liver damage in embryos and posthatching juveniles (Jacquet, Thermes et al. 2004).

As part of the long-term chronic study in mice of *Microcystis* extract in drinking water, male and female toxin-exposed mice were mated and the number, viability, and body weight of the young measured. There were no differences in the number of young or in the weight, viability, or gender of young between chronically exposed females and control females mated with chronically exposed males and control males respectively (Falconer, Smith et al. 1988). There is no current evidence of endocrine-disrupting effects of microcystin on reproduction in mammals.

7.18 CONCLUSION

It is concluded that microcystin has well-demonstrated acute toxic effects on cells, exerted primarily through inhibition of phosphatase enzymes. The sensitivity of cells in particular organs depends on the presence or absence of the toxin transporter, which is present in the cell membrane of intestinal lining cells and hepatocytes but largely absent elsewhere. As a consequence of the widespread control functions of phosphatase enzymes in cells, their inhibition causes damage to cell structure, cell metabolism, and cell cycle control; it also causes cell death by apoptosis and necrosis.

Longer-term adverse effects include tumor promotion in liver and intestine and possible carcinogenesis. At present no clear conclusion can be drawn on teratogenic effects on mammals, though the potential for damage exists. No endocrine-disrupting effects of microcystin have yet been demonstrated.

REFERENCES

Ashworth, C. T. and M. F. Mason (1946). Observations on the pathological changes produced by a toxic substance present in blue-green algae (*Microcystis aeruginosa*). *American Journal of Pathology* 22: 369–383.

Barford, D. (1996). Molecular mechanisms of the protein serine/threonine phosphatases. *Trends in Biochemical Sciences* 21(11): 407–412.

Botha, N., M. M. Gehringer, et al. (2004). The role of microcystin-LR in the induction of apoptosis and oxidative stress in CaCo2 cells. *Toxicon* 43: 85–92.

Brodsky, W. Y. and I. V. Uryvaeva (1977). Cell polyploidy: Its relation to tissue growth and function. *International Review of Cytology* 50: 275–332.

Carbis, C. R., J. A. Simons, et al. (1994). A biochemical profile for predicting the chronic exposure of sheep to *Microcystis aeruginosa*, an hepatotoxic species of blue-green alga. *Research in Veterinary Science* 57: 310–316.

Carbis, C. R., D. L. Waldron, et al. (1995). Recovery of hepatic function and latent mortalities in sheep exposed to the blue-green alga *Microcystis aeruginosa*. *Veterinary Record* 137: 12–15.

Chernoff, N., E. S. Hunter, et al. (2002). Lack of teratogenicity of microcystin-LR in the mouse and toad. *Journal of Applied Toxicology* 22(1): 13–17.

Cohen, P. and P. T. W. Cohen (1989). Protein phosphatases come of age. *Journal of Biological Chemistry* 264(36): 21435–21438.

Cohen, P., C. F. B. Holmes, et al. (1990). Okadaic acid: A new probe for the study of cellular regulation. *Trends in Biochemical Sciences* 15: 98–102.

Ding, W. X. and C. N. Ong (2003). Role of oxidative stress and mitochondrial changes in cyanobacteria-induced apoptosis and hepatotoxicity. *FEMS Microbiology Letters* 220(1): 1–7.

Ding, W. X., H.-S. Shen, et al. (2000). *Microcystis* cyanobacteria extract induces cytoskeletal disruption and intracellular glutathione alteration in hepatocytes. *Environmental Health Perspectives* 108(7): 605–609.

Ding, W. X., H.-M. Shen, et al. (2001). Pivotal role of mitochondrial Ca^{2+} in microcystin-induced mitochondrial permeability transition in rat hepatocytes. *Biochemical and Biophysical Research Communications* 285: 1155–1161.

Ding, W. X., H. M. Shen, et al. (2002). Calpain activation after mitochondrial permeability transit in microcystin-induced cell death in rat hepatocytes. *Biochemistry and Biophysics Research Communications* 291(2): 321–331.

Dvorakova, D., K. Dvorakova, et al. (2002). Effects of cyanobacterial biomass and purified microcystins on malformations in *Xenopus laevis*: Teratogenesis assay (FETAX). *Environmental Toxicology* 17(6): 547–555.

Elleman, T. C., I. R. Falconer, et al. (1978). Isolation, characterization and pathology of the toxin from a *Microcystis aeruginosa* (*Anacystis cyanea*) bloom. *Australian Journal of Biological Science* 31: 209–218.

Eriksson, J., J. Meriluoto, et al. (1990). Hepatocyte deformation induced by cyanobacterial toxins reflects inhibitions of protein phosphatases. *Biochemical and Biophysical Research Communications* 173(3): 1347–1353.

Eriksson, J., G. I. L. Paatero, et al. (1989). Rapid microfilament reorganization induced in isolated rat hepatocytes by microcystin-LR, a cyclic peptide toxin. *Experimental Cell Research* 185: 86–100.

Falconer, I. R. (1991). Tumor promotion and liver injury caused by oral consumption of cyanobacteria. *Environmental Toxicology and Water Quality* 6: 177–184.

Falconer, I. R. (1993). *Algal Toxins in Seafood and Drinking Water.* London, Academic Press.

Falconer, I. R., T. Buckley, et al. (1986). Biological half-life, organ distribution and excretion of 125-I–labelled toxic peptide from the blue-green alga *Microcystis aeruginosa*. *Australian Journal of Biological Science* 39(1): 17–21.

Falconer, I. R. and T. H. Buckley (1989). Tumour promotion by *Microcystis* sp., a blue-green alga occurring in water supplies. *Medical Journal of Australia* 150: 351.

Falconer, I. R., M. D. Burch, et al. (1994). Toxicity of the blue-green alga (cyanobacterium) *Microcystis aeruginosa* in drinking water to growing pigs, as an animal model for human injury and risk assessment. *Environmental Toxicology and Water Quality* 9: 131–139.

Falconer, I. R., M. Dornbusch, et al. (1992). Effect of the cyanobacterial (blue-green algal) toxins from *Microcystis aeruginosa* on isolated enterocytes from the chicken small intestine. *Toxicon* 30(7): 790–793.

Falconer, I. R. and A. R. Humpage (1996). Tumour promotion by cyanobacterial toxins. *Phycologia* 35(6 suppl): 74–79.

Falconer, I. R., A. R. B. Jackson, et al. (1981). Liver pathology in mice in poisoning by the blue-green alga in *Microcystis aeruginosa*. *Australian Journal of Biological Science* 34: 179–187.

Falconer, I. R. and M. T. Runnegar (1987). Effects of the peptide toxin from *Microcystis aeruginosa* on intracellular calcium, pH and membrane integrity in mammalian cells. *Chemico-Biological Interactions* 63(3): 215–225.

Falconer, I. R., J. V. Smith, et al. (1988). Oral toxicity of a bloom of the cyanobacterium *Microcystis aeruginosa* administered to mice over periods up to 1 year. *Journal of Toxicology and Environmental Health* 24(3): 291–305.

Falconer, I. R. and D. S. K. Yeung (1992). Cytoskeletal changes in hepatocytes induced by *Microcystis* toxins and their relation to hyperphosphorylation of cell proteins. *Chemico-Biological Interactions* 81: 181–196.

Fawell, J. K., C. P. James, et al. (1994). Toxins from blue-green algae: Toxicological assessment of microcystin-LR and a method for its determination in water. Medmenham, U.K., Water Research Centre plc.

Fischer, W. J., B. C. Hitzfeld, et al. (2000). Microcystin-LR toxicodynamics, induced pathology, and immunohistochemical localization in livers of blue-green algae exposed rainbow trout (Oncorhynchus mykiss). *Toxicological Sciences* 54(2): 365–373.

Fladmark, K. E., O. T. Brustugun, et al. (1999). Ultrarapid apoptosis induction by serine/threonine phosphatase inhibitors. Unpublished manuscript.

Fladmark, K. E., O. T. Brustugun, et al. (2002). Ca^{2+}/calmodulin-dependent protein kinase II is required for microcystin-induced apoptosis. *Journal of Biological Chemistry* 277(4): 2804–2811.

Fleming, L. E., C. Rivero, et al. (2002). Blue green algal (cyanobacterial) toxins, surface drinking water, and liver cancer in Florida. *Harmful Algae* 1: 157–168.

Gehringer, M. M. (2004). Microcystin-LR and okadaic acid-induced cellular effects: a dualistic response. *FEBS Letters* 557(1–3): 1–8.

Gehringer, M. M., E. G. Shephard, et al. (2004). An investigation into the detoxification of microcystin-LR by glutathione pathway in Balb/c mice. *International Journal of Biochemistry and Cell Biology* 36(5): 931–941.

Gottifredi, V., S. Y. Shieh, et al. (2000). Regulation of p53 after different forms of stress and at different cell cycle stages. *Cold Spring Harbour Symposia in Quantitative Biology* 65: 483–488.

Guzman, R. E. and P. F. Solter (2002). Characterization of sublethal microcystin-LR exposure in mice. *Veterinary Pathology* 39(1): 17–26.

Guzman, R. E., P. F. Solter, et al. (2003). Inhibition of nuclear protein phosphatase activity in mouse hepatocytes by the cyanobacterial toxin microcystin-LR. *Toxicon* 41(7): 773–781.

Holmes, C. F. B., J. T. Maynes, et al. (2002). Molecular enzymology underlying regulation of protein phosphatase-1 by natural toxins [Review]. *Current Medicinal Chemistry* 9(22): 1981–1989.

Honkanen, R. E., J. Zwiller, et al. (1990). Characterization of microcystin-LR, a potent inhibitor of type 1 and type 2a protein phosphatases. *Journal of Biological Chemistry* 265(32): 19401–19404.

Hooser, S. B., V. R. Beasley, et al. (1989). Toxicity of microcystin LR, a cyclic heptapeptide hepatotoxin from *Microcystis aeruginosa*, to rats and mice. *Veterinary Pathology* 26: 246–252.

Hooser, S. B., V. R. Beasley, et al. (1990). Microcystin-LR–induced ultrastructural changes in rats. *Veterinary Pathology* 27: 9–15.

Hooser, S. B., M. S. Kuhlenschmidt, et al. (1991). Uptake and subcellular localization of tritiated dihydro-microcystin-LR in rat liver. *Toxicon* 29(6): 589–601.

Humpage, A. R. (1997). Tumour Promotion by the Cyanobacterial Toxin Microcystin. Ph.D. thesis, Adelaide, Australia, University of Adelaide.

Humpage, A. R. and I. R. Falconer (1999). Microcystin-LR and liver tumour promotion: Effects on cytokinesis, ploidy and apoptosis in cultured hepatocytes. *Environmental Toxicology* 14(1): 61–75.

Humpage, A. R., S. J. Hardy, et al. (2000). Microcystins (cyanobacterial toxins) in drinking water enhance the growth of aberrant crypt foci in the mouse colon. *Journal of Toxicology and Environmental Health, Part A* 61: 155–165.

Ito, E., F. Kondo, et al. (1997). Neoplastic nodular formation in mouse liver induced by repeated intraperitoneal injections of microcystin-LR. *Toxicon* 35(9): 1453–1457.

Ito, E., A. Takai, et al. (2002). Comparison of protein phosphatase inhibitory activity and apparent toxicity of microcystins and related compounds. *Toxicon* 40(7): 1017–1025.

Jackson, A. R. B., A. McInnes, et al. (1984). Clinical and pathological changes in sheep experimentally poisoned by the blue-green alga *Microcystis aeruginosa*. *Veterinary Pathology* 21(1): 102–113.

Jacquet, C., V. Thermes, et al. (2004). Effects of microcystin-LR on development of medaka fish embryos (Oryzias latipes). *Toxicon* 42: 141–147.

Katoh, F., D. J. Fitzgerald, et al. (1990). Okadaic acid and phorbol esters: comparative effects of these tumor promoters on cell transformation, intracellular communication and differentiation *in vitro*. *Japanese Journal of Cancer Research (Gann)* 81: 590–597.

Kirpenko, Y. A., L. A. Sirenko, et al. (1981). Some aspects concerning remote after-effects of blue-green algal toxin impact on warm-blooded animals. *The Water Environment. Algal Toxins and Health*. W. W. Carmichael, ed. New York, Plenum Press: 257–269.

Kondo, F., H. Matsumoto, et al. (1996). Detection and identification of metabolites of microcystins formed *in vivo* in mouse and rat livers. *Chemical Research in Toxicology* 9(8): 1355–1359.

Kuiper-Goodman, T., I. Falconer, et al. (1999). Human health aspects. *Toxic Cyanobacteria in Water: A Guide to Their Public Health Consequences, Monitoring and Management*. I. Chorus and J. Bartram, eds. London, E & FN Spon (on behalf of WHO): 113–153.

Lankoff, A., A. Banasik, et al. (2003). Effect of microcystin-LR and cyanobacterial extract from Polish reservoir of drinking water on cell cycle progression, mitotic spindle, and apoptosis in CHO-K1 cells. *Toxicology and Applied Pharmacology* 189(3): 204–213.

Leithe, E., V. Cruciani, et al. (2003). Recovery of gap junctional intercellular communication after phorbol ester treatment requires proteasomal degradation of protein kinase C. *Carcinogenesis* 24(7): 1239–1245.

Levine, A. J. (1997). P53, the cellular gatekeeper for growth and division. *Cell* 88: 323–331.

Lockshin, R. A. and Z. Zakeri. (2001). Programmed cell death and apoptosis: Origins of the theory. *Nature Reviews in Molecular Biology* 2(7): 545–550.

Lorca, T., J. C. Labbe, et al. (1992). Dephosphorylation of cdc2 on threonine 161 is required for cdc2 kinase inactivation and normal anaphase. *EMBO Journal* 11(7): 2381–2390.

Mackintosh, C., K. A. Beattie, et al. (1990). Cyanobacterial microcystin-LR is a potent and specific inhibitor of protein phosphatases 1 and 2A from both mammals and higher plants. *FEBS Letters* 264(2): 187–192.

Mackintosh, R. W., K. N. Dalby, et al. (1995). The cyanobacterial toxin microcystin binds covalently to cysteine-273 on protein phosphatase 1. *FEBS Letters* 371(3): 236–240.

Mankiewicz, J., M. Tarczynska, et al. (2001). Apoptotic effect of cyanobacterial extract on rat hepatocytes and human lymphocytes. *Environmental Toxicology* 16(3): 225–233.

Matsushima, R., S. Yoshizawa, et al. (1990). *In vitro* and *in vivo* effects of protein phosphatase inhibitors, microcystins and nodularin, on mouse skin and fibroplasts. *Biochemical and Biophysical Research Communications* 171(2): 867–874.

Meriluoto, J., S. Nygard, et al. (1990). Synthesis, organotropism and hepatocellular uptake of two tritium-labeled epimers of dihydromicrocystin-LR, a cyanobacterial peptide toxin analog. *Toxicon* 28, 12: 1439–1446.

Milczarek, G. J., W. Chen, et al. (1999). Okadaic acid mediates p53 hyperphosphorylation and growth arrest in cells with wild-type p53 but increases aberrant mitoses in cells with non-functional p53. *Carcinogenesis* 20: 1043–1048.

Nishiwaki-Matsushima, R., S. Nishiwaki, et al. (1991). Structure–function relationships of microcystins, liver tumor promoters in interaction with protein phosphatase. *Japanese Journal of Cancer Research (Gann)* 82: 993–996.

Nishiwaki-Matsushima, R., T. Ohta, et al. (1992). Liver tumor promotion by the cyanobacterial cyclic peptide toxin microcystin-LR. *Journal of Cancer Research and Clinical Oncology* 118: 420–424.

Oberemm, A., J. Becker, et al. (1999). Effects of cyanobacterial toxins and aqueous crude extracts of cyanobacteria on the development of fish and amphibians. *Environmental Toxicology* 14(1): 77–88.

Oberemm, A., J. Fastner, et al. (1997). Effects of microcystin-LR and cyanobacterial crude extracts on embryo-larval development of zebrafish (*Danio rerio*). *Water Research* 31: 2918–2921.

Ohta, T., R. Nishiwaki, et al. (1992). Hyperphosphorylation of cytokeratins 8 and 18 by microcystin-LR, a new liver tumor promoter, in primary cultured rat hepatocytes. *Carcinogenesis* 13(12): 2443–2447.

Ohta, T., E. Sueoka, et al. (1994). Nodularin, a potent inhibitor of protein phosphatases 1 and 2a, is a new environmental carcinogen in male f344 rat liver. *Cancer Research* 54: 6402–6406.

Petzinger, E. and M. Frimmer (1980). Comparative studies on the uptake of ^{14}C bile acids and ^{3}H-demethylphalloin in isolated liver cells. *Archives of Toxicology* 44: 127–135.

Robinson, N. A., C. F. Matson, et al. (1991). Association of microcystin-LR and its biotransformation product with a hepatic-cytosolic protein. *Journal of Biochemical Toxicology* 6: 171–180.

Robinson, N. A., G. A. Miura, et al. (1989). Characterization of chemically tritiated microcystin-LR and its distribution in mice. *Toxicon* 27(9): 1035–1042.

Robinson, N. A., J. G. Pace, et al. (1991). Tissue distribution, excretion and hepatic biotransformation of microcystin-LR in mice. *Journal of Pharmacology and Experimental Therapeutics* 256(1): 176–182.

Runnegar, M. T., J. Andrews, et al. (1987). Injury to hepatocytes induced by a peptide toxin from the cyanobacterium *Microcystis aeruginosa. Toxicon* 25(11): 1235–1239.

Runnegar, M., N. Berndt, et al. (1995a). Microcystin uptake and inhibition of protein phosphatases: Effects of chemoprotectants and self-inhibition in relation to known hepatic transporters. *Toxicology and Applied Pharmacology* 134(2): 264–272.

Runnegar, M., N. Berndt, et al. (1995b). *In vivo* and *in vitro* binding of microcystin to protein phosphatases 1 and 2A. *Biochemistry and Biophysics Research Communications* 216(1): 162–169.

Runnegar, M. T. and I. R. Falconer (1982). Similarities and differences in the effects of microcystis hepatotoxin. *Proceedings of the Vth International IUPAC Symposium on Mycotoxins and Phycotoxins*, Vienna, 212–215.

Runnegar, M. T. and I. R. Falconer (1986). Effect of toxin from the cyanobacterium *Microcystis aeruginosa* on ultrastructural morphology and actin polymerization in isolated hepatocytes. *Toxicon* 24(2): 109–115.

Runnegar, M. T., I. R. Falconer, et al. (1986). Lethal potency and tissue distribution of ^{125}I-labelled toxic peptides from the blue-green alga *Microcystis aeruginosa. Toxicon* 24(5): 506–9.

Runnegar, M. T., I. R. Falconer, et al. (1981). Deformation of isolated rat hepatocytes by a peptide hepatotoxin from the blue-green alga *Microcystis aeruginosa. Naunyn-Schmiedebergs Archives of Pharmacology* 317(3): 268–272.

Runnegar, M. T., R. G. Gerdes, et al. (1991). The uptake of the cyanobacterial hepatotoxin microcystin by isolated rat hepatocytes. *Toxicon* 29(1): 43–51.

Runnegar, M. T., S. Kong, et al. (1993). Protein phosphatase inhibition and *in vivo* hepatotoxicity of microcystins. *American Journal of Physiology* 265(2 Pt 1): G224–G230.

Sekijima, M., T. Tsutsumi, et al. (1999). Enhancement of glutathione S-transferase placental form positive liver cell foci development by microcystin-LR in aflatoxin B1-initiated rats. *Carcinogenesis* 20(1): 161–165.

Severin, E., E. M. Meier, et al. (1984). Flow cytometric analysis of mouse hepatocyte ploidy. I. Preparative and mathematical protocol. *Cell Tissue Research* 238: 643–647.

Shen, P. P., S. W. Zhao, et al. (2003). Effects of cyanobacterial bloom extract on some parameters of immune function in mice. *Toxicology Letters* 143: 27–36.

Sivonen, K. and G. Jones (1999). Cyanobacterial toxins. *Toxic Cyanobacteria in Water. A Guide to Their Public Health Consequences, Monitoring and Management.* I. Chorus and J. Bartram, eds. London, E & FN Spon (on behalf of WHO): 41–111.

Stotts, R. R., A. R. Twardock, et al. (1997a). Distribution of tritiated dihydromicrocystin in swine. *Toxicon* 35: 937–953.

Stotts, R. R., A. R. Twardock, et al. (1997b). Toxicokinetics of tritiated dihydromicrocystin-LR in swine. *Toxicon* 35: 455–465.

Suganuma, M., H. Fujiki, et al. (1988). Okadaic acid: An additional non-phorbol-12-tetradecanonate-13-acetate-type tumor promoter. *Proceedings of the National Academy of Sciences of the USA* 85: 1768–1771.

Teixera, M. G. L. C., M. C. N. Costa, et al. (1993). Gastroenteritis epidemic in the area of the Itaparica Dam, Bahia, Brazil. *Bulletin of the Pan-American Health Organization* 27(3): 244–253.

Tiwari, S., R. Jamal, et al. (1996). Protein kinases and phosphatases in cell cycle control. *Protein Phosphorylation in Cell Growth Regulation*. M. Clemens, ed. Amsterdam, Harwood Academic Publishers: 255–282.

Toivola, D. M., R. D. Goldman, et al. (1997). Protein phosphatases maintain the organization and structural interactions of hepatic keratin intermediate filaments. *Journal of Cell Science* 110(Pt 1): 23–33.

Turner, P. C., A. J. Gammie, et al. (1990). Pneumonia associated with contact with cyanobacteria. *British Medical Journal* 300: 1440–1441.

Ueno, Y., S. Nagata, et al. (1996). Detection of microcystins, a blue-green algal hepatotoxin, in drinking water sampled in Haimen and Fusui, endemic areas of primary liver cancer in China, by highly sensitive immunoassay. *Carcinogenesis* 17: 1317–1321.

Wang, Y. L. (1987). Mobility of filamentous actin in living cytoplasm. *Journal of Cell Biology* 105: 2811–2816.

Williams, C. D., J. Burns, et al. (2001). Assessment of cyanotoxins in Florida's lakes, reservoirs, and rivers. *Cyanobacteria Survey Project*. Harmful Algal Bloom Task Force, St. John's River Water Management District, Palatka, FL.

Yoshida, T., Y. Makita, et al. (1998). Immunohistochemical localization of microcystin-LR in the liver of mice: A study of the pathogenesis of microcystin-LR–induced hepatotoxicity. *Toxicologic Pathology* 26(3): 411–418.

Yoshizawa, S., R. Matsushima, et al. (1990). Inhibition of protein phosphatases by microcystin and nodularin associated with hepatotoxicity. *Journal of Cancer Research and Clinical Oncology* 116: 609–614.

Yu, S. Z. (1989). Drinking water and primary liver cancer. *Primary Liver Cancer*. T. Zhao-You, W. Meng-Chao, and X. Sui-Sheng, eds. Beijing, China Academic Publishers/Springer-Verlag: 30–37.

Yu, S. Z. (1995). Primary prevention of hepatocellular carcinoma. *Journal of Gastroenterology and Hepatology* 10(6): 674–682.

Zhang, Z., M. Lian, et al. (2002). Teratosis and damage of viscera induced by microcystin in SD rat fetuses. *Zhonghua Yi Xue Za Zhi* 82(5): 345–347.

Zhou, L., H. Yu, et al. (2002). Relationship between microcystin in drinking water and colorectal cancer. *Biomedical and Environmental Science* 15: 166–171.

Zilberg, B. (1966). Gastroenteritis in Salisbury European children: A five-year study. *The Central African Journal of Medicine* 12, 9: 164–168.

8 Risk and Safety of Drinking Water: Are Cyanobacterial Toxins in Drinking Water a Health Risk?

We are all naturally concerned about our own health and the health of others around us. The main focus of our concerns will, however, be different depending on whether we live in a developed nation or in a less developed part of the world. In the relatively recent past, communicable gastrointestinal diseases were major causes of infant mortality worldwide and were often transmitted through drinking water. This disease source has been combated with success by the construction of sewage systems and the provision of clean, disinfected drinking water supplies. Epidemics of the more lethal gastrointestinal diseases such as cholera still occur in rural populations with no clean drinking water and in towns where drinking water disinfection has failed.

An example of failure of effective disinfection of a town drinking water supply leading to severe illness and deaths in the population is the recent dramatic instance at Walkerton, Ontario, Canada. Enteric disease organisms coming from a farm were washed into a shallow well by heavy rain and distributed in the town drinking water. Illness occurred in 2300 people out of a population of 4800, and 7 deaths resulted. Other severely affected patients had lasting organ damage (O'Connor 2002; Hrudey, Payment et al. 2003).

A primary responsibility of the drinking water supply industry is therefore to prevent the transmission of disease through the drinking water, and the regulations governing drinking water have a necessary focus on disease organisms. Of lesser importance are turbidity, taste, odor, and chemical contaminants. As the availability of disinfected, microbiologically safe drinking water has increased, attention has focused on these other issues. Consumers are inevitably concerned about turbidity, taste, and odor, which are immediately discernible and underlie most of the complaints that drinking water utilities receive. Changes in the apparent quality of drinking water are interpreted by consumers to reflect lack of adequate treatment and to be associated with a health risk.

More subtle, yet likely to present a more real risk to health, are chemical contaminants in drinking water. These may be natural constituents of the water, chemicals resulting from water treatment, or contaminants such as agricultural pesticides or sex hormones. An example of a natural constituent of water is arsenic, which may be present in considerable concentration in groundwater. In the U.S., extensive discussion in the recent past has been stimulated by revision of the safe level for arsenic in drinking water (the Maximum Contaminant Level), arising in part from increased evidence of human poisoning and cancer in areas where natural arsenic in groundwater is high (Yang, Chang et al. 2003).

The majority of water treatment worldwide uses chlorine, chloramine, or chlorine dioxide as a disinfectant. New treatment plants increasingly use ozone. The use of all of these oxidants results in reaction with naturally occurring organic matter in the water, leading to a range of compounds collectively referred to as disinfection by-products. The presence of these disinfection by-products in drinking water, some of which are carcinogens in experimental animals, has also led to controversy and a move away from chlorine as a disinfectant. A large amount of epidemiological research is currently directed toward establishing the possible relationship between human health and the chlorinated and brominated compounds in drinking water (Hwang, Magnus et al. 2002; Windham, Waller et al. 2003).

Pesticide contamination has long been known to be a risk in drinking water due to the widespread use of these chemicals in agriculture. In response to the potential risks involved in the consumption of pesticides, the World Health Organization (WHO) and national regulatory bodies have specified Guideline Values, Maximum Contaminant Levels, or Reference Doses for safe drinking water based on lifetime exposure to the chemical (WHO 1996; USEPA 2004) (Table 8.1). These drinking water concentrations are calculated in two quite different ways, depending on whether the chemical contaminant is carcinogenic or noncarcinogenic. Later in this chapter the implications of this difference are explored in the context of the cyanobacterial toxins, cylindrospermopsins, microcystins, and nodularins.

Examples of chemicals for which Guideline Values are determined on the basis of carcinogenicity are benzene (formerly a component of gasoline) and bromate (a disinfection by-product), which have been shown to be carcinogenic in animal testing and are likely to be carcinogenic in humans. Examples of chemicals determined on the basis of toxicity are atrazine (herbicide) and copper, for which there is no good evidence of a carcinogenic risk to humans but that are demonstrably toxic (WHO 1996).

8.1 RISK ASSESSMENT AND LEGISLATION

Because of the perceived risks to the population of chemical contaminants in food, water, and air, the majority of countries have legislated the maximum concentration of a potentially hazardous contaminant that can be present in these three sources of human exposure. Legislation for safe food generally preceded that for safe water, and both are in a process of continuous evolution and refinement. The major changes in approach to chemical contamination of drinking water occurred in the 1970s and 1980s as a consequence of the activities of the WHO and the U.S. Environmental Protection Agency (USEPA) in trying to quantitate the adverse effects of individual

TABLE 8.1
Drinking Water Guideline Values for Toxic Contaminants, for Lifetime Safe Consumption, as Listed by the WHO, 1996

Contaminant	Guideline Value, μg/L
Nitrite	3000
Copper	2000
Lead	10
Arsenic	10
Mercury	1
Trichloroethane	2000
Xylene	500
Dichloromethane	20
Carbon tetrachloride	2
DDT	2
Atrazine	2
Chlordane	0.2
Aldrin	0.03

From WHO 1996.

chemicals. The outcome of these efforts was a series of Guideline Values for contaminants in drinking water that could be implemented by legislation.

In the U.S., the Safe Drinking Water Act of 1974 established the responsibility of the USEPA for determining the safe levels of water contaminants. To quote the House Report to Congress:

"The purpose of this legislation is to assure that water supply systems serving the public meet minimum national standards for protection of public health."

The USEPA was to identify contaminants "which have an adverse effect on the health of persons" and to protect the public "to the maximum extent feasible."

This broad brief can be interpreted with varying amounts of rigor, and Congress appreciated the problems of proof for adverse effects on public health. Even at present, more than a quarter of a century later, there is little consensus on the evidence, for example, of the effects of steroid hormone contamination of drinking water on human reproduction. To ensure that the U.S. legislation was as comprehensive in its application as possible, the following clarification was recorded: "The Committee did not intend to require conclusive proof that any contaminant *will* cause adverse effects as a condition for regulation of a specific contaminant, rather, all that is required is that the administrator make a reasoned and plausible judgment that a contaminant *may* have such an effect."

To support this approach, the House Report stated that the USEPA administrator was to carry out the following procedures:

"The known adverse health effects of contaminants are to be compiled."

"The Administrator must decide whether any adverse effects can reasonably be anticipated, even though not proved to exist. It is at this point that the Administrator must consider the possible impact of synergistic effects, long-term and multi-media exposures, and the existence of more susceptible groups in the population."

"The recommended maximum contaminant level must be set to prevent the occurrence of any known or anticipated health event."

However, the technical capability to measure the contaminant and the cost of removal of the contaminant in water treatment were realized to be major constraints on the practicality of any particular Maximum Contaminant Level. This issue was left to the USEPA to resolve: "Economic and technological feasibility [is] to be considered by the USEPA and then only for the purpose of determining how soon it is possible to reach recommended maximum contaminant levels and how much protection of the public health is feasible until then" (all quotations from Robertson 1988).

The regulatory and enforcement responsibility under the Safe Drinking Water Act was left to the USEPA until the individual states had legislation, monitoring, and enforcement processes in place. This proceeded reasonably quickly, with the states progressively assuming control of implementation of the act.

During the early 1980s, the WHO set up expert groups to assess microbiological, radiological, and chemical contaminant risks in drinking water. The existing International Program on Chemical Safety (IPCS) and the International Agency for Research on Cancer (IARC) played major roles. The outcome was the publication by the WHO of *Guidelines for Drinking Water Quality* in three volumes in 1984 and 1985 (WHO 1984). These volumes provided a large amount of background on contaminants, for which actual numerical Guideline Values could not be set, as well as recommended values for major harmful contaminants (Table 8.1).

In many countries, national health agencies set up safe drinking water guidelines for contaminants in a manner parallel to the USEPA. The WHO's *Guidelines for Drinking Water Quality* were generally followed as a basis for national decisions, though each country used local criteria to determine the relevance of particular contaminants and the actual numerical value for the chemical. For implementation of these contaminant levels in drinking water supplies, the relevant national, state, or provincial legislature then passed acts that brought into force regulations for the Maximum Contaminant Level or equivalent concentration of chemical.

By 1986, the U.S. Senate and Congress were not satisfied with the progress that the USEPA had made in setting Maximum Contaminant Levels for drinking water, in particular the few chemicals that had been finally set as regulated contaminants. The amendments of 1986 required a substantial advance in progress, with 83 specified contaminants to be regulated within 3 years. In this legislation, the definition of a contaminant was broadened to state the following: "The term *contaminant* means any physical, chemical, biological or radiological substance or matter in water." Thus "natural" biological toxins in drinking water were included. This definition is highly relevant for the inclusion of cyanobacterial toxins among regulated contaminants once the assessment of adverse health effects has been undertaken.

8.2 WHAT IS A RISK, AND HOW CAN IT BE ASSESSED?

To reach a definition of risk and an assessment of risk that can be applied widely, the terms and procedures must, to a considerable extent, be formalized. Even the definition of a risk has been codified, so that there is a common understanding of what is meant. The IPCS together with the Organization for Economic Cooperation and Development (OECD) have defined risk as "the probability of adverse effects caused under specified circumstances by an agent in an organism, a population or an ecological system."

This immediately identifies risk as a quantitative term, which can be calculated by statistical analysis of observational, experimental, or epidemiological data and expressed as a probability. The other related term, *hazard*, is a qualitative expression of potential for harm. *Hazard* is defined as "an inherent property of an agent or situation capable of having adverse effects on something" (in the case in point, the drinking water consumer).

Having stated these basic definitions of key terms, there are a further set of terms that describe processes used in risk assessment. The joint publication of the WHO/Food and Agriculture Organization (1995) on risk analysis for food contaminants provided these definitions:

Risk assessment: The scientific evaluation of known or potentially adverse health effects resulting from (in this context waterborne) hazards. The process consists of the following steps: (1) hazard identification, (2) hazard characterization, (3) exposure assessment, and (4) risk characterization. The definition includes quantitative risk assessment, which emphasizes reliance on numerical expressions of risk, as well as an indication of attendant uncertainties.

Hazard identification: The identification of known or potential health effects associated with a particular agent.

Hazard characterization (hazard assessment/dose–response assessment): The quantitative and/or qualitative evaluation of the nature of adverse effects associated with biological, chemical, and physical agents (which may be present in water). For chemical agents, a dose–response assessment should be performed if the data are available.

Exposure assessment: The quantitative and/or qualitative evaluation of the degree of intake likely to occur.

Risk characterization: Integration of hazard identification, hazard characterization, and exposure assessment into an estimation of the adverse effects likely to occur in a given population, including attendant uncertainties.

Risk management: The process of weighing policy alternatives to accept, minimize, or reduce assessed risks and to select and implement appropriate options.

8.3 RISK MANAGEMENT

The last of these definitions is different in character from the others, as it encompasses political, social, and economic factors as well as the available science-based data in resolving the appropriate actions to be taken. The area of risk management is in a phase of rapid change, as a rebound from the complex and costly regulatory approach to contaminants in drinking water. A practical consequence of defining Maximum Contaminant Levels or regulated Guideline Values for an increasing list of chemicals is the cost and futility of repeatedly analyzing for large numbers of chemicals that are below the limits of detection and highly unlikely to occur in that water supply.

The food industry has developed a different approach, called Hazard Analysis and Critical Control Point (HACCP). This is based on an initial analysis that first identifies hazards and their severity and likelihood of occurrence, and, second, identifies critical control points and their monitoring criteria to establish controls that will reduce, prevent, or eliminate the identified hazards. This has been modified from the food industry for use in the drinking water industry and is currently under development in Australia and Europe as a safe and practical approach to the prevention of adverse health effects from contaminants (National Health and Medical Research Council of Australia 2004, under approval). Hazard identification and risk assessment are integral parts of this process, with measures of likelihood of occurrence of the hazard as well as of severity of consequences from the hazard. The approach encourages the development of preventive strategies, in particular the multibarrier design of catchment management and water treatment, discussed in Chapter 11.

8.4 RISK AND CHEMICAL SAFETY IN DRINKING WATER — CYANOBACTERIAL TOXINS AS TOXIC CHEMICALS

This approach to determining the safe concentration of the cyanobacterial toxins cylindrospermopsin and microcystin in drinking water makes the basic assumption that these are noncarcinogenic. In this case the normal detoxification processes in the liver (in particular) are assumed to remove the compounds from the body via oxidation and conjugation up to a threshold dose, which overcomes the metabolic capacity to render the toxins inactive. The biochemical pathways for detoxification and excretion of these cyanobacterial toxins have been described earlier and reflect similar mechanisms for other ingested xenobiotics. Thus the dose–response curve of injury from microcystin and cylindrospermopsin has a threshold below which no adverse effects can be observed. It was therefore possible to experimentally determine the minimum dose that would cause an adverse effect and the maximum dose that could be administered without ill effect, which are the experimental doses lying on either side of the actual threshold dose.

Above this dose or concentration a log dose/linear injury response was seen, up to the point at which the cells or animals died (Figure 8.1). The concentration of toxin resulting in 50% cell death is stated as the effective concentration 50% (EC_{50}).

FIGURE 2.1 Stromatolites exposed at low tide in a hypersaline bay, Shark Bay, Western Australia.

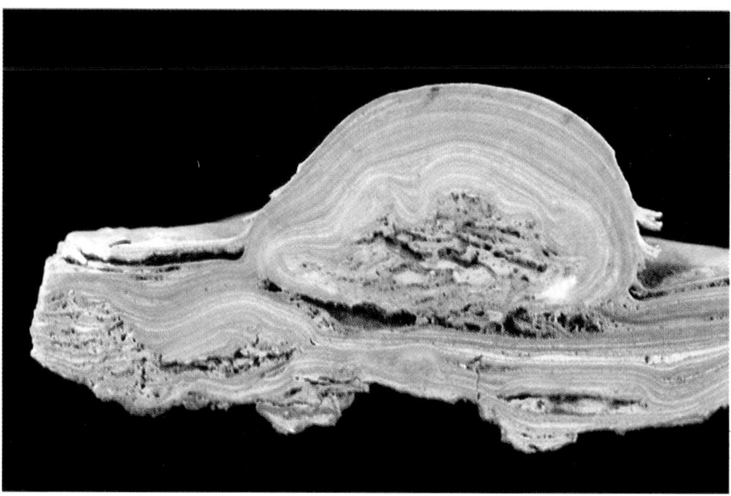

FIGURE 2.2 Section of stromatolite from a saline lake in Innes National Park, South Australia, showing cyanobacterial layers.

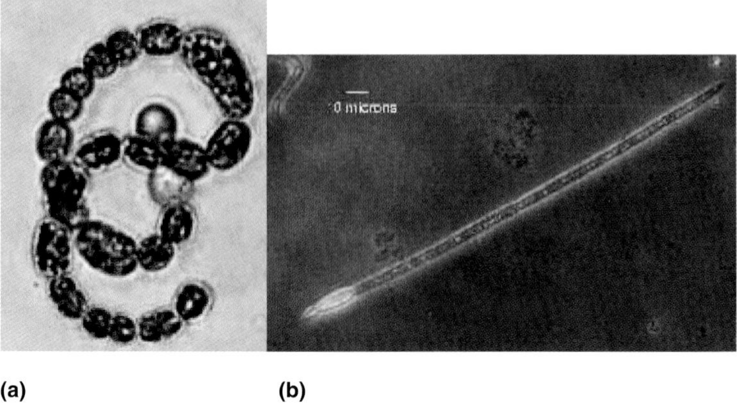

FIGURE 2.3 (**a**) *Anabaena circinalis* showing akinetes (large dense oval cells) and heterocysts (translucent spherical cells); (**b**) *Cylindrospermopsis raciborskii* showing akinete (large oval cell) and terminal heterocyst. (Images from Roger Burks, University of California at Riverside; Mark Schneegurt, Wichita State University; and Cyanosite, www.cyanosite.bio.purdue.edu. With permission.)

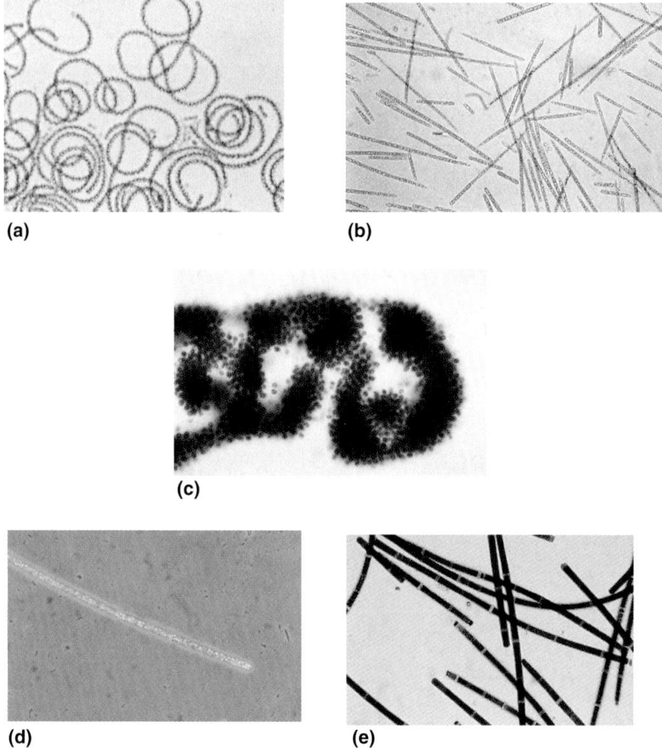

FIGURE 2.4 Photomicrographs of toxic species of cyanobacteria: (**a**) *Anabaena circinalis*; (**b**) *Cylindrospermopsis raciborskii*; (**c**) *Microcystis aeruginosa*; (**d**) *Planktothrix* sp.; (**e**) *Nodularia spumigena*. (Images (b), (c), and (e) from Cyanobacteria-toxins in drinking water, Ian R. Falconer, *Encyclopedia of Microbiology*, p. 985. With permission from Wiley. Image (d) from Dr. B. Ernst. With permission.)

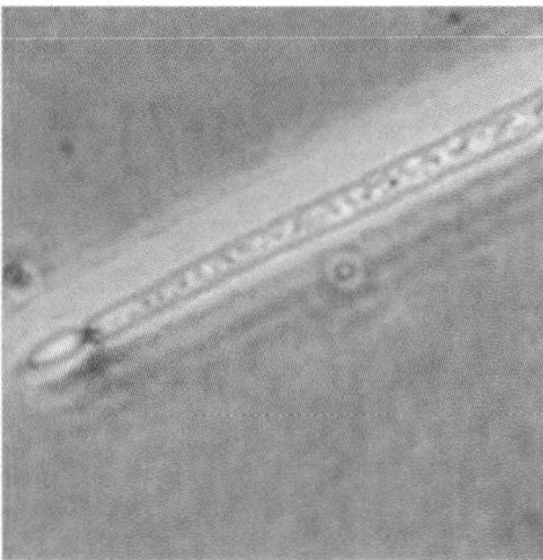
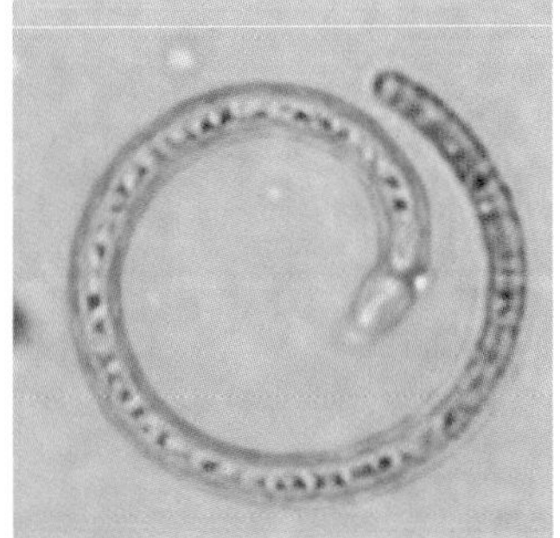

FIGURE 4.5 Straight and coiled forms of *Cylindrospermopsis raciborskii*. (From Fabbro and Andersen 2003. With permission.)

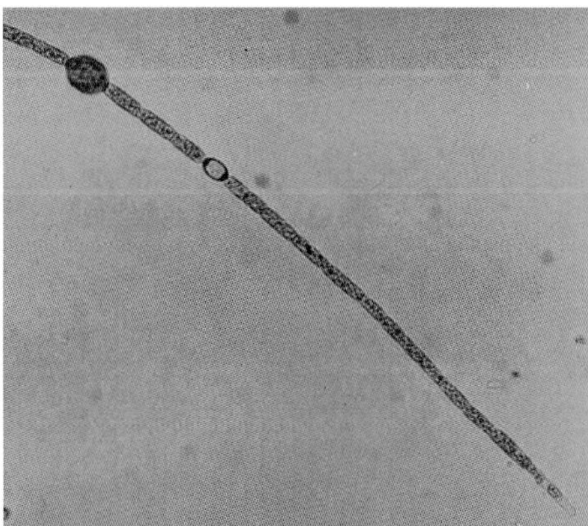

FIGURE 4.6 *Aphanizomenon ovalisporum*. (From Peter Baker, Cooperative Research Centre for Water Quality and Treatment, Australia. With permission.)

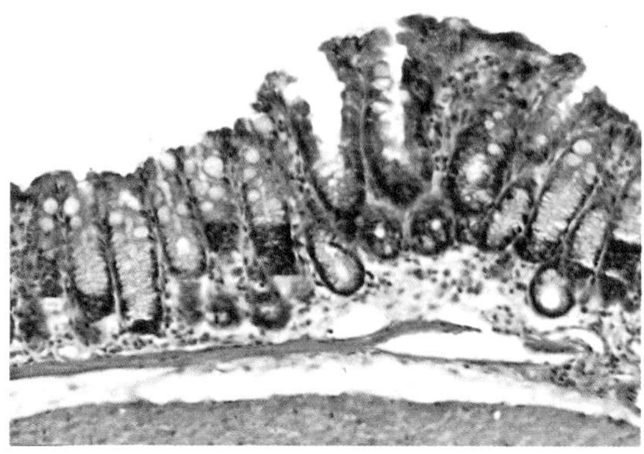

FIGURE 7.3 Precancerous colon crypt in mouse dosed with azoxymethane to initiate hypertrophic crypt formation and then exposed to microcystin in drinking water for 212 days. (From Humpage, Hardy et al. 2000. With permission.)

FIGURE 11.1 Distribution of copper sulfate to a drinking water reservoir. An airplane, normally used to distribute superphosphate fertilizer, is dropping copper sulfate on a reservoir with a *Microcystis* bloom. (Photograph courtesy of M. Choice, Armidale, Australia. With permission.)

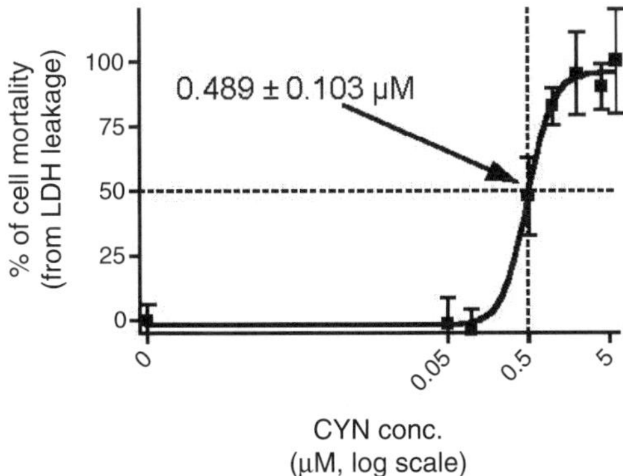

FIGURE 8.1 Death of cultured hepatocytes as a result of incubation with increasing concentrations of the cyanobacterial toxin cylindrospermopsin. Death was measured by leakage of lactate dehydrogenase from the cells.

For acute measurement of toxicity in whole animals, the lethal dose killing 50% of the animals (LD_{50}) over a fixed period of time can be calculated following administration of a single dose. In order to be able to compare different toxic chemicals, the standard procedure for experimental determination of LD_{50} is to inject young mice or rats with measured doses of the toxin into the peritoneal cavity. The doses cover the range between no observed effect and complete mortality over 24 h. The LD_{50} is expressed as milligrams per kilogram of body weight. This approach provides a basis for assessing comparative toxicities, which can be applied to any toxic chemical. Of more value to understanding of toxicity in drinking water is the oral LD_{50}, which is determined by dosing by mouth. Table 8.2 provides examples of oral toxicities. The much higher doses needed for toxicity by mouth are due to the barrier provided by the gastrointestinal tract and the destruction of chemicals in the intestine by enteric enzymes and bacteria.

The threshold concept applies with even more effect when chronic exposure to a toxic chemical occurs. In this case the bodily defense mechanisms may be activated to induce increased levels of detoxifying enzymes in the hepatocytes. These cells are then able to remove xenobiotics at a greater rate than unprepared cells. To establish experimentally the dose just below and that just above the threshold when given for an extended period, experimental animals are orally dosed for at least 10 weeks. The most commonly used period of dosing is 13 weeks for a subchronic exposure experiment and for the whole lifetime of the animal for chronic exposure.

In order to minimize the number of animals exposed, a range-finding experiment is often conducted with a minimum number of animals dosed orally for 14 days over a wide range of concentrations. After experimentally determining a dose range

TABLE 8.2
Comparative Toxicities to Rodents of Possible Drinking Water Contaminants — Oral LD_{50} (oral dose causing 50% mortality over 24 h) mg/kg

Compound	Oral LD_{50}
Atrazine	850
Copper	400
Acrylamide	100–270
Chlorpyriphos	60
Parathion	5
Microcystin-LR	5
Cylindrospermopsin	6 (at 7 days)
Saxitoxin	0.12

that causes limited toxicological symptoms at the upper dose and none at the lowest dose, a dose regime is set that brackets the threshold dose. This is followed by a 13-week oral dosing of groups of at least 15 animals of each gender at each dose, with controls and a minimum of three toxin dose rates. At the end of the dosing period, the animals are clinically examined, euthanized, and examined postmortem for biochemical and histopathological injury (Fawell, James et al. 1994).

From these data are found the highest dose, expressed in micrograms or milligrams per kilogram of body weight, causing no injury to the animals [termed the No Observed Adverse Effect Level (NOAEL)] and the lowest dose causing injury to the animals [termed the Lowest Observed Adverse Effect Level (LOAEL)]. These doses are often a factor of 5 or 10 apart, limiting the accuracy of the final values. A Tolerable Daily Intake (TDI) or Reference Dose (RfD) can then be calculated, by the incorporation of a series of safety or uncertainty factors (WHO 1996).

These factors are aimed at providing a safe and conservative adjustment to the data derived from rodent experiments when applied to human health. The most valuable data for safety calculations for the population is that from accidental human exposure to the toxin, with clinical injury to individuals and accurate exposure data. Fortunately such data are very rare, so that experimental animal data must be substituted.

The safety factors are standardized, so as to provide comparability between methodologies and results. To allow for the range of sensitivity within the human population to a particular toxin, a reduction factor of 10 is applied to the NOAEL (intraspecies uncertainty). To allow for the possible differences in toxin sensitivity between rodent and human populations, a further factor of 10 is applied (interspecies uncertainty). As the majority of the studies are performed over 10 to 13 weeks of toxin exposure and the desired outcome is a safe level of toxin over the lifetime of the consumer, an additional safety factor is required. Often there is a lack of data on teratogenicity, reproductive injury, or tumor promotion, and the uncertainties from these are incorporated with the lack of lifetime data to give an additional factor of 10 (data uncertainty). This provides a combined safety or uncertainty factor of 1000, which is the most commonly applied factor to data from rodent experiments.

Each of these factors can be reduced if the source and quality of the data are suitable. For example, the interspecies factor is not used if human epidemiological data are the source of the dose information. Similarly, if the experiment was done using primates or animals with metabolic processes similar to those of humans, such as pigs, the interspecies factor is lessened. As the overall quality and comprehensiveness of the data improve, further reduction can be made in the data uncertainty. There is one additional factor that can be applied if the toxin under consideration has particularly severe and lasting effects — for example the dioxins — and particular care must be taken in determining safe exposures. If the injury seen at the lowest dose is a teratogenic or potentially carcinogenic response, this additional factor, which can range from 1 to 10, applies (WHO 1996).

8.5 THE TOLERABLE DAILY INTAKE

This terminology is adopted by WHO for the estimation of the amount of a substance that can be ingested from food or drinking water or by inhalation daily over a lifetime without an appreciable health risk. The term has been criticized on the basis that no toxin intake is tolerable; however, it is less vulnerable to this criticism than the term that preceded it, the *Acceptable Daily Intake*. In the U.S., the term *Reference Dose*, calculated on a similar basis, is employed. The TDI is expressed in micrograms or milligrams of toxin per kilogram of body weight, as are the NOAEL or LOAEL data.

TDI is therefore calculated as

$$\text{TDI} = \frac{\text{NOAEL(or LOAEL)}}{\text{Uncertainty factors}}$$

where the combined uncertainty factors for experimental data can range from 100 to (exceptionally) 10,000, with the majority of data employing an uncertainty of 1000. The WHO considers that the combined factors should not exceed 10,000, as the resulting TDI would be so imprecise as to lack meaning.

Once the TDI for a particular toxic compound has been calculated, this information can be used to set safety guidelines for food, air, or water. In all cases the relative proportion of the dose derived from each of these exposure sources must be assessed.

For nonvolatile compounds, air is not a major environmental source and can be omitted. Thus the contribution from food and from drinking water must be determined. For the majority of metals, industrial contaminants, and pesticides, food is likely to be a significant source. However, groundwater and surface water are also liable to contamination and will contribute to the intake.

In the particular case of the cyanobacterial toxins, surface water will be the major source unless the individual is consuming toxic cyanobacteria in a health food. An arbitrary allocation of 80 or 90% of cyanobacterial toxin intake from drinking water has been applied. This is quite different from the normal situation for toxic contaminants, where food is the main source. In such cases, unless there are data that can be used to improve the accuracy of the percentage, the WHO suggests that an arbitrary value of 10% of the intake of a contaminant arising from drinking water is applicable.

The Guideline Value (GV) for a noncarcinogenic toxicant in drinking water is therefore

$$GV = \frac{\text{TDI} \times \text{Body weight} \times \text{Proportion of intake from drinking water}}{\text{Daily drinking water consumption}}$$

where body weight is 60 kg for adults, 10 kg for children, and 5 kg for infants and daily water consumption is 2 L for adults, 1 L for children, and 0.75 L for infants.

A wide range of toxic contaminants have now been assessed to determine Guideline Values; a few examples are shown in Table 8.2. These compounds have not been identified as human carcinogens, though in some cases an additional or increased uncertainty factor has been incorporated to account for tumor promotion or suspected carcinogenesis in nonhuman mammals.

In the U.S., the maximum concentration of a contaminant allowed in drinking water is defined as the Maximum Contaminant Level (MCL, also based on toxicological trials in experimental animals, with the incorporation of safety factors to determine the RfD. Up to the present, no MCLs have been set for cyanobacterial toxins in the U.S.

In Canada, the equivalent of the GV, calculated similarly, has been defined as the Maximum Acceptable Concentration (MAC), and a concentration for microcystin-LR has been determined.

8.6 DETERMINATION OF A GUIDELINE VALUE FOR CYLINDROSPERMOPSIN

There have been several published accounts of the oral toxicity of cylindrospermopsin, the majority of studies using a single dose (Falconer, Hardy et al. 1999; Seawright, Nolan et al. 1999; Shaw, Seawright et al. 2000). Repeat oral dosing after a 2-week interval showed unexpectedly enhanced toxicity, indicating residual damage to the animals from the first dose (Falconer and Humpage 2001).

A recent study, following the protocols set out by the OECD for subchronic oral toxicity assessment in rodents, used male Swiss albino mice exposed to cylindrospermopsin through drinking water and through gavage (dosing by mouth) (OECD 1998). The first trial used a cylindrospermopsin-containing extract from cultured *Cylindrospermopsis raciborskii,* supplied in drinking water for 10 weeks. The dose ranged from 0 to 657µg/kg/day, at four levels. The animals were examined clinically during the trial and showed no ill effects other than a small dose-related decrease in body weight compared to controls after 10 weeks. Liver and kidney weights were significantly higher with increasing dose.

The biochemical indicators of liver function showed dose-related changes. Serum total bilirubin and albumin increased while serum bile acids decreased. Liver enzyme changes in the serum showed a quite different pattern to those seen with acute liver poisoning or hepatitis, as only a small increase in serum alanine aminotransferase, a larger increase in alkaline phosphatase, and a decrease in aspartate aminotransferase were observed. The most substantial change was in the urine protein/creatinine concentration, which decreased sharply with dose. This was interpreted as reflecting decreased protein synthesis in the kidney through inhibition by the toxin.

Histopathological examination of all internal organs showed changes only in the liver and kidney. Dose-related hepatocyte damage and renal proximal tubular necrosis were observed (Humpage and Falconer 2003).

It was apparent from these results that lower oral doses were required to find the NOAEL, and a second trial was carried out in which mice were dosed by gavage over 11 weeks with 0, 30, 60, 120, and 240 μg/kg/day of purified cylindrospermopsin. The same trends in serum parameters were seen, but with no statistically significant changes. Organ weights showed more sensitivity to these low doses, with significant increases in body weight, and, as a percentage of body weight, in liver, kidney, adrenal glands, and testis.

Minor histopathological damage was seen in liver at the two upper dose levels and in kidney proximal tubules at the highest dose. Urine protein/creatinine decreased progressively with dose, reaching significance at 120 μ/kg/day of oral cylindrospermopsin (see Figure 6.2).

At very low dose levels of toxins, compensatory changes occur in metabolism to restore homeostasis. The increases in organ weight can be expected to compensate for reductions in function, as seen in the liver and kidneys, and compensation for stresses resulting from the toxin — for example, in the adrenal glands. It therefore becomes subjective to decide where the NOAEL occurs, depending on which effect is considered adverse. From these data (Figure 6.2), it is clear that the NOAEL is below120 μg/kg/day. However, statistically significant change in kidney weight occurred at 60 μg/kg/day. Thus, to adopt the conservative viewpoint that the most sensitive response should be considered as the indicator of adverse effect, the dose of 30 μg/kg/day is accepted as the NOAEL from these trials (Humpage and Falconer 2003).

From this value, the TDI for cylindrospermopsin in drinking water can be calculated:

$$\text{TDI} = \frac{30}{\text{Uncertainty factors}} = \frac{30}{1000} = 0.03\ \mu\text{g/kg/day}$$

Uncertainty factors are as follows: 10 intraspecies (human variability); 10 interspecies (rodent compared to human); 10 limitations in data, including subchronic, not lifetime, exposure; use only of male mice; possibility of mutagenicity or carcinogenicity; and lack of data for teratogenicity or reproductive toxicity, which gives an overall uncertainty of 1000.

The GV for safe drinking water is

$$GV = \frac{0.03 \times 60(\text{kg}) \times 0.9(\text{proportion in water})}{2\ \text{L/day}} = 0.81\ \mu\text{g/L}$$

Or, for practical purposes, the GV for cylindrospermopsin is 1 μg/L.

The need for a GV for cylindrospermopsin is currently under consideration by the WHO Chemical Safety in Drinking Water committee, together with the available data from which the Guideline Value can be determined.

8.7 THE TOLERABLE DAILY INTAKE AND DRINKING WATER GUIDELINE VALUE FOR MICROCYSTIN

Microcystin has been the most thoroughly investigated cyanobacterial toxin and is still the major toxin under investigation. As described in Chapter 7, the research has included studies of acute, subchronic, and chronic oral exposure to microcystins in several species of animal and humans. The criteria set out for oral exposure studies by the OECD, contributing to TDI calculations, have, however, been completely met only by Fawell, James et al. (1994) in their study of mouse exposure. This met the criteria for duration of exposure, both genders of animal, and experimental design. The data are discussed in Chapter 7. The conclusion was drawn that the NOAEL for microcystin-LR was 40 μg/kg/day. This was supported by the oral toxicity study carried out in pigs, which resulted in a LOAEL of 100 μg/kg/day of microcystin-LR equivalents (Kuiper-Goodman, Falconer et al. 1999). Therefore,

$$TDI = \frac{40}{\text{Uncertainty factors}} = \frac{40}{1000} = 0.04\ \mu\text{g/kg/day}$$

In this case the uncertainty factors were the same as those used for cylindrospermopsin, the limitations in data including evidence of tumor promotion, suspicion of carcinogenesis, conflicting data in teratogenesis, and less than lifetime exposure studies.

From this TDI, the GV for drinking water was calculated as

$$GV = \frac{0.04 \times 60\ \text{kg} \times 0.8\ (\text{proportion in drinking water})}{2\ \text{L}} = 0.96\ \mu\text{g/L}$$

Thus the GV for safe drinking water for microcystin-LR is 1 μg/L.

This was adopted as a provisional guideline by the WHO in 1998 as applying only to microcystin-LR (WHO 1997). Since that time Australia, Brazil, Canada, the European Union, and New Zealand have incorporated guideline levels or concentration standards for microcystins in their national drinking water supplies. Because microcystin-LR is not the only common microcystin in water supply reservoirs, consideration must be given to toxicity arising from other microcystins. In particular instances reservoirs and lakes have carried heavy water blooms of

TABLE 8.3
Toxicity of Microcystin Variants with Different L-Amino Acids in the Peptide Ring — Absence of Methyl Groups from Methylated Amino Acids Reduces Toxicity in Des-Methyl Variants

Microcystin	LD_{50}
MCYST-LA	50
MCYST-YM	56
MCYST-LR	60
MCYST-YR	70
MCYST-LY	90
MCYST-WR	150–200
MCYST-FR	250
MCYST-AR	250
MCYST-RR	600

From Sivonen and Jones 1999. With permission.

Microcystis aeruginosa that contained predominantly microcystin-LA, others microcystin-LY, others microcystin-YM, and yet others microcystin-RR. Almost all blooms have a mixture of microcystins present.

In the case of provision of safe drinking water, specifying a concentration for a single microcystin may be quite inappropriate. Even worse would be chemical or immunochemical analysis for microcystin-LR alone, which may miss high toxic concentrations of other microcystins. National guidelines have adapted the WHO guideline by using the concept of total microcystins expressed as equivalent toxicity to microcystin-LR. The toxicities of many microcystins are known, and others can be presumed equal to microcystin-LR as a safe default value (Table 8.3).

The most commonly used analytical methods will identify the range of microcystins, as discussed in Chapter 9. By converting the quantitative chemical data for separate microcystins to toxicity equivalents on the basis of comparative toxicity to microcystin-LR, a total toxicity can be determined equivalent to microcystin-LR, and applied to the Guideline Value of 1 μg/L. This will provide the level of safety for drinking water intended by the WHO guideline.

8.8 CYLINDROSPERMOPSINS AND MICROCYSTINS AS CARCINOGENS?

Carcinogens present a well-recognized hazard to the human population. The risk of getting cancer from substances in the environment is the topic of much controversy and has led to considerable research. The early recognition of a connection between the inhalation of substances later found to be carcinogens and cancer in the exposed workers was one major motivation for the establishment in 1948 in the U.S. of the Environmental Cancer Section of the National Cancer Institute. Through the work of Wilhelm Heuper, occupational exposure to β-napthylamine by dye industry workers was shown to result in bladder cancer (discussed in Hrudey 1998). More recently, exposure of miners and building workers to asbestos fiber has been shown to result in a particular type of lung cancer — mesothelioma — with damages cases currently before the law courts. Because of these and other demonstrated cancers resulting from occupational exposure, the risk of cancer from environmental contaminants has become increasingly apparent.

What is much more difficult to achieve than the qualitative identification of a hazard is to accurately determine risk from environmental exposures. The results of human epidemiology studies are strongest when the amount of exposure to a potential carcinogen can be related to the subsequent rate of cancer in the population. This has been done for some occupational exposures to carcinogens but is very difficult for environmental exposures. An example of the difficulty of relating human exposure to outcomes, including cancer, can be seen in the current debate and research into endocrine-disrupting compounds. No clear consensus has emerged on the risk to the population of environmental exposures, whereas clear evidence exists for both pharmacological and occupational exposure (WHO/IPCS 2002).

As cancer is such a considerable component of total mortality, with one-quarter to one-third of western populations dying of the disease, the identification of "extrinsic" or external factors resulting in cancer is of great importance. The WHO suggested in 1964 that three-quarters of all cancers were of extrinsic origin, as compared with only one-quarter from internal genetic or biochemical origins (WHO 1964). It is of value to identify what proportion of cancers due to these extrinsic factors can be attributed to food or water, so that modifications to diet, or food and water contaminant regulations, can be used to reduce cancer rates. It was estimated in 1981 that the proportion of cancer deaths that could be attributed to diet was 35%, higher than tobacco at 30% and much higher than alcohol at 3% or pollution at 2% (Doll and Peto 1981). In particular, it was found that voluntary modifications to diet can substantially reduce cancer risk without any regulatory involvement (Thomas and Hrudey 1997).

In the recent past, one of the biggest avoidable causes of death from disease was smoking. In Canada, the 1991 data showed 26% of all male deaths and 15% of all female deaths attributed to smoking (Thomas and Hrudey 1997). Of these deaths, 40% were due to cancer. Thus the risk of death from smoking-related cancer in the overall male population was roughly 10%, or 0.1.

Estimation of the additional risk, or additional deaths, that can be attributed to a particular environmental contaminant is best done from data from human epidemiology if available. As cancer risk is proportional to carcinogen dose, accurate human risk calculations require exposure data that are almost never available. For environmental exposures, this is particularly difficult, as individuals have multiple sources of contaminants, normally at very low doses.

To provide estimates of cancer risk for the variety of carcinogens from industrial and natural sources that occur in food or drinking water, experimental animal models have been widely applied. The basic data is obtained from a dose–response trial in which a range of doses are applied for a lifetime to a large number of experimental animals and cancers recorded. The highest dose is arbitrarily aimed at being the maximum dose that can be given orally with a weight loss of less than 10% compared to controls and with no overt signs of toxicity, which is termed the Maximum Tolerated Dose (MTD). The lower doses are simple proportions of this, such as one-half and one-quarter. Thus a set of dose–response data is generated, with cancer rates at three dose levels and zero dose, which reflects the underlying cancer rate of untreated animals.

The concept that there is no threshold dose of a carcinogen has been widely adopted on the basis that a single mutational event may lead to cancer and that the increased cancer incidence will be directly proportional to increased dose. This appears to fit well with radiation-induced cancers and also with what evidence is available for chemical carcinogenesis (McMichael 1991). On the strength of this assumption, several curve-fitting models have been developed, all projecting back to zero dose, at which there is no increased cancer probability. The most widely applied is the linear multistage model. This model simplifies to

$$A(d) = q_1^* d$$

where $A(d)$ is the additional risk (probability) of cancer from exposure to dose d, q_1^* is the slope of the probability/dose plot, and d is the dose in question.

This model can be extrapolated back to a point at which an arbitrary risk probability is reached, providing a dose for that risk level, or alternatively extrapolated to provide a figure for the risk probability at any specified dose (USEPA 1996). Because of the inherent variability of biological data, the 95% upper confidence limit of the slope estimate is used for the probability estimate to give a conservative figure.

Experimental data from animal cancer trials is likely to give moderate percentages of affected animals, at doses of carcinogen vastly higher than likely to be encountered in the environment. To determine the dose level that provides an acceptable level of excess risk, the line from the experimental data is extrapolated down to low doses. The level of risk used by the WHO for the determination of Guideline Values is a probability of 10^{-5} additional cancers — i.e., 1 in 100,000 (WHO 1996). As the experimental data will be likely to require 1 (or more) in 10 excess cancers to meet statistical significance, the downward extrapolation of dose is considerable. This can lead to an overestimation of the risk or underestimation of the dose. In practice, the major factor determining the slope of the dose–response line is the toxicity of the compound, which sets the doses used in the trial. Highly toxic carcinogens will have a steep slope, compared with less toxic carcinogens, irrespective of carcinogenicity (Lovell and Thomas 1996; Hrudey 1998). This is illustrated in Figure 8.2 from Hrudey (1998). The outcome of this effect is that the slope value, q_1^*, of risk against dose shows a strong negative correlation with MTD. A range of slope factors and drinking water Guideline Values calculated by carcinogen risk assessment are shown in Table 8.4. There has been considerable discussion on the continued use of no-threshold models and their lack of consideration of many factors affecting carcinogenesis in humans and experimental animals. This has resulted in proposals for alternative models. One of these is the Benchmark Dose (BMD), which

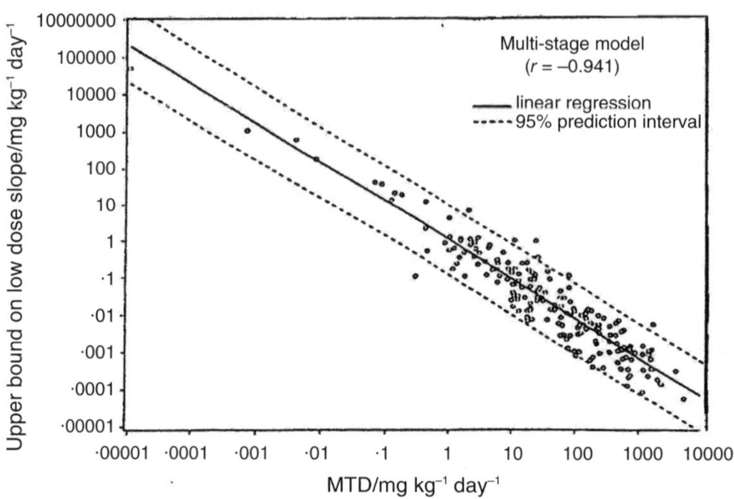

FIGURE 8.2 Association between the upper bounds on low-dose slope estimates and the maximum dose used in rodent carcinogen bioassays. (From Hrudey 1998. With permission.)

is also calculated from the additional cancers resulting from a range of doses of carcinogen in rodents. This model required less extrapolation, as it defines the probability of excess cancers — i.e., the excess risk, at 1, 5, or 10% — as the starting point from which the BMD was calculated. This risk level is likely to be close to or within the experimental results. It used the upper probability of the 95% confidence interval to account for statistical variation. The curve-fitting model may be sigmoid or whatever model best fits the experimental data. Figure 8.3 (Di Marco, Anderssen et al. 1999) illustrates this approach.

The BMD is thus directly related to risk, as the probability of a particular level of additional cancers is decided in advance and the dose providing the risk obtained from the experimental data. To this dose is then applied a series of uncertainty factors

TABLE 8.4
Slope Factors for Carcinogens in Drinking Water (mg/kg/day) and Their Guideline Values for Drinking Water (μg/L), Based on 1 In 100,000 Risk Probability of Excess Cancers

Compound	Slope Factor	Guideline Value
Acrylamide	4.5	0.08
Hexachlorobenzene	1.6	0.2
Arsenic	1.5	0.2
Bromate	0.7	0.5
Benzene	0.015–0.055	10–100

From the USEPA Integrated Risk Information System database (IRIS).

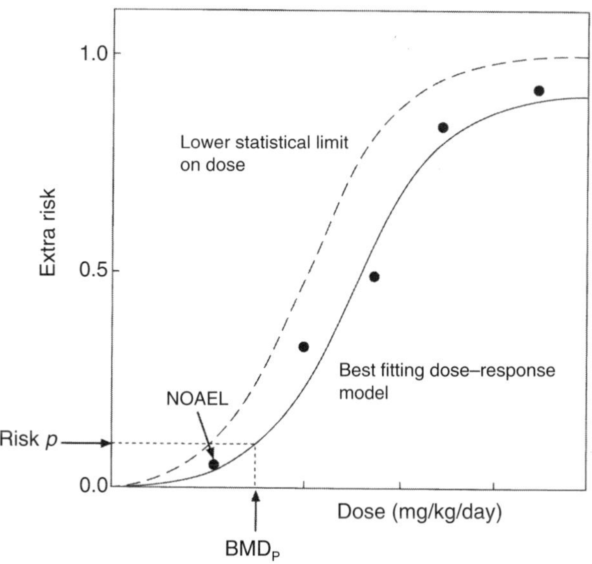

FIGURE 8.3 Hypothetical data for the determination of the benchmark dose for a carcinogen for a specified level of rise, using a best-fitting curvilinear dose–response model. At risk *p*, a horizontal line meeting the dose–response curve determines the Benchmark Dose (BMDp). (From DiMarco, Anderssen et al. 1999. With permission.)

and modifying factors to obtain the Guideline Value, which is an estimate of the dose giving no increased risk for lifetime exposure. This has been developed further into the *modified* BMD, which is based on the dose of the substance which produces a 5% increase in cancer incidence, using the central estimate of the dose–response relationship. The Guideline Value is then obtained after applying four safety or uncertainty factors. Two are the same as applied in the TDI calculation — that is, 10 for intraspecific variability and 10 for interspecific variability. The third is for quality of information, ranging from 1 to 10 on the basis of uncertainty of data. This is comparable to the third factor in TDI calculations. The fourth factor is for extent of malignancy, organ susceptibility, and genotoxicity; together, these are assigned an uncertainty factor from 1 to 50. Thus the range of overall uncertainty factors is from 100 to 50,000 for rodent data, giving considerable room for subjective assessments (Di Marco, Anderssen et al. 1999).

To resolve whether a compound should be regarded as a human carcinogen and Guideline Values determined using a no-threshold approach, a set of standardized criteria have been applied (USEPA 1986).

These identify the following:

A human carcinogen as a substance for which sufficient evidence has been provided from epidemiological studies to support a causal association between exposure to the agent and cancer.

A probable human carcinogen is a substance for which limited evidence is available from epidemiological studies for human carcinogenesis, or sufficient evidence is available from animal studies and no evidence available from epidemiology.

A possible human carcinogen is a substance for which there is limited evidence of carcinogenicity in animals and an absence of evidence from human epidemiology.

A substance not classifiable as to human carcinogenicity for which there is inadequate human or animal data for carcinogenicity.

A noncarcinogenic substance for which there is negative evidence in at least two adequate animal tests in two species or negative evidence in both adequate epidemiological and animal studies.

Those substances classified as human carcinogens and those classified as probable human carcinogens are assessed for risk and guideline levels on the basis of the no-threshold model. The other groups of substances are assessed by experimentally determining the NOAEL for calculation of the TDI and the Guideline Value, as described earlier.

8.9 CYLINDROSPERMOPSIN — IS IT A CARCINOGEN?

To answer this question, the present experimental and epidemiological data must be examined in the light of the USEPA criteria set out above.

The first and strongest criterion for a human carcinogen is that of human epidemiology, establishing a cause–effect relationship between exposure and cancer. There are no published data on this for cylindrospermopsin. Very preliminary data with small numbers of excess cancers of the liver and gastrointestinal tract have been recorded in the Palm Island population, who were exposed to cylindrospermopsin poisoning in 1979 (unpublished personal data). Geographically based analysis of cancer rates in Florida showed a significantly increased risk of liver cancer in populations located in areas supplied with surface water for drinking compared with those in contiguous areas (Fleming, Rivero et al. 2002). Earlier, a survey of microcystins and cylindrospermopsin in tap water in Florida had shown appreciable concentrations, especially of cylindrospermopsin, in reservoirs and finished water in some localities supplied from surface water sources (Williams, Burns et al. 2001). None of these data meet the requirements for an established dose–effect relationship.

The second criterion for a probable human carcinogen accepts data from experimental studies of carcinogenesis as well as epidemiology. At present there are several studies that can be considered, as well as the nature of the molecule itself. Cylindrospermopsin is a substituted pyrimidine, with potential to intercalate into the DNA double helix. The clearest experimental data on genotoxicity is the study of the effect of cylindrospermopsin on a well-understood human white cell line in culture. This demonstrated both a clastogenic (chromosome breakage) and aneugenic (whole chromosome loss) action of the toxin on dividing cells (Humpage, Fenech et al. 2000). These data show DNA damage of a major type, which in the experiments led to micronucleus formation and hence defective cells through DNA loss. Other evidence of potential DNA damage by cylindrospermopsin was shown by data suggesting DNA–cylindrospermopsin adduct formation in hepatocytes (Shaw, Seawright et al. 2000). The absence of DNA strand breaks in Chinese hamster ovary cells incubated with cylindrospermopsin may indicate that a metabolite of cylindrospermopsin is responsible for genotoxicity, rather than the parent compound (Fessard and Bernard 2003). Further studies of the mutagenic activity of cylindrospermopsin are currently in progress in Australia and Europe.

The only published whole-animal study investigating carcinogenesis after cylindrospermopsin was supplied orally to mice reported a relative risk in dosed mice of 6.6, calculated from 5 tumors in 53 dosed mice compared to none in 27 control mice. These numbers of experimental animals did not give statistical significance for the increased risk (Falconer and Humpage 2001). The data are, however, indicative that the potential for carcinogenesis from cylindrospermopsin requires urgent investigation. Cylindrospermopsin is currently a "candidate" toxin for the U.S. National Toxicology Program, which is at present exploring the feasibility of a standard toxicological assessment.

The IARC is the WHO group that resolves whether the data for a particular chemical are strong enough for a determination of a substance as a probable human carcinogen. On the basis of the present data, it is unlikely that such a determination can be made. The IARC may wish to await the results from the U.S. National Toxicology Program prior to review of the data for cylindrospermopsin as a carcinogen.

In the absence of adequate data and the likelihood of several years' delay in obtaining carcinogenicity data from standard protocol experiments, it is of interest to try to model the possible situation for cancer risk from cylindrospermopsin. Assuming that the toxin is classed as a probable human carcinogen, then assessment of its carcinogenicity can be presumed to fall within the considerable body of present data for carcinogens. As discussed earlier, the major component that determines the slope factor for a carcinogen is its MTD. A linear relationship for slope factor (obtained from the multistage model) against MTD has a negative correlation of $r = -0.941$, demonstrating the high correlation of toxicity to slope (Figure 8.2) (Krewski, Gaylor et al. 1993; Hrudey 1998). Applying an oral toxicity for cylindrospermopsin in mice of approximate MTD of 500 μg/kg/day to the regression above, the slope factor (the upper bound on the low-dose slope) is 1.6 mg/kg/day. This is comparable to arsenic at 1.5 and hexachlorobenzene at 1.6 mg/kg/day (WHO 1996). The calculated risk is then the slope factor multiplied by the exposure, so that a lifetime exposure of 1.0 μg of cylindrospermopsin per liter of drinking water (equal to 0.03 μg/kg/day of cylindrospermopsin in a 60-kg adult drinking 2 L of water) will result in a theoretical risk of 1 in 20,000 excess cancers.

This is appreciably higher than the standard accepted risk for carcinogens in drinking water of 1 in 100,000 used by the WHO. To generate a risk estimate of 1 in 100,000 for cylindrospermopsin in drinking water by this approach, the Guideline Value would be reduced to approximately 0.2 μg/L. The range for the Guideline Value for cylindrospermopsin in drinking water therefore appears to fall between 0.2 μg/L from the carcinogenicity approach to 1.0 μg/L from NOAEL data. This relative closeness in outcomes is not uncommon for toxins, irrespective of the use of the threshold model or the linear multistage model for calculation. Both approaches use the precautionary principle, with the safety factors designed to provide a wide margin of safety.

8.10 MICROCYSTINS AND NODULARINS — ARE THEY CARCINOGENS?

This question must by approached in the same way as the comparable question for cylindrospermopsin.

First, is there epidemiological evidence of human carcinogenesis, which may place these toxins in the human carcinogen category? There is considerable evidence from China that the consumption of surface water is associated with an increased risk of liver cancer (Yu 1995). It has been suggested that microcystins present in the surface water are responsible. What has not been established is the microcystin exposure data for these populations with sufficient accuracy to ensure that the effect is due to microcystins and not other carcinogenic substances in the water. The microcystin concentrations that have been measured appear low compared with concentrations in Australian or northern European surface waters (Zhang, Carmichael et al. 1991; Ueno, Nagata et al. 1996).

A recent epidemiological study of colorectal cancer in an area of China has shown an association between drinking surface water and these cancers. Measurement of microcystins in the drinking water sources show a positive correlation of colorectal cancer with microcystin content in the water (Zhou, Yu et al. 2002). This is stronger epidemiological evidence than the data for liver cancer, as a result of the assessments of exposure of the population.

Without further epidemiological data for a dose–response relationship between cancer rate and microcystin concentrations in drinking water, it is not possible to classify microcystins as human carcinogens.

For classification as probable human carcinogens, the case is stronger. For microcystins and nodularins, there is a large body of evidence from animal studies that is relevant and can be considered with the epidemiological data. Thus the classification of these toxins as probable human carcinogens requires careful consideration. Whole-animal studies by researchers in several laboratories have clearly shown that microcystins and nodularin are active tumor promoters in liver, colon, and skin, as discussed earlier. This does not imply that they are carcinogenic but leaves open the possibility that they may be nongenotoxic carcinogens. Observation of liver tumor growth following repeated high doses of microcystin-LR without prior dosing with carcinogen has been regarded as evidence for carcinogenesis (Ito, Kondo et al. 1997). Similarly, induction of precancerous foci in liver, caused by nodularin in the absence of prior carcinogen treatment, has been interpreted as implying direct carcinogenesis (Ohta, Sueoka et al. 1994). The difficulty with this interpretation is that a tumor promoter will stimulate cells mutated by prior exposure to dietary carcinogens, radiation, or natural errors in chromosome replication into precancerous foci or, with extended high doses, into cancers. Thus a range of evidence for carcinogenesis by microcystins or nodularins using differing experimental designs is required before a finite conclusion can be drawn.

Evidence for genotoxicity of microcystins and nodularin is similarly inconclusive, as these toxins cause apoptosis and necrosis of hepatocytes and other cells. This results in DNA damage, which is observed in experimental systems such as the Comet assay and in other *in vitro* and *in vivo* tests (Rao and Bhattacharya 1996). Some of the genotoxicity research has used *Microcystis* cell extracts containing microcystins and potentially a range of other bioactive components, making it difficult to ascertain the cause of any effects seen (Ding, Shen et al. 1999; Mankiewicz, Walter et al. 2002).

On balance, the available data are not strong enough to support classification of microcystins or nodularins as probable human carcinogens, though the definitive answer to this lies with the IARC, which has not yet reviewed the data.

Evidence for tumor promotion by these toxins is strong and unambiguous. Together with the epidemiological data and the possibility of carcinogenesis discussed above, the evidence supports the classification of these compounds as possible human carcinogens. On this basis, the threshold hypothesis is the most applicable to risk assessment. In the absence of dose–response data for cancers, the BMD method for carcinogens cannot be applied. However, experimental measurement of the NOAEL, data for which is available, can be used to calculate a TDI and Guideline Value for microcystins in drinking water. This does not depend on the outcome of carcinogenicity trials; however, the determination incorporates a combined uncertainty factor including tumor promotion and is the basis for the present WHO Guideline Value of 1 μg/L microcystin-LR.

8.11 CHRONIC LIFETIME DOSE, INTERMITTENT ACUTE DOSES, AND RECREATIONAL EXPOSURES

Among the issues that arise from the very fluctuating concentration of toxic cyanobacterial cells in water sources is the interpretation of Guideline Values in the case of short times where the value is exceeded in drinking water. This issue of possible intermittent exposure to high concentrations of toxin also arises in recreational exposure to cyanobacterial toxins while swimming or participating in other body-immersion water sports. The WHO Guideline Values are conservative figures aimed at providing safety over a lifetime of consumption at this concentration and therefore are not directly applicable to brief exposures to toxins.

How, then, should an acute rise in cyanobacterial toxin concentration in drinking water to above the Guideline Value be regarded? Clearly the risk associated with toxin in drinking water is directly related to the concentration and also, but less directly, to the duration of exposure. It has been argued that the logical approach to the concentration question can be seen from scrutiny of the safety factors used in calculating the Guideline Value. If, for example, the trial for ascertaining the NOAEL was done by gavage for a subchronic duration, then an additional uncertainty factor of 10 may have been applied in the calculation of the TDI for a lifetime duration. For calculating a safe dose from a single exposure or a short duration, this factor would not be required (Fitzgerald, Cunliffe et al. 1999). Hence an increase in the Guideline Value concentration by a factor of 10 may provide an estimate of the toxin concentration unlikely to cause harm from an acute exposure.

This approach has been used to develop Alert Levels for microcystins in water supplies. These Alert Levels can be legislated, so that the drinking water supplier must notify the health authorities if they are reached. The health authority then has the responsibility to determine further action — for example, discontinuance of a particular water source. The Alert Level proposed for South Australia for both total microcystins and for nodularin in drinking water is 10 μg/L (Fitzgerald, Cunliffe et al. 1999). This may be converted into a cell concentration of 20,000 cells per milliliter by using the cell content of microcystins determined from highly toxic blooms (WHO 2003). It may be considered that these levels are insufficiently conservative if the likelihood of toxin contamination at this level occurs several times a year. In this case notification of the health authority may be more appropriate at 5 μg/L. State and provincial legislatures should consider local circumstances when setting regulated levels of cyanobacterial toxins in drinking water, both as Guideline Values and as Alert Levels.

For recreational waters, the toxin concentration is not the most practical measure to determine safety, as it can be known only after analysis, which would delay action by responsible authorities. Cyanobacterial cell concentrations vary quickly, especially in situations where scums can form on bathing beaches. Cell numbers form a reasonable approximation to toxin concentration provided that the toxic species is identified. There are extensive data on the toxin content of cells, so it is possible to base recommendations on the highest toxicity seen in natural samples. This approach has been described by Chorus, Falconer et al. (2001), who set out a decision structure for the control of recreational exposure — considered in more detail in Chapter 9 (Chorus and Bartram 1999). The WHO has published *Guidelines for Safe Recreational Water Environments* (WHO 2003), which discusses the approach to safety in the presence of cyanobacterial blooms, similarly based on species identification and cell numbers in the water. The WHO classification of "moderate probability of adverse health effects" (WHO 2003, p. 149) is set at 100,000 cells per milliliter. This may be associated with toxin concentrations up to 100 μg/L, though more probably 20 to 40 μg/L if the bloom is *Microcystis*, *Planktothrix*, or *Cylindrospermopsis.* This can be used as the basis for legislated Alert Levels for recreational waters. How it is interpreted will depend on local circumstances. A conservative approach would be to designate 2,000 cells per milliliter as the first Alert Level, with increased scrutiny of the water body. At 20,000 cells per milliliter as the second Alert Level, warning signs could be posted but the area left open to bathing. At 100,000 cells per milliliter as the third Alert Level, the area is closed to body-contact water sports, including water skiing, sailboarding, jet skiing, and other sports in which there is a likelihood of toxin inhalation. At this cell concentration, there is a high chance of scum formation on bathing beaches, with associated high probabilities of adverse health effects.

REFERENCES

Chorus, I. and J. Bartram (1999). *Toxic Cyanobacteria in Water: A Guide to Their Public Health Consequences, Monitoring and Management.* London, E & FN Spon (on behalf of WHO).

Chorus, I., I. R. Falconer, et al. (2001). Health risks caused by freshwater cyanobacteria in recreational water. *Journal of Toxicology and Environmental Health* Part B, 3: 323–347.

Di Marco, P., R. Anderssen, et al. (1999). *Toxicity Assessment for Carcinogenic Soil Contaminants.* Canberra, National Health and Medical Research Council: 90.

Ding, W. X., H. M. Shen, et al. (1999). Genotoxicity of microcystic cyanobacteria extract of a water source in China. *Mutation Research* 442: 69–77.

Doll, R. and R. Peto (1981). The causes of cancer: Quantitative estimates of avoidable risks of cancer in the United States today. *Journal of the National Cancer Institute* 66(6): 1192–1308.

Falconer, I. R., S. J. Hardy, et al. (1999). Hepatic and renal toxicity of the blue-green alga (cyanobacterium) *Cylindrospermopsis raciborskii* in male Swiss Albino mice. *Environmental Toxicology* 14(1): 143–150.

Falconer, I. R. and A. R. Humpage (2001). Preliminary evidence for *in-vivo* tumour initiation by oral administration of extracts of the blue-green alga *Cylindrospermopsis raciborskii* containing the toxin cylindrospermopsin. *Environmental Toxicology* 16(2): 192–195.

Fawell, J. K., C. P. James, et al. (1994). *Toxins from Blue-Green Algae: Toxicological Assessment of Microcystin-LR and a Method for Its Determination in Water.* Medmenham, U.K., Water Research Centre, plc.

Fessard, V. and C. Bernard (2003). Cell alterations but no strand breaks induced *in vitro* by cylindrospermopsin in CHO K1 cells. *Environmental Toxicology* 18(5): 353–359.

Fitzgerald, D. J., D. A. Cunliffe, et al. (1999). Development of health alerts for cyanobacteria and related toxins in drinking-water in South Australia. *Environmental Toxicology* 14(1): 203–209.

Fleming, L. E., C. Rivero, et al. (2002). Blue green algal (cyanobacterial) toxins, surface drinking water, and liver cancer in Florida. *Harmful Algae* 1: 157–168.

Hrudey, S. E. (1998). Quantitative cancer risk assessment: Pitfalls and progress. *Risk Assessment and Risk Management.* R. E. Hester and R. M. Harrison, eds. Cambridge, U.K., The Royal Society of Chemistry: 57–90.

Hrudey, S. E., P. Payment, et al. (2003). A fatal waterborne disease epidemic in Walkerton, Ontario: Comparison with other waterborne outbreaks in the developed world. *Water Science and Technology* 47(3): 7–14.

Humpage, A. R. and I. R. Falconer (2003). Oral toxicity of the cyanobacterial toxin cylindrospermopsin in male Swiss albino mice: Determination of No Observed Adverse Effect Level for deriving a drinking water Guideline Value. *Environmental Toxicology* 18: 94–103.

Humpage, A. R., M. Fenech, et al. (2000). Micronucleus induction and chromosome loss in WIL2-NS cells exposed to the cyanobacterial toxin, cylindrospermopsin. *Mutation Research* 472: 155–161.

Hwang, B. F., P. Magnus, et al. (2002). Risk of specific birth defects in relation to chlorination and the amount of natural organic matter in the water supply. *American Journal of Epidemiology* 156(4): 374–382.

Ito, E., F. Kondo, et al. (1997). Neoplastic nodular formation in mouse liver induced by repeated intraperitoneal injections of microcystin-LR. *Toxicon* 35(9): 1453–1457.

Krewski, D., D. W. Gaylor, et al. (1993). An overview of the report. Correlation between carcinogenic potency and maximum tolerated dose: Implications for risk assessment. *Risk Analysis* 13(4): 383–398.

Kuiper-Goodman, T., I. Falconer, et al. (1999). Human health aspects. *Toxic Cyanobacteria in Water: A Guide to Their Public Health Consequences, Monitoring and Management*. I. Chorus and J. Bartram, eds. London, E & FN Spon (on behalf of WHO): 113–153.

Lovell, D. P. and G. Thomas (1996). Quantative risk assessment and the limitations of the linearised multistage model. *Human and Experimental Toxicology* 15: 87–104.

Mankiewicz, J., Z. Walter, et al. (2002). Genotoxicity of cyanobacteria extracts containing microcystins from Polish water reservoirs as determined by SOS Chromotest and comet assay. *Environmental Toxicology* 17(4): 341–350.

McMichael, A. J. (1991). Setting environmental exposures standards: Current concepts and controversies. *International Journal of Environmental Health Research* 1: 2–13.

National Health and Medical Research Council (2004). Discussion document on the *Framework for Management of Drinking Water Quality.* www.nhmrc.gov.au/advice/pdf/watergly/pdf.

O'Connor, D. R. (2002). *Reports of the Walkerton Commission of Inquiry, Ontario Ministry of the Attorney General*. 2003.

OECD (1998). *OECD Guideline for the Testing of Chemicals*. Paris, Organization for European Cooperation and Development: 10.

Ohta, T., E. Sueoka, et al. (1994). Nodularin, a potent inhibitor of protein phosphatases 1 and 2a, is a new environmental carcinogen in male f344 rat liver. *Cancer Research* 54: 6402–6406.

Pan-American Health Organization (1984). Guidelines for Drinking Water Quality: Vol. 1. Recommendations. Washington, D.C., PAHO.

Rao, P. V. and R. Bhattacharya (1996). The cyanobacterial toxin microcystin-LR induced DNA damage in mouse liver *in vivo*. *Toxicology* 114(1): 29–36.

Robertson, P. (1988). Development of maximum contaminant levels under the Safe Drinking Water Act. Washington, D.C., U.S. Environmental Protection Agency: 23.

Seawright, A. A., C. C. Nolan, et al. (1999). The oral toxicity for mice of the tropical cyanobacterium *Cylindrospermopsis raciborskii* (Woloszynska). *Environmental Toxicology* 14(1): 135–142.

Shaw, G. R., A. A. Seawright, et al. (2000). Cylindrospermopsin, a cyanobacterial alkaloid: Evaluation of its toxicologic activity. *Therapeutic Drug Monitoring* 22(1): 89–92.

Sivonen, K. and G. Jones (1999). Cyanobacterial toxins. *Toxic Cyanobacteria in Water. A Guide to Their Public Health Consequences, Monitoring and Management.* I. Chorus and J. Bartram, eds. London, E & FN Spon (on behalf of WHO): 41–111.

Thomas, S. P. and S. E. Hrudey (1997). *Risk of Death in Canada*. Edmonton, University of Alberta Press.

USEPA (1986). Cancer risk assessment guidelines. *Federal Register* 51: 33992–34005.

USEPA (1996). *Draft Revision to the Guidelines for Carcinogenic Risk Assessment*. Washington, D.C., Office of Health and Environmental Assessment, Office of Research and Development, United States Environmental Protection Agency.

USEPA (2004). *Integrated Risk Information System*. United States Environmental Protection Agency electronic database, www.epa.gov/iris/

Ueno, Y., S. Nagata, et al. (1996). Detection of microcystins, a blue-green algal hepatotoxin, in drinking water sampled in Haimen and Fusui, endemic areas of primary liver cancer in China, by highly sensitive immunoassay. *Carcinogenesis* 17: 1317–1321.

WHO (1964). *Prevention of Cancer.* Geneva, World Health Organization.

WHO (1984). *Guidelines for Drinking Water Quality*. First edition. Geneva, World Health Organization.

WHO (1996). *Guidelines for Drinking Water Quality*. Second edition. Geneva, World Health Organization.

WHO (1997). Microcystin-LR. From the Report of the World Health Organization Working Group Meeting on Chemical Substances in Drinking Water, 22–26 April 1997. Geneva, World Health Organization.

WHO (2003). Algae and cyanobacteria in fresh water. *Guidelines for Safe Recreational Water Environments*. Geneva, World Health Organization: 1: 136–158.

WHO/Food and Agriculture Organization (1995). *Application of Risk Analysis to Food Standards Issues*. WHO/FNU/FOS/95.3. Geneva, World Health Organization.

WHO/IPCS (2002). *Global Assessment of the State-of-the-Science of Endocrine Disruptors*. Geneva, World Health Organization/International Program on Chemical Safety: 180.

Williams, C. D., J. Burns, et al. (2001). Assessment of cyanotoxins in Florida's lakes, reservoirs, and rivers. *Cyanobacteria Survey Project*. Harmful Algal Bloom Task Force, St. John's River Water Management District, Palatka, FL.

Windham, G. C., K. Waller, et al. (2003). Chlorination by-products in drinking water and menstrual cycle function. *Environmental Health Perspectives* 111: 935–941.

Yang, C. Y., C. C. Chang, et al. (2003). Arsenic in drinking water and adverse pregnancy outcome in an arseniasis-endemic area in northeastern Taiwan. *Environmental Research* 91(1): 29–34.

Yu, S. Z. (1995). Primary prevention of hepatocellular carcinoma. *Journal of Gastroenterology and Hepatology* 10(6): 674–682.

Zhang, Q. X., W. W. Carmichael, et al. (1991). Cyclic hepatotoxins from freshwater cyanobacterial (blue-green algae) waterblooms collected in central China. *Environmental Toxicology and Chemistry* 10: 313–321.

Zhou, L., H. Yu, et al. (2002). Relationship between microcystin in drinking water and colorectal cancer. *Biomedical and Environmental Science* 15: 166–171.

9 Monitoring of Reservoirs for Toxic Cyanobacteria and Analysis of Nutrients in Water

The need for monitoring of drinking water sources for toxic cyanobacteria derives from their identification as a potential hazard to consumers. For the assessment of risk from cyanobacteria in water supplies, hazard characterization is required, which includes data for occurrence of the hazard as well as the extent of the hazard. To provide this information, the monitoring of water sources is essential. Monitoring supplies the key data from which management decisions are made to avoid or minimize risk to consumers. Effective monitoring will provide predictive information that can be used to guide remedial measures in advance of any formation of a toxic bloom. When water blooms occur, monitoring will assist in assessing the extent of the hazard to consumers and guide the response of the supply operator. Without monitoring, the public health risk increases sharply, to the point where adverse health effects occur.

Monitoring of water supplies coupled with remediation and contingency planning can be more cost-effective than emergency responses to health risks caused by toxic cyanobacteria. The costs of loss of drinking water supply and provision of alternative supplies in an emergency situation can be very substantial, indeed the overall costs of cyanobacterial blooms to urban water supplies can be considerable. In Australia the estimated increased cost caused by cyanobacterial blooms in urban water supplies is $U.S. 25 million per year (Atech Group, Land and Water Resources, 1999).

Most common species of toxic cyanobacteria are well described and can be identified by normal light microscopy to the level of genera by trained field and laboratory staff. Identity at species level is more difficult, but it is not required for field monitoring purposes.

Developing a monitoring program for a water supply requires knowledge of the local situation, as each water body differs and the geography and hydrology of the lake or river will substantially determine the potential for cyanobacterial occurrence. The history of cyanobacterial blooms in an established reservoir will provide a useful guide to designing a cost-effective monitoring program. If no such data exist, much more extensive monitoring will be needed to supply a database for cyanobacterial bloom prediction.

Even with a long historical record of cyanobacterial populations in a particular water body, unpredicted bloom events occur, sometimes with species that had not previously been recorded. For example, *Aphanizomenon ovalisporum* formed a water bloom in Lake Kinneret (Sea of Galilee), Israel, for the first time in 1994. This lake provides 30% of Israel's water supply and had been continuously monitored for more than 30 years. The seasonal abundance of phytoplankton had been recorded, and only once before had a toxic cyanobacterial bloom occurred, in 1964–1965, and this comprised *Microcystis aeruginosa* and *M. flos-aquae*. However in 1994, *A. ovalisporum* appeared for the first time in high concentration in the summer and reached 3000 trichomes per milliliter by October. The organism was evenly distributed in the epilimnion of this large lake.

Toxicity testing of the *Aphanizomenon* by intraperitoneal injection into mice showed lethal toxicity at 465 mg dry weight of cells per kilogram of mouse body weight. Extraction and characterization of the toxin identified cylindrospermopsin, found for the first time in this organism and in this lake (Banker, Carmeli et al. 1997; Sukenik, Rosin et al. 1998).

This is a salutary example of the need for monitoring of drinking water supplies and of the unpredictability of the occurrence and proliferation of cyanobacterial species. More commonly, however, dominance by a particular species recurs, with a reasonable level of year-to-year similarity. Global warming may have an impact on the abundance of species with capacity to thrive in warmer waters, as illustrated by the northward distribution of *Cylindrospermopsis raciborskii* into Europe (Padisak 1997). This may result in species that formerly were abundant only in the tropics becoming substantial components of temperate lake phytoplankton.

9.1 MONITORING SITES

The location of monitoring sites in a reservoir must be determined on the basis of surface and underwater topography, location of inflow streams, offtake location, and direction of prevailing winds. If the monitoring program is substantially aimed at providing data relevant to cyanobacterial problems, the characteristics of the water body will have a major role in the design of monitoring sites.

Factors that will directly affect the value of a monitoring program and its cost effectiveness include a number of issues in site selection.

For overall assessment of nutrient loading and total cyanobacterial biomass, one or more central reference sites in open water — selected on the basis of experience as characteristic of the water body as a whole — are essential. To assess the extent of the mixed zone, depth of stratification, and nutrient loading arising from anoxic solubilization of phosphorus and nitrogen from the sediments, depth profile samples will be needed. Additionally, some species of cyanobacteria form high cell densities at the deeper end of the photic zone, which may be up to 12 m below the surface in clear lakes. This is the case for *Planktothrix* in cool temperate lakes (Mur, Skulberg et al. 1999) and *Cylindrospermopsis* in subtropical lakes and rivers (Fabbro and Duivenvoorden 1996). To identify subsurface blooms of these organisms that cannot necessarily be seen from the surface, a depth profile is required. The techniques applicable to these samples are described by Utkilen (Utkilen, Fastner et al. 1999).

FIGURE 9.1 Water offtake tower in Malpas Reservoir, New South Wales, Australia. Prevailing winds drive surface blooms of cyanobacteria into the narrow channel in which the tower is located.

For assessment of nutrient inputs, sites in incoming streams — at their entry points to the reservoir and in the reservoir adjacent to major erosion sources — are needed.

Sites where cyanobacterial blooms initiate can often be identified. These may be coincident with locations for major nutrient inflow or in shallow bays, where light penetrates down to the sediments. Monitoring at these locations can provide advance warning of local cyanobacterial proliferation, which may then extend further into the water body and lead to a more extensive bloom. Scums forming in these areas can be drifted downwind into more vulnerable locations, as around water offtake towers adjacent to the dam wall (Figure 9.1).

Sites prone to scum formation will usually be bays or inlets downwind of the prevailing air movement in calm weather. In the presence of moderate winds, scums will mix into the epilimnion, to re-form when calm, warm conditions recur. If the reservoir is used for recreational purposes, the shoreline formation of cyanobacterial scums can become a significant public health issue. This is discussed in detail in the World Health Organization's *Guidelines for Safe Recreational Water Environments,* published in 2003.

A depth profile for cyanobacterial concentration at the water extraction tower or offtake is essential to obtain the profile of organisms in the water column likely to interfere with drinking water quality. This has immediate implications for selection of the depth for water abstraction, where a strategy for minimizing cyanobacterial cell concentration in the raw water for the treatment plant relies on accurate information on the cell concentrations down the water column.

9.2 MONITORING FREQUENCY

For cyanobacterial monitoring, it is preferable to monitor at high frequency during the periods when historical records indicate bloom formation and at lower frequencies at times when blooms have not been observed. In temperate and subtropical climates, cyanobacterial blooms occur most frequently in mid- to late summer, when stratification occurs. During winter and spring, many genera — including *Microcystis* and *Cylindrospermopsis* — are quiescent, with gradually increasing numbers with warming and stratification of the surface water. In locations where these are the predominant toxic species, monthly monitoring is sufficient for the winter, with an increase to weekly monitoring during spring, summer, and autumn. During an actual water bloom, monitoring on the basis of daily visual inspection of the reservoir may be necessary.

In water bodies where the cool-temperate genus *Planktothrix* is predominant, a depth profile of cyanobacteria may be required during winter to allow selection of optimal water depth for drinking water abstraction. This genus can form substantial cell density deep in the water column or under ice in winter.

9.3 PARAMETERS FOR MONITORING — PREDICTIVE PARAMETERS

On the reasonable working assumptions that (1) most species of planktonic cyanobacteria can occur at low populations in most lakes and rivers worldwide and (2) the relative proliferation of species is dependent on the immediate environment of the cells, a number of hydrological measures of lakes become useful for general predictive purposes.

One of the physical measures that is informative in cyanobacterial bloom prediction is hydraulic retention time of the water body in a reservoir. Cyanobacteria are relatively slow-growing species compared with eukaryotic algae and diatoms. As a result, they cannot compete successfully in reservoirs, lakes, or rivers in which the retention time is short or the river flowing. Thermal stratification of water bodies substantially assists cyanobacterial growth, allowing diatoms and green algae to settle into deeper layers, whereas the buoyant cyanobacteria can rise into the zone of optimal light availability. Therefore reservoirs with extended retention times and summer stratification are highly liable to cyanobacterial blooms, and the monitoring of stratification is informative in predicting blooms.

The other determining features are nutrient availability and light penetration into the water. The effects of phosphorus and nitrogen concentrations in lakes and reservoirs have been particularly studied.

The most comprehensive model for the relationship between phytoplankton biomass and total phosphorus in water was established by a cooperative study for the Organization for Economic Cooperation and Development (OECD) (Vollenweider and Kerekes 1982). Phytoplankton biomass was approximated by measurement of chlorophyll-*a* per unit volume of water. This does not discriminate between eukaryotic and prokaryotic plankton, so that it is not a measure of cyanobacteria but rather of total planktonic photosystems in the water. In hypereutrophic lakes, eukaryotic phytoplankton predominate; whereas at low phosphorus concentrations in mesotrophic lakes, cyanobacteria often predominate. There will always be a mixture, though it is possible that high cyanobacterial densities suppress growth of green algae.

Total phosphorus was measured, rather than soluble orthophosphate, as phosphorus cycling between planktonic organisms is fast; with a flourishing phytoplankton community in a lake, the actual soluble phosphate content in the water may be negligible. For further discussion of this point see Chorus and Mur (1999).

Measurements of annual mean values and maximum values for chlorophyll-*a* in a large number of phosphorus-limited lakes were plotted against annual mean concentrations of total phosphorus. This demonstrated a general positive regression between chlorophyll-*a* concentration and total phosphorus (correlation coefficient *r* approximately 0.9). The slope approximated to 1 and the intercept for the biomass was 0.64 mg/m^3 chlorophyll-*a* at 1.0 mg/m^3 phosphorus, with both parameters in the same units (equal to micrograms per liter). Thus a reasonable approximation is that 10 µg/L of total phosphorus is equivalent to a maximum potential for cyanobacterial growth of 10 µg/L of chlorophyll-*a*. If this chlorophyll-*a* is from toxic cyanobacteria, then some estimations of possible toxin concentration can be made.

One of the most studied and commonly found toxic cyanobacteria is *M. aeruginosa*; for this organism, the toxin content per cell mass and per cell has been measured, as has the chlorophyll-*a* content. If the concentration of microcystins in highly toxic natural blooms of this species is related to chlorophyll-*a*, an approximate equivalence results. That is, in a bloom sample, the chlorophyll-*a* content and microcystins content can be approximately equal (Chorus and Mur 1999; Fastner, Neumann et al. 1999). Thus a lake with a total phosphorus concentration of 10 µg/L has the potential for (say) a *Microcystis* bloom containing 10 µg/L chlorophyll-*a* and 10 µg/L of microcystins. Increased phosphorus concentrations will result in the prediction of increased cyanobacterial cell mass and increased toxins in the same proportions.

The approximations employed above use the maximum case for cyanobacterial development at this phosphorus concentration and the higher range of toxin concentrations found in the cyanobacteria. In most instances, cyanobacterial blooms do not occur at high density at this low concentration of phosphorus in water, nor do most *Microcystis* blooms have this concentration of toxin. Thus only in exceptional circumstances would both maximal organism growth occur and the cyanobacterial strain contain this level of toxin and the predicted toxin concentration result. Unfortunately, scum-forming cyanobacteria like *Microcystis* will accumulate in bays and channels in reservoirs, developing very high concentrations in the vicinity of water offtakes. Thus low cell concentrations in the bulk water of the epilimnion in a lake can result in potentially hazardous concentrations at water intakes.

The conclusion arising from these considerations is that monitoring of total phosphorus concentration in reservoirs and rivers used as drinking water sources is essential for prediction of cyanobacterial bloom-forming potential. Below 10 µg/L of total phosphorus, cyanobacterial blooms of significant magnitude are unlikely. Above 20 µg/L of phosphorus in a reservoir, the potential for blooms rises sharply. The development of blooms and the species forming the bloom will depend on other environmental factors.

Analytical techniques for phosphorus measurement have been standardized by the International Standards Organization (ISO) in Geneva. By varying the filtration protocol, soluble reactive phosphorus (orthophosphate), dissolved organic phosphorus, particulate phosphorus, and total phosphorus can be determined. The process involves the hydrolysis and oxidation of organic phosphorus to orthophosphate, as well as solubilization of particle-bound phosphorus, in sealed tubes. The product is soluble orthophosphate, which is reacted with acid molybdate reagent and ascorbic acid to give a bright blue color, read at 880 nm wavelength in a spectrophotometer (ISO 6878, in ISO 1998).

Monitoring of inorganic nitrogen in lakes can help in predicting the potential biomass of phytoplankton that can develop and its composition. Under conditions in which the phosphorus concentration is not limiting, overall phytoplankton content of a reservoir will be controlled by nitrogen availability. To maintain substantial growth of cyanobacteria during the summer, a minimum concentration of 100 μg/L of dissolved inorganic nitrogen was required (Reynolds 1992). Non-nitrogen-fixing species will be advantaged by raised concentrations of ammonia in the hypolimnion and nitrate in the epilimnion. This is particularly beneficial for *Microcystis*, which lacks heterocysts to fix atmospheric nitrogen and has buoyancy control, which allows the cells to sink deeply to take up nutrients below the epilimnion. Inorganic nitrogen in reservoirs and rivers can arise from agricultural fertilizer use, seepage, or runoff from feedlots or manure spreading, or discharge of treated sewage from plants not employing nitrification/denitrification technologies. Cyanobacteria can utilize ammonia, nitrite, or nitrate, with a preference for ammonia (see Chapter 4). All cyanobacterial genera will use dissolved inorganic nitrogen if it is available.

Methods for nitrate and ammonia analysis are available from the ISO. Two spectrophotometric methods, using phenolic reagents, are available for nitrate. These are methods ISO 7890-1 and 7890-2, using a standard spectrophotometer (ISO 1986a,b). One method involves distillation, which makes it ponderous and requires extra spaces and extraction hoods. Ammonia must be analyzed separately from nitrate. The older approach required distillation of ammonia gas; however, new techniques can be automated and are based on spectrophotometric detection (ISO 7150-2, in ISO1986c).

Several toxic cyanobacterial genera can fix nitrogen gas dissolved in water, using the specialized heterocysts present in the trichomes. Examples of these toxic genera — which produce microcystins, nodularins, and/or cylindrospermopsins — are *Anabaena*, *Nostoc*, *Nodularia*, *Cylindrospermopsis*, and *Aphanizomenon*. This nitrogen-fixing ability will provide for growth under conditions where nitrogen availability is limiting, providing the cells with some independence of dissolved nitrate or ammonia. Nitrogen fixation is, however, expensive in terms of energy and will reduce the cell growth otherwise available from photosynthetic activity. Thus it is not straightforward to predict cyanobacterial growth from water concentrations of dissolved inorganic nitrogen. The ratio of nitrogen to phosphorus in phytoplankton biomass is approximately 7:1, so the overall capability of a water body to support algal and cyanobacterial growth can be approximated from a knowledge of dissolved inorganic nitrogen in circumstances when phosphorus concentration is not limiting

(Round 1965). However, because of the nitrogen-fixing capability of some cyanobacterial genera at very low dissolved inorganic nitrogen concentrations with phosphorus available, these species will have a substantial competitive advantage and form water blooms.

Measurement of turbidity, which is necessary for treatment plant operation, can also have value for understanding cyanobacterial blooms. Turbidity coming from organic matter or suspended solids will suppress phytoplankton, including cyanobacterial growth, by reducing light penetration into the water. Cyanobacteria, however, include species that will grow at light intensities appreciably lower than are optimum for green algae; at moderate turbidity, therefore, cyanobacteria may be advantaged. Similarly, turbidity from high cell concentrations of eukaryotic algae will "shadow" phytoplankton growth, which will assist the growth of species of cyanobacteria adapted to low light.

For maximum cyanobacterial growth, light penetration down to the bottom of the surface mixed zone is needed. This is expressed as the euphotic depth — the depth at which the intensity of photosynthetically active radiation will support growth. This is directly related to turbidity and water coloration and is approximated as two or three times the Secchi depth (Mur, Skulberg et al. 1999).

9.4 PARAMETERS FOR MONITORING — IDENTITY AND NUMBER OF CYANOBACTERIAL CELLS

The most cost-effective method of direct monitoring for toxic cyanobacteria in drinking water sources is to monitor cells in the water storage site. Cyanobacterial cells are readily observed using a light microscope at magnifications of 200 to 600 times, and initial screening can be done on fresh specimens at the waterside if necessary. Cyanobacterial genera can be identified with relatively short training, and tend to recur in the same water body. There are only about 20 common toxic genera, which are sufficiently distinctive for initial identification in the field. Several toxic species are illustrated in Figure 2.4, others are illustrated in Skulberg, Carmichael et al. (1993).

Identification of species is much more difficult, as many look very similar and quantitative measures, such as cell diameter and the shape of trichomes of cells, may vary within a species. Good identification keys are available (Anagnostidis and Komarek 1988; Baker 1991; Baker 1992; Baker, Humpage et al. 1993; Baker and Fabbro 2002), but even with the help of preserved specimens it is sometimes impossible to arrive at a firm conclusion. Genetic analysis is a considerable step forward in species assignment, as discussed in Chapter 2. Fortunately for practical water monitoring purposes, it is rarely necessary to arrive at a firm species identity; assignment to a genus is sufficient to determine whether the organism is likely to be toxic.

Table 9.1 lists the more abundant genera and species of planktonic cyanobacteria identified as synthesizing microcystins, nodularins, and cylindrospermopsins, but this is only the present position, as further additions to the list occur annually.

TABLE 9.1
Cyanobacterial Genera or Species Containing Microcystins, Nodularins or Cylindrospermopsins[a]

Toxic Genus or Species	Toxin	Geographic Location	Reference
Anabaena flos-aquae	Microcystins	Canada	Krishnamurthy, Carmichael et al. 1986
Anabaena circinalis	Microcystins	France	Vezie, Brient et al. 1998
Anabaenopsis millerii	Microcystins	Greece	Lanaris and Cook 1994
Microcystis aeruginosa	Microcystins	South Africa	Botes, Viljoen et al. 1982
Microcystis botrys	Microcystins	Denmark	Henriksen 1996
Microcystis viridis	Microcystins	Japan	Watanabe and Oishi 1986
Nostoc sp.	Microcystins	Finland	Sivonen, Carmichael et al. 1990
Planktothrix agardhii	Microcystins	Finland	Sivonen 1990
Planktothrix rubescens	Microcystins	Germany	Fastner, Neumann et al. 1999
Planktothrix mougeotii	Microcystins	Denmark	Henriksen 1996
Snowella lacustris	Microcystins	Norway, Patagonia	Personal communication
Nodularia spumigena	Nodularins	Baltic Sea, Australia	Sivonen, Kononen et al. 1989; Jones, Blackburn et al. 1994
Anabaena bergii	Cylindrospermopsins	Australia	Fergusson and Saint 2003
Aphanizomenon ovalisporum	Cylindrospermopsins	Israel	Banker, Carmeli et al. 1997
Cylindrospermopsis raciborskii	Cylindrospermopsins	Australia	Hawkins, Runnegar et al. 1985
Raphidiopsis curvata	Cylindrospermopsins	China	Li, Carmichael et al. 2001
Umezakia natans	Cylindrospermopsins	Japan	Harada, Ohtani et al. 1994

[a] The references and locations are chosen arbitrarily to show the diversity of data. A more extensive listing can be found in Sivonen and Jones 1999.

9.5 SAMPLING

Sampling is of considerable importance, as cyanobacteria are rarely distributed evenly across or through a water body. In the surface layers, depth will often determine cell concentration on calm days; hence a common practice is to sample with a 5-m pipe lowered slowly into the water, collecting a representative vertical section of the water column. This is then mixed, providing an average cell concentration over that depth. In shallower areas or for obtaining a cell distribution with depth, samples can be collected at different depths by using bottles that can be opened after submerging (Utkilen, Fastner et al. 1999).

Because of the uneven distribution of cyanobacteria in reservoirs, general scrutiny of the location and abundance of the colonies in the water should be undertaken and used as a guide to sites for additional sampling. This will result in more samples at times when cyanobacterial blooms are occurring than from the routine sampling at set sites used for other hydrological data. Sampling frequency should be modified to ensure that peak periods for cyanobacterial growth are followed closely — if necessary, sampling weekly — with visual examination of the water on a daily basis during major bloom events. This will inform action for reservoir management and remediation as well as water treatment.

In addition to sample collection at the previously determined reference sites and sites of visible cyanobacterial colonies, it will be helpful to collect a concentrated scum sample from an area of cyanobacterial concentration so that toxin analysis can be undertaken more easily. If there are no accessible locations of cyanobacterial concentration, use of a plankton net in areas where cyanobacterial colonies can be seen in the water will enable a concentrated cell sample to be collected. For total toxin analysis, cooling after collection and freezing the sample at –18ºC as soon as possible will prevent microbial degradation.

If free and cell-bound toxins are to be measured separately, it is necessary to filter the sample on collection through glass-fiber papers, followed by freezing the papers and freezing a sample of the filtrate. This is rarely needed for water safety monitoring, as cell lysis will redistribute toxin from cells to water during travel through offtake pipes and pumps and during prechlorination, so that total toxin content is more informative.

For field identification of cyanobacteria to genus level, freshly collected samples without preservation are suitable. This reduces the effects of colony breakup or filament dissociation, giving a clearer opportunity for identification. When samples of cyanobacterial suspensions are carried for hours at ambient temperatures, cyanobacteria are affected to varying extents. *Anabaena,* for example, readily disintegrates, whereas *Microcystis* is more resistant to damage. *Cylindrospermopsis* may lose the terminal heterocysts from the trichomes, which are essential for identification. Carrying samples chilled to 0ºC has limited protective effect. For laboratory identification of species, samples are preserved in Lugol's iodine solution (Utkilen, Fastner et al. 1999). This enables specimens to be handled at ambient temperatures and examined under laboratory conditions, where more detailed information can be obtained than is possible in the field.

9.6 CELL COUNTING, MEASUREMENT, AND CHLOROPHYLL-*a* ANALYSIS

Quantification of cyanobacterial cell concentrations in water samples is necessary to obtain data on the increase or decrease of organisms with time, so as to assess the potential hazard to consumers of drinking water taken from that source and to plan remediation and water treatment. Three approaches can be used for the quantitation of cyanobacteria in water samples. All require attention to detail and are time-consuming. These approaches are as follows:

1. Counting cells and colonies to provide a cell number per unit volume, usually quoted as cells per milliliter
2. Measuring the biomass or biovolume of organisms, usually quoted as micrograms or cubic microliters per milliliter
3. Determining chlorophyll-*a* concentration, quoted as micrograms per liter or milligrams per cubic meter

The relative advantages and disadvantages of these methods depend on the samples being quantified. If the water contains a mix of green algae, diatoms, and cyanobacteria, the only effective approach is to count the cyanobacteria microscopically. This will provide data on both the number and genus or species of the cyanobacteria present, and samples collected over time will show trends in species composition as well as the numbers of cells of each species or genus present at a particular time. This information can provide advance warning of potential for bloom formation as well as an indication of which toxins are likely to be present. The cyanobacterial cell numbers can be used as an information base for health warnings or treatment modifications and are of value for recreational water use as well as drinking water (WHO 1998). The disadvantage is the time-consuming task of cell counting, especially with a mixed population of organisms.

Because there are large differences in the sizes of cells and filaments of different cyanobacterial species, a knowledge of the biomass or biovolume of cyanobacteria in the water can be of value. This is particularly true when the filamentous cyanobacteria are considered, as genera such as *Planktothrix* and *Nodularia* may have 10 times or more biomass per filament or per cell than occurs in *Cylindrospermopsis*. Similarly, the unicellular *Microcystis* has much smaller mass per cell than the cells in a *Planktothrix* filament. Therefore, unless the biomass is also known, cell numbers are only an indication of the potential concentration of toxin in the water. The disadvantage is that the cell measurements as well as the cell numbers must be taken for biovolume. Because cell size in a single species varies between samples and locations, tables of standard cell biomass will potentially introduce an error in the determination.

More indirect than cell counting and measurement are analyses for chlorophyll-*a,* as this measures total photosystems present in the water without differentiating between algae, diatoms, and cyanobacteria. Thus, in the presence of a mixed population of phytoplankton, the data cannot be used to quantify cyanobacterial numbers. Only in the case of water blooms of almost pure cyanobacteria can the measurement of chlorophyll-*a* be informative. In this case the content of chlorophyll will be a good guide to biomass and potential toxin concentration, as biomass is directly related to chlorophyll content and the effects of different sizes of organism are eliminated. As discussed earlier in this chapter, for a highly toxic unialgal *Microcystis* bloom there is approximate equivalence between chlorophyll-*a* content and total microcystins concentration in water; thus a bloom containing 20 μg/L of chlorophyll-*a* may contain 20 μg/L of microcystins. The analytical procedure is simple and requires only standard laboratory equipment. Thus many samples can be processed rapidly, at relatively low cost in time or materials. When a cyanobacterial bloom occurs in the absence of other phytoplankton in a reservoir, a combination of microscopic counting at less frequent intervals or with a smaller proportion of samples, with chlorophyll-*a* analysis at more locations and more frequently, will give effective, lower-cost data.

Methods of cell counting have been systematized but are not uniform worldwide. Looking down into a water bloom of cyanobacteria may give the impression that there are great numbers of colonies or filaments in the water; however, when the sample is taken, surprisingly few may be collected. As a result, there is often a need for concentration, which can be achieved by sedimentation, centrifugation, or filtration of bulk water samples. The older approach was to allow colonies to sediment down a cylinder with a counting chamber at the base, which could be fitted to an inverted microscope (Lawton, Marsalek et al. 1999). The colonies or filaments were then counted and related back to the volume of water sedimented.

A second and faster approach is to filter a measured quantity of water through a suitable filter paper or membrane. Colonies, filaments, and individual cells can be counted directly on the filter. Use of a 10- or 20-mL hypodermic syringe with a membrane filter of 0.45-μm pore size will enable easy preparation. After the water is slowly pressed through the filter, the filter holder is disassembled and the membrane, with the colonies uppermost, air-dried on a microscope slide. When it is dry, a few drops of microscope immersion lens oil are added to the membrane, which is covered with a cover slip. The cyanobacteria are then clearly visible on a transparent membrane.

To obtain an accurate estimate of the number of cyanobacterial cells, colonies, or filaments under the microscope, transects of the area of the sediment or filter or randomly selected subareas of the sediment or filter, must be counted. This requires a counting grid. It can be a simple cross-hatched grid, a grid marked with parallel lines, or a grid with random circular fields (Lawton, Marsalek et al. 1999). In every case the area that is to be counted must be known, as well as the area over which the cyanobacteria are spread and the volume of the original sample. Counting grids of the hemocytometer type are made as part of the slide chamber; other grids for upright and inverted microscopes fit into the eyepiece lens. The area of the grid that is to be used for counting can be determined by use of a calibrated slide, from which the length and hence area of the grid can be measured.

As filaments and colonies will project out from the area of grid to be counted, an arbitrary but accurate method for correcting the count is to disregard the filaments projecting to the left out of the counting field and count those projecting from the right. For square or round counting fields, a similar approach can be used.

To convert filament counts to cell numbers requires counting the number of cells in (say) 20 to 40 representative filaments to obtain an average number, which is then multiplied by the filament count. Nonfilamentous cyanobacteria forming globular colonies are almost impossible to count without disrupting the colony. This can be achieved by judicious ultrasonification, mild alkaline hydrolysis, or storage and strong shaking in Lugol's iodine solution. Clumps of cells and individual cells can be counted much more readily than colonies. If this quantification of cell numbers in colonies is done on 20 to 40 colonies, the average cell number can then be applied to colony counts for the calculation of cell numbers. As cell numbers in colonies vary widely with the growth of the bloom and hence size of the colony, it is necessary to repeat the cell counting each sampling period or when visual inspection or colony measurement shows a changed colony size.

Biomass determination can be carried out by a count of the cell numbers and measurement of cell dimensions or, in the case of monocultures of cyanobacteria, by chlorophyll-*a* analysis. Cyanobacteria have approximately 1 mg of cell mass for 1 mm^3 of cell volume, so that measures of cell dimensions can be directly converted to mass. Colonial cyanobacteria such as *Microcystis* have spherical cells, whose diameter can be measured and volume calculated. Measurement of 20 cells should provide an average that is sufficiently accurate provided that the cells are of similar size.

Filaments are more difficult, as they may have tapering ends, many heterocysts or akinetes, and smaller cells nearer the tips than those in the center of the filament. A working assumption must be made that the cells are cylindrical; however, the differing length of cells results in the need to measure a greater number in order to provide an accurate average. The average lengths and diameters can be used to calculate the cell volume. These methods will provide measures of biomass in the form of "wet weight" of cyanobacteria in the water body, which together with toxin analysis on a wet-weight basis can estimate the toxin per unit volume of water.

An alternative approach to the same end is to take measured volumes of water containing cyanobacteria, centrifuge the cells down to a packed sediment, and then air-dry or dry by oven at a low temperature — say 40°C — and measure dry weight. This can then be used with toxin analysis on a dry-weight basis to again calculate the toxin content of the water. This approach is of limited use in mixed phytoplankton samples, as a substantial proportion of the dry weight can then arise from the eukaryotic organisms. In a cyanobacterial bloom with few other organisms present, it is quicker and less demanding than microscopic techniques for measuring biomass. The presence of noncyanobacterial mass does not, however, affect the outcome of calculation of toxin concentration in the reservoir provided that the toxin analysis is also done on the dry weight of similar samples.

9.7 CHLOROPHYLL-*a* ANALYSIS

Because of the ease of the analytical technique and its rapidity compared with cell counting and measurement, chlorophyll-*a* analysis is the most widely used method for determining cyanobacterial or phytoplankton biomass. It provides a cost- and labor-effective mechanism for quantitative assessment of phytoplankton. However, the relative sizes of cyanobacteria and green algae, with green algal cells one or more orders of magnitude larger than cyanobacterial cells, mean that water samples containing both types of organism will be predominantly measuring eukaryotic algae. Unless the water sample is very largely cyanobacteria, the results will grossly overestimate cyanobacterial biomass. Many cyanobacterial blooms are in practice almost monospecific, so that chlorophyll-*a* becomes an effective tool for following growth.

The chlorophyll-*a* content of cyanobacterial cells varies with the light availability, lower light intensity resulting in a higher proportion of chlorophyll. Overall this does not have a large impact on the results, as the proportion generally lies between 0.5 and 1% of the organic content of the cells' dry weight (Lawton, Marsalek et al. 1999).

There is an ISO standard protocol for chlorophyll-*a* measurement (ISO 1992); a detailed account of the procedure can be found in Lawton et al. (1999). The basic technique is to filter off the cyanobacteria from the bulk water sample, extract the filter with boiling 90% ethanol, grind the filter and cells, centrifuge, collect the clear supernatant, and read absorbance of the chlorophyll-*a* spectrum at 665-nm wavelength. Corrections are need for turbidity and for the chlorophyll-*a* breakdown product, phaeophytin-*a*.

9.8 FLUORESCENCE MEASUREMENT OF CYANOBACTERIAL CONCENTRATION IN RESERVOIRS

An alternative to the measurement of light absorbance for the quantitation of phytoplankton pigments is the determination of fluorescence. This can be used for the quantitation of the biomass of phytoplankton without the need for extraction, as it is effective on whole live cells. It has been used for chlorophyll-*a* measurement using the excitation peak at 440 nm and the emission at 680 nm, which will provide information on the overall phytoplankton biomass (Heaney 1978; Maxwell and Johnson 2000).

In cyanobacteria, the fluorescent pigment phycocyanin provides a major component of the overall fluorescence spectrum, with excitation at 550 to 650 nm and emission at 645 nm. The fluorescence from phycocyanin provides about 100-fold more sensitivity than fluorescence from chlorophyll-*a* for measurement of cyanobacteria (Lee, Tsuzuki et al. 1994). Because of the differences in overall fluorescence spectra between green algae and cyanobacteria, the species composition of a mixed plankton sample can be assessed (Gerhardt and Bodemer 2000).

The phycocyanin fluorescence can therefore be used as an analytical tool to determine cyanobacterial biomass in a water bloom (Otsuki, Omi et al. 1994). This can be applied to samples taken in the course of normal monitoring procedures and has also been developed as an online assay method. Application of online techniques to assessment of the concentration of *Microcystis* blooms in a drinking water reservoir supplying Lodz, Poland, has demonstrated a statistically significant correlation between phycocyanin fluorescence and cyanobacterial biomass and also between fluorescence and total microcystin concentration (personal communication, K. Izydorczyk, Lodz University). Monitoring online enabled the cyanobacterial concentration in water drawn into the water offtake to the treatment plant to be monitored at 15-min intervals, which showed an approximately 10-fold variation in cell concentration over a 24-h period. This technique has valuable implications for water treatment, discussed in Chapter 11.

9.9 MONITORING BY GENETIC METHODS

With advances in the genetic identification of genes for toxin synthesis in cyanobacterial populations, it has become possible to screen naturally occurring water blooms for the distribution of potentially toxic organisms (Vaitomaa, Repka et al. 2002; Hisbergues, Christiansen et al. 2003; Kurmayer and Kutzenberger 2003). In populations containing mixed species several toxigenic genes may be present, and their abundance and nature vary with time and location. This approach has been used on a drinking water supply reservoir subject to frequent cyanobacterial blooms. Within this population, genes for both microcystin synthesis and saxitoxin synthesis were identified and shown to change over time (Baker, Entsch et al. 2002). It is apparent that this technique has considerable potential, both in identifying possible toxicity and in quantitating the occurrence of toxigenic species.

A recent review and description of the molecular detection methods available for the genes responsible for production of microcystins, nodularin, and cylindrospermopsin has set out the value of these approaches (Burns, Saker et al. 2004). With the range of genetic identification available, the potential of cyanobacterial blooms to form the key toxins likely to cause harm when present in drinking water can be identified. With these gene detection methods there is no requirement to identify toxic species of cyanobacteria, which removes one of the potential obstacles in monitoring. The methods can also identify nontoxigenic blooms of otherwise toxic species. As these methodologies develop into either laboratory-based detection systems for routine use or portable kit systems that can be carried by vehicle to a lake side, genetic monitoring is likely to become a valuable tool for management of drinking water reservoirs.

Monitoring of water supply reservoirs for cyanobacterial cells is very widely undertaken by supply authorities as well as authorities controlling recreational water use. Operational decisions are regularly made on the basis of this monitoring, in conjunction with the water quality guidelines and Alert Levels recommended by the WHO and national regulatory agencies.

REFERENCES

Anagnostidis, K. and J. Komarek (1988). Modern approach to the classification of cyanophytes. *Archiv für Hydrobiologie* 80: 327–472.

Atech Group (1999). *Cost of Algal Blooms*. Canberra, Australia, Land and Water Resources Research and Development Corporation: 41.

Baker, J. A., B. Entsch, et al. (2002). Monitoring changing toxigenicity of cyanobacterial bloom by molecular methods. *Applied Environmental Microbiology* 68(12): 6070–6076.

Baker, P. (1991). *Identification of Common Noxious Cyanobacteria: Part 1 — Nostocales*. Melbourne, Urban Water Research Association of Australia.

Baker, P. (1992). *Identification of Common Noxious Cyanobacteria: Part II — Chroococcales, Oscillatoriales*. Melbourne, Australia, Urban Water Research Association of Australia.

Baker, P. and L. D. Fabbro (2002). *A Guide to the Identification of Common Blue-Green Algae (Cyanoprokaryotes) in Australian Freshwaters*. Thurgoona, Australia, Cooperative Research Centre for Freshwater Ecology.

Baker, P. D., A. R. Humpage, et al. (1993). *Cyanobacterial Blooms in the Murray Darling Basin: Their Taxonomy and Toxicity*. Adelaide, Australia, Australian Centre for Water Quality Research: 159.

Banker, P., S. Carmeli, et al. (1997). Identification of cylindrospermopsin in *Aphanizomenon ovalisporum* (Cyanophyceae) isolated from Lake Kinneret, Israel. *Journal of Applied Phycology* 33: 613–616.

Botes, D. P., C. C. Viljoen, et al. (1982). Structure of toxins of the blue-green alga *Microcystis aeruginosa*. *South African Journal of Science* 78: 378–379.

Burns, B. P., M. L. Saker, et al. (2004). Molecular detection of genes responsible for cyanobacterial toxin production in the genera *Microcystis*, *Nodularia* and *Cylindrospermopsis*. *Methods in Molecular Biology* 268: 213–222.

Chorus, I. and L. Mur (1999). Preventative measures. *Toxic Cyanobacteria in Water: A Guide to Their Public Health Consequences, Monitoring and Management*. I. Chorus and J. Bartram, eds. London, E & FN Spon (on behalf of WHO): 236–273.

Fabbro, L. D. and L. J. Duivenvoorden (1996). Profile of a bloom of the cyanobacterium *Cylindrospermopsis raciborskii* (Woloszynska) Seenaya and Subba Raju in the Fitzroy River in tropical central Queensland. *Marine and Freshwater Research* 47(5): 685–694.

Fastner, J., U. Neumann, et al. (1999). Microcystins (hepatotoxic heptapeptides) in German fresh water bodies. *Environmental Toxicology* 14(1): 13–22.

Fergusson, K. M. and C. P. Saint (2003). Multiplex PCR assay for *Cylindrospermopsis raciborskii* and cylindrospermopsin-producing cyanobacteria. *Environmental Toxicology* 18(2): 120–125.

Gerhardt, V. and U. Bodemer (2000). Delayed fluorescence spectroscopy: A method for automatic determination of phytoplankton composition of freshwaters and sediments and the algal composition of benthos. *Limnologica* 28: 313–322.

Harada, K., I. Ohtani, et al. (1994). Isolation of cylindrospermopsin from a cyanobacterium *Umezakia natans* and its screening method. *Toxicon* 32: 73–84.

Hawkins, P. R., M. T. C. Runnegar, et al. (1985). Severe hepatotoxicity caused by the tropical cyanobacterium (blue-green alga) *Cylindrospermopsis raciborskii* (Woloszynska) Seenaya and Subba Raju isolated from a domestic supply reservoir. *Applied and Environmental Microbiology* 50(5): 1292–1295.

Heaney, S. I. (1978). Some observations on the use of the *in-vivo* fluorescence technique to determine chlorophyll-*a* in natural populations and cultures of freshwater phytoplankton. *Freshwater Biology* 8: 115–126.

Henriksen, P. (1996). Toxic Cyanobacteria/Blue-Green Algae in Danish Fresh Waters. Ph.D. thesis, Copenhagen, University of Copenhagen.

Hisbergues, M., G. Christiansen, et al. (2003). PCR-based identification of microcystin-producing genotypes of different cyanobacterial genera. *Archives of Microbiology* 180: 402–410.

ISO (1986a). Water quality — determination of nitrate — part 1: 2,6-Dimethylphenol spectrometric method. *ISO 7890-1*. Geneva, International Organization for Standardization.

ISO (1986b). Water quality — determination of nitrate — part 2: 4-Fluorophenol spectrometric method after distillation. *ISO 7890-2*. Geneva, International Organization for Standardization.

ISO (1986c). Water quality — determination of ammonium — part 2. Automated spectrometric method. *ISO 7150-2*. Geneva, International Organization for Standardization.

ISO (1992). Water quality — measurement of biochemical parameters: Spectrometric determination of the chlorophyll-a concentrations. *ISO-10260*. Geneva, International Organization for Standardization.

ISO (1998). Water quality — spectrophotometric determination of phosphorus using ammonium molybdate. *ISO/FDIS 6878*. Geneva, International Organization for Standardization.

Jones, G. J., S. I. Blackburn, et al. (1994). A toxic bloom of *Nodularia spumigena* Mertens in Orielton Lagoon, Tasmania. *Australian Journal of Marine and Freshwater Research* 45: 787–800.

Krishnamurthy, T., W. W. Carmichael, et al. (1986). Toxic peptides from freshwater cyanobacteria (blue-green algae). I. Isolation, purification and characterization of peptides from *Microcystis aeruginosa* and *Anabaena flos-aquae*. *Toxicon* 24(9): 865–873.

Kurmayer, R. and T. Kutzenberger (2003). Application of real-time PCR for quantification of microcystin genotypes in a population of the toxic cyanobacterium *Microcystis* sp. *Applied Environmental Microbiology* 69(11): 1258–1262.

Lanaris, T. and C. M. Cook (1994). Toxin extraction from an *Anabaenopsis millerii*–dominated bloom. *The Science of the Total Environment* 142: 163–169.

Lawton, L., B. Marsalek, et al. (1999). Determination of cyanobacteria in the laboratory. *Toxic Cyanobacteria in Water: A Guide to Their Public Health Consequences, Monitoring and Management.* I. Chorus and J. Bartram, eds. London, E & FN Spon (on behalf of WHO): 347–367.

Lee, T., M. Tsuzuki, et al. (1994). *In–vivo* fluorimetric method for early detection of cyanobacterial waterblooms. *Journal of Applied Phycology* 6: 489–495.

Li, R. H., W. W. Carmichael, et al. (2001). The first report of the cyanotoxins cylindrospermopsin and deoxycylindrospermopsin from *Raphidiopsis curvata* (cyanobacteria). *Journal of Phycology* 37(6): 1121–1126.

Maxwell, K. and G. N. Johnson (2000). Chlorophyll fluorescence — a practical guide. *Journal of Experimental Botany* 51: 659–668.

Mur, L. R., O. M. Skulberg, et al. (1999). Cyanobacteria in the environment. *Toxic Cyanobacteria in Water: A Guide to Their Public Health Consequences, Monitoring and Management*. I. Chorus and J. Bartram, eds. London, E & FN Spon (on behalf of WHO): 15–40.

Otsuki, A., T. Omi, et al. (1994). HPLC fluorometric determination of natural phytoplankton phycocyanin and its usefulness as cyanobacterial biomass in a highly eutrophic shallow lake. *Water, Air and Soil Pollution* 76: 383–396.

Padisak, J. (1997). *Cylindrospermopsis raciborskii* (Woloszynska) Seenayya et Subba Raju, an expanding, highly adaptive cyanobacterium: worldwide distribution and review of its ecology. *Archiv für Hydrobiologie* 107 (suppl): 563–593.

Reynolds, C. S. (1992). Eutrophication and the management of planktonic algae: what Vollenweider couldn't tell us. *Eutrophication: Research and Application to Water Supply.* J. G. Jones and D. W. Sutcliffe, eds. Ambleside, U.K., Freshwater Biological Association.

Round, F. E. (1965). *The Biology of the Algae*. London, Edward Arnold.

Sivonen, K. (1990). Effects of light, temperature, nitrate, orthophosphate, and bacteria on growth of and hepatotoxin production by *Oscillatoria agardhii* strains. *Applied and Environmental Microbiology* 56 (9): 2658–2666.

Sivonen, K., W. W. Carmichael, et al. (1990). Isolation and characterization of hepatotoxic microcystin homologs from the filamentous freshwater cyanobacterium *Nostoc* sp. strain 152. *Applied and Environmental Microbiology* 56(9): 2650–2657.

Sivonen, K. and G. Jones (1999). Cyanobacterial toxins. *Toxic Cyanobacteria in Water: A Guide to Their Public Health Consequences, Monitoring and Management.* I. Chorus and J. Bartram, eds. London, E & FN Spon (on behalf of WHO): 41–111.

Sivonen, K., K. Kononen, et al. (1989). Occurrence of the hepatotoxic cyanobacterium *Nodularia spumigena* in the Baltic Sea and structure of the toxin. *Applied and Environmental Microbiology* 55(8): 1990–1995.

Skulberg, O. M., W. W. Carmichael, et al. (1993). Taxonomy of toxic Cyanophyceae (cyanobacteria). *Algal Toxins in Seafood and Drinking Water.* I. R. Falconer, ed. London, Academic Press: 145–164.

Sukenik, A., C. Rosin, et al. (1998). Toxins from cyanobacteria and their potential impact on water quality of Lake Kinneret, Israel. *Israel Journal of Plant Sciences* 46(2): 109–115.

Utkilen, H., J. Fastner, et al. (1999). Fieldwork: Site inspection and sampling. *Toxic Cyanobacteria in Water: A Guide to Their Public Health Consequences, Monitoring and Management.* I. Chorus and J. Bartram, eds. London, E & FN Spon (on behalf of WHO): 329–345.

Vaitomaa, J., S. Repka, et al. (2002). Aminopeptidase and phosphatse activities in cyanobacteria dominated basins at Lake Hiidevesi and in laboratory grown *Anabaena. Freshwater Biology* 47: 1582–1593.

Vezie, C., L. Brient, et al. (1998). Variation of microcystin content of cyanobacterial blooms and isolated strains in Lake Gand-lieu (France). *Microbial Ecology* 35(2): 126–135.

Vollenweider, R. and J. Kerekes (1982). *Eutrophication of Waters, Monitoring, Assessment, Control.* Paris, Organization for Economic Cooperation and Development.

Watanabe, M. F. and S. Oishi (1986). Strong probability of lethal toxicity in the blue-green alga *Microcystis viridis* Lemmermann. *Journal of Phycology* 22: 552–556.

WHO (1998). Freshwater algae and cyanobacteria. *Guidelines for Safe Recreational Water Environments. Volume 1: Coastal and Fresh-Waters.* Geneva, WHO.

10 Detection and Analysis of Cylindrospermopsins and Microcystins

To assure public safety of drinking water supplies, harmful organisms and toxic contaminants must be reduced to harmless levels. In order to be able to provide this assurance when toxic cyanobacterial water blooms occur on supply reservoirs, analytical techniques are required of sufficient sensitivity to characterize any hazard.

The early approach to assessing potential hazard from a cyanobacterial bloom was the toxicity testing of scum or concentrate samples by injection into mice (Falconer 1993). This method has limitations through lack of sensitivity and specificity and ethical difficulties due to subjecting animals to potentially painful treatment. In the last decade, a series of alternative approaches have been developed, including microbiotests using invertebrates, enzyme inhibition assays, enzyme-linked immunosorbent assays (ELISAs), and a range of chemical analytical techniques. This chapter considers these methods and also evaluates the remaining role for *in vivo* mouse assays.

Cyanobacterial toxins are synthesized within the cells of the organisms and largely remain within the cells during growth. However, when the cells senesce or are killed, there may be high concentrations of toxin that are free in the water. Blooms senesce and lyse (die) naturally, so that a water body with a *Microcystis* or *Planktothrix* bloom that is forming a decaying scum will have appreciable quantities of free toxin in the water. Copper treatment of reservoirs kills the cyanobacterial cells and releases toxins into the water. During water treatment, the early addition of chlorine will lyse the cells, similarly releasing toxin into the water. Thus, for accurate measurement of a cyanobacterial toxin in a drinking water supply, it is important to measure total toxin in the water — that is, toxin in cells plus free toxin in the water — for a reliable assessment of potential hazard.

The majority of bioassay and analytical techniques are insufficiently sensitive to directly measure the low concentrations of microcystins, nodularins, and cylindrospermopsins that occur in the bulk water in reservoirs. To provide sufficient concentration of toxins, several methods have been developed that will selectively concentrate toxin for analysis. These are described in this chapter.

It has recently become possible to measure the total toxins in water without a concentration step, and these methods are currently being validated. As the toxin levels are frequently in the submicrogram-per-liter range, great sensitivity is required. With the World Health Organization's (WHO) determination of a provisional

Guideline Value of 1 μg/L for microcystin-LR in drinking water and a similar concentration recommended for cylindrospermopsin, analytical techniques for tap water must be accurate down to concentrations of 0.1 μg/L (approximately 0.1 nM for microcystins and 0.2 nM for cylindrospermopsin). Immunoassays and analytical techniques of suitable sensitivity are becoming available and others are under development. Several based on ELISAs for microcystins and nodularin are available commercially as kits.

10.1 TOXIN CONCENTRATION

There are two quite different approaches to concentrating cyanobacterial toxins present in a water bloom in a reservoir so that an effective analysis can be undertaken. Which approach is selected depends on the genus of toxic cyanobacterium and the growth phase of the water bloom. The microcystin- and nodularin-containing genera, such as *Microcystis*, *Planktothrix*, and *Nodularia*, retain the toxin within the cells in growing, healthy colonies and filaments (Welker, Steinberg et al. 2001). The simplest method of concentrating toxin for analysis in water blooms of these organisms is to concentrate the cells, as minimal amounts of free toxin will be present in the water. This approach is discussed later in Section 10.7.

The other approach applies when a water bloom is naturally lysing, has been dosed with copper sulfate, or is of a genus that "leaks" toxin into the free water. *Cylindrospermopsis* is such a genus, with a considerable quantity of toxin in the water even in healthy, growing cultures and natural blooms (Hawkins, Putt et al. 2001). In these cases a deceptive underestimate of total toxin in the water will result from measuring cell toxicity only.

To ensure that all the toxin content of a sample of water containing cyanobacteria is available for concentration and assay, it is essential to lyse the intact cyanobacterial cells. Freeze-thawing of samples is effective for some cyanobacterial species, but *Microcystis* is particularly difficult to disrupt. Repeated cycles of sonication and freeze-thawing may be required. Freeze-drying followed by resuspension and sonication are also effective. The cell debris can be removed by filtration or centrifugation (Falconer 1993; Lawton, Beattie et al. 1994).

Concentration of toxin from lysed cyanobacteria can be undertaken by two methods. One is to freeze-dry, or evaporate off, the bulk of the water prior to assay. The resulting solution may require pH adjustment. The other (preferable) approach relies on adsorption of toxins onto a solid phase in an appropriate cartridge. As the technique differs between toxins, it is dealt with separately in Sections 10.3 and 10.7.

10.2 *IN VIVO* RODENT TOXICITY ASSAYS

The basis of toxicology is the adverse effect of the toxic compound on mammals, which provides evidence for potential toxicity to people. The identification of toxicity in cyanobacteria followed poisonings of domestic animals and of people and was initially investigated in rodents and domestic animals. Concentrated scum samples were the preferred material for toxicity investigation, as sufficient concentration of toxin for observable pathological changes was required. Direct measurement of low concentrations of toxins in water was not feasible with the *in vivo* rodent assays, as their low sensitivity required about 3 μg of toxin per mouse for lethality.

To provide effective concentrations of toxin for rodent assays, most samples require cell or toxin concentration; for fresh bloom or cell-culture samples, this has been done by cell concentration followed by *in vivo* toxicity measurement. This method has provided the basic data for the investigations of the toxic species, the mechanisms of toxicity, and the development of more sensitive assays.

The use of whole-animal assays is currently opposed on ethical grounds and is being replaced by the variety of microbiotests and *in vitro* test systems available. Only in the case of investigation of a possible public health risk from a newly discovered toxic cyanobacterial species or verification of the cause of livestock poisoning by an uncharacterized cyanobacterial bloom is it essential to use live animals. When a cyanobacterial species that has not been investigated for toxicity appears in a drinking water supply, especially in the posttreatment distribution system, it is still necessary to undertake mouse toxicity tests. In a recent case of this problem, *Phormidium*, a normally benthic cyanobacterium, detached from the sediment of a distribution reservoir and entered the public water supply. Immediate toxicity testing using mice identified the organism as poisonous, and the users of the supply were alerted and provided with alternative drinking water (Baker et al. 2001). In this case the toxin has not been identified and is not any of the presently described toxins. For investigation of the toxicity of cyanobacterial blooms of genera known to produce particular toxins, microbiotests or *in vitro* tests are appropriate and sufficiently rapid.

10.2.1 Methods for Mouse Tests — Intact Cells

Toxicity testing, using rodents, of an uncharacterized cyanobacterial bloom posing a health risk can be undertaken by the following procedure. It is first necessary to collect a sample of as dense bloom material as possible in order to obtain sufficient toxin. This can be done by collecting from a scum concentration along the edge of the water or by use of a plankton net. The cells can be processed to remove water by filtration at the lakeside or transported to the laboratory with minimal heating or shaking. This material can be concentrated by allowing the sample to stand overnight and collecting the buoyant cells, or by centrifuging a sample at sufficient speed to collapse the gas vacuoles and sediment the cells, or by filtering the sample through a paper, glass-fiber, or membrane filter. The cell paste can then be used directly, stored refrigerated, or freeze- or air-dried for more extended storage.

For *in vivo* toxicity testing, resuspension of the cells is done in physiological or phosphate-buffered saline (pH 7.5, 0.05 M), at 200-mg dry weight in 10 mL of saline. Concentrated suspensions of fresh cyanobacterial cells can also be used directly, without drying, using about 1 g of cell paste in 10 mL of saline. In this case a dry-weight determination is required for the suspension. Cell rupture of fresh or dried cells is essential for the rodent assay and can be performed by sonication and freeze-thawing of the suspension (Falconer 1993; Lawton, Beattie et al. 1994).

The assay is carried out by intraperitoneal injection into test animals. This requires bacteriologically sterile solution to avoid infection. The suspension can be filtered through a bacterial filter prior to injection or, if the toxicity is anticipated to be due to peptide or other heat-stable toxins, the suspension can be held in a boiling water bath for 10 min and then filtered through a sterile filter for injection. On the basis of the dry weight of the solids in the suspension, doses from 50 to 500 mg/kg body weight can be used to determine lethal dose, with doses of 0.1 to 1.0 mL injected. To minimize the number of animals used, four dose rates (say 50, 100, 250, and 500 mg/kg) administered to pairs of animals will provide basic information. If no pathological changes are seen at the highest dose, it is unlikely that the bloom is of significance as a health risk. Any animals showing distress should be euthanized immediately, and all animals should be euthanized at 24 h after dosing. Clinical observation should include respiratory rate, motor activity, piloerection, salivation and lacrymation, and blanching of extremities.

Postmortem examination of internal organs is informative, as hepatotoxins, such as microcystins and nodularins, and cytotoxins, such as cylindrospermopsins, will cause changes in liver weight and appearance and show hepatocyte damage on histopathological examination. Microcystins and nodularins at acutely toxic doses will cause swelling and darkening of the liver without substantial damage to other organs (Falconer, Jackson et al. 1981).

Cylindrospermopsins will cause damage to several organs, including lymphocyte necrosis in the spleen and damage to the proximal tubule in the kidney, which can be seen on histopathological examination (Falconer, Hardy et al. 1999; Seawright, Nolan et al. 1999). Neurotoxins cause neurological symptoms and death in the absence of visually observable postmortem changes, and hence can be differentiated from other toxins on postmortem examination (Falconer 1993).

An assessment of relative toxicity, given by Harada, Kondo et al. (1999) as LD_{50} in milligrams of dry cyanobacterial cells per kilogram of mouse body weight (the dose at which 50% of the mice are killed within 24 h by toxin), is as follows:

Greater than 1000 mg/kg body weight: Nontoxic
500 to 1000 mg/kg: Low toxicity
100 to 500 mg/kg: Medium toxicity
Less than 100 mg/kg: High toxicity

The same scale of toxicity can be applied to the minimal lethal dose (the lowest dose at which death occurred), which can be expected to be below but close to the LD_{50}.

10.2.2 Senescent or Lysed Samples

The method described above assesses *in vivo* toxicity of cell-bound toxins and is applicable only to healthy bloom material collected and handled without cell lysis. If cell lysis has occurred in the bloom or after collection, then total lysis of the bloom sample should be undertaken by freeze-thawing and sonication. The extent of lysis in collected bloom material depends substantially on the genus of cyanobacterium as well as handling conditions. *Microcystis* is very resistant to lysis, whereas the filamentous cyanobacteria are much more sensitive, especially *Anabaena.* After lysis, the suspension should be centrifuged and filtered to obtain a particle-free solution of toxins. Toxin concentration from lysed samples can be undertaken as described later for the individual types of toxin.

For injection into mice, the toxin extract is redissolved in physiological saline or phosphate-buffered saline for intraperitoneal injection. Dose rate can be approximated by assuming a lethal dose of (say) 100 μg/kg and administering doses of 50, 100, 250, and 500 μg/kg. Clinical observation, euthanasia, and postmortem examination follow the procedures described above unless cylindrospermopsin is suspected. In this case, if no mortality is seen at 24 h, it will be necessary to extend the time of observation to 7 days, as this toxin is slow acting. All animals should then be euthanized at 7 days for postmortem examination and histopathology.

10.2.3 Ethics Permission

In most countries permission is required from an ethics committee within the institution prior to any toxicity testing in mammals. Some jurisdictions require the experimenter to also have a personal license, which certifies the individual as being competent to carry out the tests specified by the license. In general, standard LD_{50} determination is not permitted unless rigorous evaluation of a potential pharmaceutical product or pesticide is being undertaken. The aim of this restriction is to minimize the number of animals used and reduce the potential suffering of lethal toxicity testing. For assessment of toxicity of water or scum samples, an approximate minimal lethal dose is a satisfactory substitute for an LD_{50} determination. This can be carried out with four doses of cyanobacterial extract administered over a 10-fold concentration range to pairs of Swiss albino mice weighing 25 to 30 g.

10.3 CYLINDROSPERMOPSIN BIOASSAY AND ANALYSIS

A *Cylindrospermopsis* bloom was responsible for the human poisoning at Palm Island, Australia, described in Chapter 5. The cyanobacterium was isolated and its toxicity investigated, initially using mice (Hawkins, Runnegar et al. 1985). Later, after isolation and identification of the toxin cylindrospermopsin (Ohtani, Moore et al. 1992), it became possible to apply cell-based, biochemical, and chemical assays. It was also possible to calibrate microbiotests for cylindrospermopsin toxicity, providing a less technological method for assay suitable for laboratories without sophisticated chemical analytical equipment.

C. raciborskii appears to release toxin into the water during growth, unlike *Microcystis*, which retains toxin in the cells until cell death (Chiswell, Shaw et al. 1999). Therefore measurement of only the cell content of toxin may substantially underestimate the overall toxin content of the water. Hence both the toxin in the cells and in the free water phase require measurement. The most direct method of assuring that total toxin is measured is to lyse/rupture the cells in the water sample by freeze-thawing and/or sonication of the bulk sample. Freeze-drying the bulk sample is also effective, with subsequent extraction of toxin in 5% concentration of acetic acid in water (Hawkins, Chandrasena et al. 1997). Dissolving cylindrospermopsin in deionized water is also possible, but methanol extraction of the dried sample may lose cylindrospermopsin while bringing into solution a range of other potentially bioactive compounds (Hiripi, Nagy et al. 1998).

If a water bloom of cyanobacteria is likely to contain cylindrospermopsins or other alkaloid cyanobacterial toxins, solid-phase adsorption cartridges can be used for toxin concentration. Polygraphite cartridges — for example, Carbograph Extract Clean — are conditioned before use by washing with methanol (5 mL) followed by high-purity water (5 mL). After the water (pH 6 to 8) containing filtered cyanobacterial lysate is run through the column, elution by 5% formic acid in methanol, 2×5 mL, will recover alkaloid toxins (Norris, Eaglesham et al. 2001). The concentrated sample that is eluted from the cartridge can be dried with nitrogen and weighed prior to assay.

Cylindrospermopsin adsorbs strongly to polyethylene, so only glass containers should be used for its extraction and analysis. It is stable in the dark between pH 4 and 10 at room temperatures but breaks down in sunlight in the presence of cell debris. It is stable to boiling at neutral pH for 15 min (Chiswell, Shaw et al. 1999).

10.3.1 Bioassays for Cylindrospermopsin

In vivo rodent assay, which has been extensively undertaken for cylindrospermopsin, describes the features of toxicity (Hawkins, Runnegar et al. 1985; Hawkins, Chandrasena et al. 1997; Falconer, Hardy et al. 1999; Seawright, Nolan et al. 1999). The mouse assay technique used is described earlier in this chapter and is not now generally required for cylindrospermopsin analysis, as a range of alternatives are available.

During recent years, the use of insects and zooplankton as assay organisms for environmental toxins has become more common, and a number of standardized kits are available for this purpose (Persoone, Janssen et al. 2000). These test systems use dehydrated cysts, eggs, or the equivalent as sources of the test organisms and a standardized protocol for measuring toxicity. Evaluation of these organisms for monitoring cyanobacterial toxins, in place of mouse tests, has shown promising results (Tarczynska, Nalecz-Jawecki et al. 2001). In a similar manner to the toxicity tests using rodents, these organisms respond to neurotoxins, hepatotoxins, and alkaloid toxins, with dose–response curves of varying sensitivity. They thus provide a general test for toxicity even when the nature of the toxin is unknown. A range of protozoa, crustacea, insecta, and eukaryotic algae have been explored for sensitivity in toxicity tests (Persoone, Janssen et al. 2000).

Larger insects have also been tried as assay systems for cyanobacterial toxins, particularly neurotoxins, and the African locust has also been shown to be sensitive to toxicity from *C. raciborskii* and *Microcystis aeruginosa* (Hiripi, Nagy et al. 1998).

The most promising organism that provides sufficient sensitivity to cyanobacterial toxins for test use is the small freshwater crustacean *Thamnocephalus platyurus*. This organism has been demonstrated to be effective for the assay of cylindrospermopsin carried out on samples cultured from a toxic water bloom of *C. raciborskii* in Lake Balaton, Hungary (Torokne 1997). The lethal concentration to 50% of the test organism (LC_{50}) for *T. platyurus* was 0.61 mg of freeze-dried cells per milliliter of assay solution, and the LD_{50} of the same sample for mice was 550 mg/kg.

On the basis of published data for *C. raciborskii* containing cylindrospermopsin, the mouse toxicity of the Lake Balaton material is equivalent to a toxin content of approximately 0.5 mg cylindrospermopsin per gram of dried cells (Hawkins, Chandrasena et al. 1997). Therefore the *T. platyurus* assay provided an LC_{50} (calculated as pure toxin) of about 0.3 μg cylindrospermopsin per milliliter of solution. This sensitivity is adequate for testing the toxicity of freeze-dried scums and cell concentrates or concentrated extracts of bulk water samples eluted from solid-phase adsorption cartridges. The LC_{50} is 300 times higher than the concentration of the proposed Guideline Value for cylindrospermopsin in drinking water of 1 μg/L (Humpage and Falconer 2003); hence this microbiotest is not suitable for the direct analysis of the toxin in water supplies.

Another organism, more widely used in toxicity tests, is the brine shrimp *Artemia salina*. This can be readily obtained in the form of eggs, which are then hatched for use. The quoted LC_{50} for purified cylindrospermopsin was 8.1 μg/mL at 24 h and 0.71 μg/mL at 72 h (Metcalf, Lindsay et al. 2002). This increased sensitivity with extended time of observation is similar to the reduction in experimental LD_{50} in mice, when the time is extended from 24 h to 7 days (see Chapter 6). It probably relates to two independent toxic mechanisms, the earlier lethal response with lower sensitivity followed by a later, more sensitive response through inhibition of protein synthesis.

The use of these microbiotests in place of the mouse bioassay has led to cost savings while also avoiding mammalian testing. These tests have similar technical advantages and disadvantages when compared to mouse tests, as they provide a general toxicity screen with some indication of toxin type. Their sensitivity is satisfactory for concentrated material but insufficient for the direct monitoring of toxin content in bulk water. Cylindrospermopsin has, however, been successfully concentrated from lake water by the use of C-18 and polygraphite cartridges in series, with 100% recovery (Metcalf, Beattie et al. 2002). The method for use of Carbograph solid-phase cartridges for the concentration of cylindrospermopsin from dilute solution in water is described earlier in this chapter.

Cylindrospermopsin is also toxic to plants, and inhibition of the growth of etiolated seedlings of *Sinapis alba* (mustard) has been used as an assay (Vasas, Gaspar et al. 2002).

10.4 CELL-BASED AND CELL-FREE TOXICITY MEASUREMENT OF CYLINDROSPERMOPSIN

Cylindrospermopsin is toxic to a wide variety of cells, though the sensitivity of different cell types to toxin is likely to vary widely through differences in xenobiotic metabolizing capability of the cells. Damage to a human lymphocyte cell line has been shown, at a concentration of 1 μg/mL (approximately 2 μM) (Humpage, Fenech et al. 2000). Primary mouse hepatocytes appear to be more sensitive, with toxicity and inhibition of protein synthesis shown at 0.25 μg/mL (approximately 0.5 μM) (Froscio, Humpage et al. 2003). These concentrations, which are toxic to isolated cells in culture, are comparable to those shown to be toxic in the microbiotests with crustaceans. The cell-culture systems do not present any substantial advantages for assay of cylindrospermopsin, due to cost, the time involved, and the low sensitivity.

The cell-free protein synthesis inhibition assay, which uses a rabbit reticulocyte lysate as a source of protein-synthesizing capacity, is appreciably more sensitive, with a detection limit of 50 nM (0.025 μg/mL) in the assay solution (Froscio, Humpage et al. 2001). This assay provided accurate quantitation of cylindrospermopsin concentrations in water of 0.2 to 1.2 μg/mL (0.5 to 3.0 μM). Water sample concentration or concentration of toxin by solid-phase adsorption is required prior to use of this assay for field samples.

10.5 ELISA OF CYLINDROSPERMOPSIN

This approach has proved successful for the analysis of microcystins, as discussed later. The method is based on the specificity of the antibody/antigen association and is therefore dependent on the characteristics of the antibodies to the toxin to be assayed. The specificity varies between antibodies according to the part of the toxin molecule that is recognized. Antibodies can be multivalent/polyclonal preparations, from animal serum, or monoclonal, derived from isolated clones of hybrid mouse lymphocytes.

Both methods of preparation require the covalent linkage of the toxin to a carrier protein, and bovine serum albumin, ovalbumin, and polylysine have been used. For antiserum production, the conjugated toxin is then administered intradermally to rabbits or other animals, together with an adjuvant to stimulate antibody formation. Repeat injections are given to boost antibody titer. For the production of polyvalent antibodies, rabbits or larger animals are used and blood samples collected for the separation of antibodies from serum (Chu, Huang et al. 1989).

For monoclonal antibody formation, mice are used, and repeat injections of the conjugated toxin are administered intraperitoneally. Spleen cells are collected and hybridized *in vitro* with a transformed cell line to form a series of clones of hybrid cells. Screening is done in culture plates using the original toxin coated onto the plate to identify cell clones secreting antibodies. A color-producing enzyme linked to an antimouse immunoglobulin is used to visualize the antibody-secreting clones. Selected clones are then multiplied in culture or injected intraperitoneally into mice to generate ascites fluid, and the antitoxin antibodies are purified (Kfir, Johannsen et al. 1986a,b).

The monoclonal or polyvalent (polyclonal) antitoxin antibodies are then used in a competitive binding assay, which can be carried out in multiwell plates. Two assay methods can be used. The direct competition assay employs competition between an unknown toxin concentration and toxin linked to an enzyme. These compete for available binding sites on the antibody, which is coated onto the plate surface. The color development arises from the enzyme linked to the toxin, which is bound on

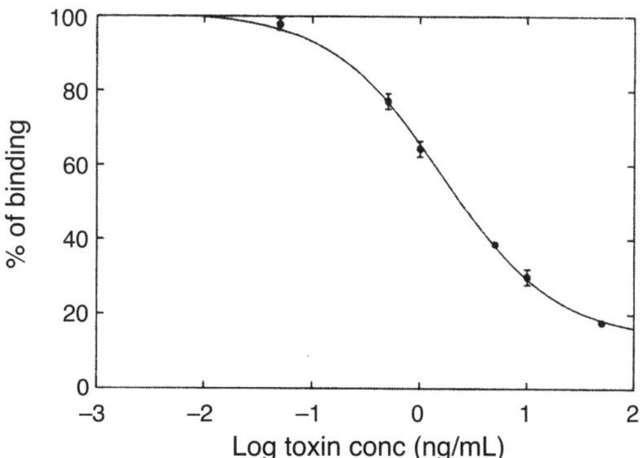

FIGURE 10.1 Standard curve for the direct competitive immunoassay of microcystin-LR using a polyclonal antibody coated onto a plate. Color development by horseradish peroxidase coupled to microcystin-LR. (From Chu, Huang et al. 1990. With permission.)

the plate. As the test toxin concentration rises, less toxin linked to enzyme is bound to the plate and the color development decreases (Figure 10.1) (Chu, Huang et al. 1990).

The alternative approach is an indirect competition assay in which the plate is coated with toxin linked to a protein and there is competition between toxin in solution and toxin on the plate for binding sites on antitoxin antibody in solution. The result is visualized by use of an anti-immunoglobulin linked to an enzyme, which will quantitatively bind to the antitoxin antibodies adhering to the toxin on the plate (Chu, Huang et al. 1989). As before, increased toxin in the test material will decrease color development. The β-galactosidase and horseradish peroxidase enzymes are often used with suitable substrates that provide the color reaction quantitating the enzyme bound to the plate.

At the time of writing, several laboratories are developing enzyme-linked immunoassays for detecting cylindrospermopsin in water supplies. These immunoassays have potential for laboratory-based quantitative assays using multiwell plates or to be developed into a "dipstick" that can be used on site at a lake for an approximate measure of toxin presence. Adequate sensitivity for direct estimation of concentrations in the range of 0.2 to 20 μg/L in bulk water will be needed in these assays for use by water supply agencies

10.6 INSTRUMENT-BASED TECHNIQUES FOR CYLINDROSPERMOPSIN

10.6.1 High-Performance Liquid Chromatography (HPLC)

Pure cylindrospermopsin has a UV absorbance maximum at 262 nm, which can be observed in a photodiode array detector. When coupled to an HPLC system fitted with an ODS column (Cosmosil $5C_{18}$-AR), a mobile phase of 5% methanol will provide separation from other cyanobacterial cell constituents (Harada, Ohtani et al.

1994). A modification of this was used by Hawkins, Chandrasena et al. (1997), employing a linear gradient of 0 to 5% methanol followed by isocratic 5% methanol with a Spherisorb ODS-2 packed column to separate cylindrospermopsin from extracts of toxic *C. raciborskii*.

Use of this approach with environmental samples has also shown advantages in gradient elution from the column. A more extensive gradient from 0 to 50% methanol containing 0.05% trifluoracetic acid was used with a sensitivity of detection from 1 to 300 ng cylindrospermopsin on the column. Some interference from peaks eluting close to cylindrospermopsin was observed (Welker, Fastner et al. 2002). Caution is required if high concentrations of organic solvents are used, as above 50% methanol or 30% acetonitrile a marked decrease in measured toxin content was observed, which may be due to self-association of the cylindrospermopsin molecule (Metcalf, Beattie et al. 2002). The majority of environmental samples will require preconcentration before application to HPLC.

Capillary electrophoresis has been used to analyze cylindrospermopsin from *Aphanizomenon ovalisporum* (Vasas, Gaspar et al. 2002), a technique that may have future potential.

The definitive analytical technique at present is HPLC followed by tandem mass spectrometry (MS/MS). The initial HPLC separation used a C-18 column at 40ºC, with a gradient of 1 to 60% methanol buffered with 5 mM ammonium acetate for 6 min, followed by 60% methanol for 1 min. The injection volume was 110 µL. Effluent splitting supplied 20% of the flow to the MS/MS interface. The original M+H ion at 416 *m/z* was fragmented to 194 *m/z*, which was measured for quantitation.

The determination was linear between 1 and 600 µg/L cylindrospermopsin in water, showing great sensitivity (Eaglesham, Norris et al. 1999). The accuracy of the assay at a concentration of 5.2 µg cylindrospermopsin per liter was 93.5%. This method is very costly for the equipment and requires highly skilled operators. It is likely to continue as the reference technique, as it has very high sensitivity, specificity and accuracy. Less costly methods will be required for routine water analysis.

10.7 MICROCYSTINS AND NODULARINS: BIOASSAY AND ANALYSIS

Microcystis blooms have caused worldwide deaths of livestock, and there are early reports of "water bloom" as a cause of poisoning of domestic animals in the U.S. (Fitch, Bishop et al. 1934); these were followed by studies of the pathology of toxicity in rats (Ashworth and Mason 1946). Many whole-animal studies followed, using laboratory and domestic animals (see Carmichael and Falconer 1993). Studies of the effects of microcystin on isolated hepatocytes (Runnegar, Falconer et al. 1981) preceded the final structural characterization of the toxins (Botes, Tuinman et al. 1984). Since that time a range of assays of varying approach have been developed for microcystins, the most widely used being ELISA, protein phosphatase inhibition assay (PPI), HPLC, and assays based on mass spectroscopy (MS).

As a consequence of the WHO's determination of a provisional Guideline Value for microcystin-LR in drinking water of 1.0 µg/L, there has been worldwide activity in refining and validating methods of analysis. Sensitivity and precision are improving, and novel approaches for online flow techniques for microcystin measurement are under development.

Nodularia first came to attention as a consequence of a large-scale poisoning of domestic livestock (Francis 1978) and later poisonings of dogs (Lundberg, Edler et al. 1983). Its toxicity was examined in mice (Runnegar, Jackson et al. 1988) and the toxin nodularin was structurally identified (Rinehart, Harada et al. 1988). The similarity of the toxic action and chemical structure of nodularin to that of microcystin has allowed modification of the assays for microcystin to be applied to nodularin assay. Nodularin has occurred in drinking water and in seafood (Sipia, Kankaanpaa et al. 2001; Sipia, Kankaanpaa et al. 2002) and has the same toxic potency as microcystin-LR. It is therefore necessary to have assays that will measure the concentrations in water and food. There are general assays for both microcystins and nodularins; a specific ELISA assay for nodularin is available (Mikhailov, Harmala-Brasken et al. 2001).

10.8 SAMPLE COLLECTION AND HANDLING FOR MICROCYSTINS

Fresh samples of reservoir water containing cyanobacteria and their toxins are subject to bacterial degradation and therefore best stored for short periods at reduced temperatures prior to concentration or analysis. Microcystins and nodularins are, however, very stable compounds, which are resistant to boiling at neutral pH; they can be stored as dried water samples or dried cell concentrates at room temperature for weeks prior to analysis. An early study used boiling to concentrate microcystins from 20-L samples collected during water treatment (Falconer, Runnegar et al. 1983). Drying in a convection oven at moderate temperature (40 to 50°C) can be used to evaporate water from cyanobacterial cell samples on filters for later extraction and analysis.

Because microcystins remain within the cyanobacterial cells until cell death, the toxicity of healthy bloom samples containing microcystins can be measured simply by collecting cells from a known volume of water for analysis. Cells can be concentrated by filtration, centrifugation, or flotation or by evaporation of water from the sample by freeze- or air-drying. The dried cells can be extracted by 75% methanol in water to quantitatively dissolve microcystins and nodularins (Fastner, Flieger et al. 1998). Wet cell concentrates can also be extracted with methanol, using 2.5 mL of cells plus 7.5 mL of methanol. In both cases a second extraction with 75% methanol will maximize toxin recovery. Filtration or centrifugation will remove particulate residue. These methanolic samples can be dried with nitrogen or by evaporation at 45°C under reduced pressure. For bioassay, they are redissolved in water, physiological saline, or culture medium; for instrumental analysis, they are redissolved in appropriate solvents. Sonication of the cell suspension, which is required for aqueous extraction of toxin, may also assist in organic solvent extraction of fresh cells or dried material.

When total toxin content is required — for example, in senescent blooms, after copper treatment of the reservoir, or in samples collected during or after water treatment — bulk water samples for microcystin or nodularin measurement are handled differently from cell concentrates. If intact cells are present in the sample, lysis is necessary to obtain all the toxin in the aqueous phase, which can be done by freeze-thawing and sonication.

Concentration of microcystins and nodularins from water and removal of interfering inorganic and organic compounds can be carried out using solid-phase adsorption cartridges. For blooms expected to contain peptide toxins, reversed-phase octadecyl (C-18) silanized silica gel (ODS) cartridges have been extensively used. There are a variety of commercially available types with different characteristics (Meriluoto 1997). SepPak, Bond Elut, and Baker cartridges, for example, have been used. The cartridge is activated by washing with 5 mL of methanol followed by 5 mL of water. The filtered cyanobacterial solution is passed through the cartridge, which is then washed with water. Elution of microcystins and nodularins can be done by 70% methanol solution, 2 × 5 mL, and the eluate dried under a stream of dry nitrogen and weighed (Lawton, Edwards et al. 1994). The product can then be dissolved in phosphate-buffered saline for intraperitoneal injection or in high-purity water or methanol/water for use in other assays.

An alternative method of concentration of microcystins and nodularin from dilute solution in water samples is the use of immunoaffinity columns. These employ a cartridge containing either Sepharose beads coated with antimicrocystin antibodies or silica beads coated with similar antibodies. The filtered water sample is passed through the affinity column, which selectively attaches microcystins and nodularins. After washing the cartridge with 25% methanol/water, the toxins are eluted with 4% acetic acid in 80% methanol/water (Aranda-Rodriguez, Kubwabo et al. 2003). The use of immunoaffinity columns to concentrate microcystins in lake water in Japan has shown that they have potential in environmental monitoring and a possibility of reuse, which minimizes cost (Kondo, Ito et al. 2002).

The most sensitive of the presently available assay techniques can be undertaken on water samples without concentration, though the effects of interfering ions and organic material on precision may be considerable. Concentration/cleanup procedures reduce these possibilities of interference in the assay results.

10.9 BIOASSAYS FOR MICROCYSTINS AND NODULARINS

Rodent assays for determination of microcystin or nodularin were of vital importance in the initial characterization of these hepatotoxins but are now not permitted in many countries, and alternatives are available. The method for rodent assay of cyanobacterial concentrates is described earlier (in Section 10.2). A range of crustacean and insect assays have been evaluated (Harada, Kondo et al. 1999), with the *A. salina* (brine shrimp) assay being among the most easily performed (Kiviranta, Sivonen et al. 1991; Campbell, Lawton et al. 1994). This assay may be carried out with commercially available eggs (sold by aquarium retailers or biological supply companies) or by purchase of a prepared kit. The protocol is described in Harada, Kondo et al. (1999). The sensitivity is moderate, with LC_{50} values of about 4 μg/mL (4mg/L) for microcystin-LR (Delaney and Wilkins 1995; Lahti, Ahtiainen et al.

1995). This is adequate for toxicity testing of scums or cell concentrates or concentrated eluates from solid-phase adsorption cartridges (see earlier in this chapter). It is not sufficiently sensitive for direct testing of bulk water samples, for which a sensitivity of 0.1 to 0.5 μg/L is necessary, to meet the WHO's Guideline Value for microcystin-LR of 1 μg/L.

10.9.1 Cell-Based Assays for Microcystins

Freshly isolated rat hepatocytes in suspension culture were shown to be sensitive to microcystins by dose-related deformation. Thirty nanograms of microcystin were shown to cause deformation of approximately 60% of 10^6 hepatocytes in 1 mL of incubation mixture (Runnegar, Falconer et al. 1981). This method was further developed as an assay tool for detection of microcystins in cyanobacterial blooms (Aune and Berg 1986). Examination of the relative potencies of microcystins-LR, -YR, and -RR to reduce hepatocyte viability showed responses that followed the *in vivo* toxicity of the three microcystins. The LD_{50} for hepatocytes after 20 h of incubation with microcystin-LR was 50 ng/mL (Heinze 1996). This is appreciably more sensitive than whole organism assays but still inadequate for unconcentrated samples. Salmon hepatocytes were found to be two- to fivefold less sensitive than rat hepatocytes to cell death in suspension culture when incubated with a range microcystin and nodularin concentrations (Fladmark, Serres et al. 1998).

10.9.2 Bacterial Luminescence Assays

Toxicity tests based on bacterial luminescence measurement have proved generally unsatisfactory for cyanobacterial toxins, as they respond to nontoxic components of cyanobacteria and not to the toxins of concern (Lawton, Beattie et al. 1994). More recently a *Vibrio fisheri* bioluminescence assay has been reevaluated for nodularin measurement, with more promising results (Dahlmann, Ruhl et al. 2001).

10.10 ELISA for Microcystins and Nodularins

10.10.1 Polyclonal Antibodies

The early ELISA methods in which antibodies to microcystin-LR were raised in rabbits proved successful and have formed the basis of commercially available kits (Chu, Huang et al. 1989, 1990). The background to this method was described in discussing ELISA techniques for cylindrospermopsin earlier in this chapter. The direct competitive assay in which the antimicrocystin antibody was coated onto the plate was used. Microcystin linked to peroxidase enzyme was used as the competitor with the standard or unknown microcystin concentration. Color development in the wells of the plate measured the amount of peroxidase activity on the washed plate. As the unknown concentration of toxin increased, so the enzyme-linked microcystin bound to the antibodies on the plate decreased (Figure 10.1). The sensitivity range was 0.5 to 10.0 ng/mL, with a detection limit of 10 pg in 50-μL volume per assay. This detection limit (of 0.2 μg/L) was sufficient to detect microcystin in unconcentrated lake water.

As there are some 60 variants of the microcystin molecule, the sensitivity of a set of polyclonal antibodies to microcystin-LR raised in rabbits to the mix of microcystins in a natural water sample has potential for considerable variability. Differences in cross-reactivity of antibodies to variants results in underestimation of natural mixtures of microcystins (Chu, Huang et al. 1990; An and Carmichael 1994). At the extreme, antibodies raised against microcystin-YR failed to detect microcystin-LA in a toxic water sample (Runnegar and Falconer, unpublished data). To minimize this variability, antibodies have been raised against synthetic ADDA, the β-linked amino acid unique to microcystins and nodularins. This eliminates the problem of antibodies specific to particular L-amino acid variants and hence missed detection of others. ADDA antibodies will not be affected by L-amino acid differences, thus providing a more general specificity to the whole range.

In one preparation, synthetic ADDA was linked to bovine serum albumin, cationized bovine serum albumin, or ovalbumin and injected intramuscularly into sheep with adjuvant. Repeat doses were given to increase antibody titers. The antisera were used together with unknown or standard toxin solutions to indirectly compete with ADDA coated onto the plates. Visualization was achieved by an antisheep immunoglobulin coupled to peroxidase enzyme, which attached to the anti-ADDA sheep antibody associated with ADDA coated onto the plate (Fischer, Garthwaite et al. 2001). The results showed a good reaction with a variety of microcystins and nodularin at concentrations relevant to direct measurement in water supplies (Figure 10.2). These anti-ADDA antibodies should detect all peptide ring variants of microcystins and nodularins and also the open-ring peptides caused by microbial degradation of the parent molecules (Bourne, Jones et al. 1996). They may have less sensitivity toward ADDA variants, which are comparatively rare. A commercial kit is available using the anti-ADDA antibodies (Abraxis kits).

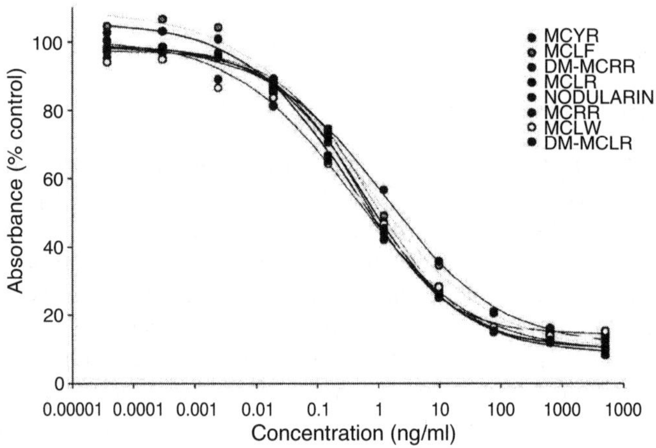

FIGURE 10.2 Competitive binding immunoassay for microcystins based on antibodies to ADDA. Closely related responses to a range of microcystin variants and to nodularin. (From Fischer, Garthwaite et al. 2001. With permission.)

10.10.2 Monoclonal Antibodies

Monoclonal antibodies to microcystins have been developed from spleen cells of mice immunized against microcystins and hybridized with a transformed recipient cell line. This was first demonstrated by Kfir and colleagues using microcystin-LA as the antigen (Kfir, Johannsen et al. 1986a,b). Since that time the approach using monoclonal antibodies has been considerably refined, the most recent using synthetic ADDA as the antigen (Zeck, Weller et al. 2001). As described for the polyclonal antibody, the monoclonal antibodies to ADDA showed cross-reactivity to a range of microcystins, thus providing the capability to determine the sum of the microcystin variants in a water sample. A direct competitive ELISA was developed using the monoclonal antibodies. The sensitivity of these antibodies was considerable, with an inhibition constant (IC_{50}) of 0.33 µg/L of microcystin-LR, and a detection limit of 0.07 µg/L. The ELISA is potentially capable of measuring microcystins in drinking water; however, at the time of writing, no commercial kit based on these antibodies is available.

A monoclonal antibody specific for nodularin with minimal cross-reactivity for microcystins has been developed and tested in an ELISA. This will be valuable when there is the possibility of both nodularin and microcystins being present in a mixed bloom or in food (Mikhailov, Harmala-Brasken et al. 2001).

10.10.3 Phage Library Antibodies

A new alternative source of antibodies against specific haptens, in this case microcystins, exists in libraries of antibody fragments carried on bacteriophage. Several unimmunized human phage libraries carrying antibodies are available. The Griffin.1 library contains 6.5×10^{10} different phage antibodies (Griffiths, Williams et al. 1994). Antibody fragments against microcystin-LR were selected from this library and cloned into an expression vector. This was introduced into *Escherichia coli* for expression of the antibodies. The recombinant antibodies so generated were characterized in a competitive ELISA. They showed cross-reactivity to microcystins-RR, -LW, and -LF as well as -LR and to nodularin. The sensitivity was adequate for direct measurement of microcystins in drinking water (McElhiney, Lawton et al. 2000; McElhiney, Drever et al. 2002; Strachan, McElhiney et al. 2002).

This approach to antibody production renders the method independent of the variability of immunized animals and the ongoing costs of sterile culture of hybrid cell clones. Simple proteins such as single-chain antibodies and antibody fragments can be produced in large amounts in recombinant *E. coli*, thus ultimately providing a relatively low-cost source of specific immunoreagents. These can be used for concentration of microcystins from dilute solution in water in immunoaffinity columns (McElhiney, Drever et al. 2002) and developed for use in online toxin detection during water treatment.

10.10.4 Immunofluorimetric Assays

All of the ELISAs are ultimately dependent on the color development generated by the enzymatic reaction that concludes the assay. An alternative detection system that may lead to greater sensitivity is based on fluorescent measurement, which is also available on microplate readers. One of the methods tested employed a europium chelate of microcystin as the competitive antigen in place of enzyme-linked microcystin. Fluorescence detection provided a sensitivity equal to the ELISA (Mehto, Ankelo et al. 2001).

A microcystin assay using a disposable immunoaffinity cartridge has been developed that employs a laser fluorescence scanner for detection. The sensitivity of this system is appropriate for direct measurement in water, with a linear range of 0.125 to 2.0 μg/L of microcystin-LR (Kim, Oh et al. 2003).

10.11 PROTEIN PHOSPHATASE INHIBITION ASSAY FOR MICROCYSTINS AND NODULARINS

Since the basic mechanism of the toxicity of microcystins and nodularins is through the inhibition of protein phosphatases 1 and 2A, an assay using this biological property should relate directly to the actual toxicity of the mixture of toxins present. Unfortunately this has not proved to be the case, as microcystin-RR has the same phosphatase inhibition potency *in vitro* as microcystin-LR (Rivasseau, Racaud et al. 1999; Heresztyn 2001), whereas *in vivo* the -RR variant is appreciably less toxic (Carmichael 1997). This discrepancy is due to the relative efficiency of transfer of the toxin variants through the plasma membrane of cells. The more hydrophobic the toxin variant, the easier penetration, and the more hydrophylic, the less (Ward and Codd 1999). A further problem that has been identified is that metabolic conjugates of microcystin, microcystin-LR-glutathione and microcystin-LR-cysteine, inhibit protein phosphatase *in vitro* similarly to microcystin-LR itself. The conjugates have about one-twelfth of the toxicity of microcystin-LR *in vivo* (Ito, Takai et al. 2002). However the errors generated by this overestimation of toxicity are not critical for public health, as they will result in conservative outcomes in safety assessment.

The phosphatase assay, which has enzyme inhibition as its basis, avoids the variations in sensitivity of immunoassays to different toxin variants and also the need to know which variants are present. It provides a cumulative toxicity, which is a problem for the chemical-based assays discussed later, as these may completely miss some toxin variants and thus underestimate overall toxicity. It therefore has inherent advantages as a toxicity assay as against a chemical-based toxin content measurement.

10.11.1 METHODOLOGY

This technique employs a target protein phosphatase enzyme, either purified from biological tissue or produced by a recombinant *E. coli* carrying the gene for the enzyme. Early assays used ^{32}P-labeled protein as the substrate and measured ^{32}P orthophosphate liberation (Mackintosh, Beattie et al. 1990; Yoshizawa, Matsushima et al. 1990). An alternative method not requiring a radioactive marker for measurement employed a phosphorylated substrate that directly or indirectly developed color when the phosphate was cleaved off by the enzyme. In both cases the reduction in the liberated phosphate (or resulting color) in the presence of microcystin or nodularin was used as the measure for the enzyme inhibition (Heresztyn and Nicholson 2001).

The microcystin-sensitive protein phosphatases have a range of natural substrates — for example, glycogen phosphorylase, in which phosphate groups are linked through serine or threonine to protein. There are also several phosphorylated proteins that will act as artificial substrates, including phosvitin from egg yolk (Heresztyn and Nicholson 2001). The phosphatases will also hydrolyze phosphate from simple inorganic substrates. One of these is *p*-nitrophenol phosphate, which when hydrolyzed yields the colored *p*-nitrophenol (An and Carmichael 1994; Ash, MacKintosh et al. 1995a,b).

The ^{32}P-phosphate assay proved effective for measuring concentrations of microcystins in water down to 0.1 μg/L, with the proviso that high false-negatives are possible due to phosphatases present in the sample analyzed, which requires checking by a chemical analytical method (Sim and Mudge 1993; Lambert, Boland et al. 1994). The practical difficulty with this method is the need to prepare radioactive protein substrate at regular intervals, as the half-life of ^{32}P is only 14 days, and the radioactivity is a high-energy β-emission. Thus a research laboratory with radioisotope facilities is required, which is not readily available in the water industry.

The techniques using color development to measure enzyme activity are less sensitive than those using radioactive markers, but they have the considerable advantage in that they can be undertaken in standard laboratories and automated by the use of reactions carried out in multiwell plates. The original assay employed a commercially available skeletal muscle protein phosphatase type 1 produced from recombinant bacterial culture. This was reacted with *p*-nitrophenol phosphate and the color read in a microwell plate reader (An and Carmichael 1994). The effective range was 0.2 to 1.0 μg/L, suitable for direct water measurement. This technique has been further developed by the use of a commercially available protein phosphatase type 2A, which is approximately 50 times more sensitive to microcystin inhibition than type 1 (Honkanen, Codispoti et al. 1994). Effects due to the ionic composition of the water being analyzed have been removed by a change in the divalent ion composition of the assay medium to include manganese chloride (Heresztyn 2001). The sensitivity of this assay is similar to that of the earlier version, and it can be employed directly on water samples.

Questions still arise over inaccuracy in protein phosphatase inhibition assays at low toxin concentrations due to interference from organic material present in cyanobacterial extracts. To eliminate these problems, it may be necessary to use a preconcentration or cleanup step in which microcystins or nodularins are initially passed into solid-phase extraction cartridges and then eluted, as described earlier. When water samples were freeze-dried and then methanol extracted, a good correlation between the phosphatase inhibition assay and HPLC analysis was observed (Wirsing, Flury et al. 1999).

10.12 HPLC FOR MICROCYSTINS AND NODULARINS

Analysis for microcystins and nodularins in cyanobacterial samples and in water blooms has essentially relied on HPLC over the last two decades, and is still the reference method against which the newer and more sensitive methods are judged. The advantages of the technique are that well-standardized procedures are available, and many laboratories undertaking analytical work have the equipment installed. The microcystins and nodularins have characteristic absorption spectra, with a peak in the UV range at 238 nm, which can easily be identified. Toxin variants containing amino acids with strong UV spectra, such as -LW with tryptophan, will give a different adsorption maximum, but the ADDA peak at 238 is still visible (Lawton, Edwards et al. 1995).

For use in microcystin or nodularin analysis, HPLC with diode array detection is valuable, as each peak is scanned for its UV spectrum as it emerges from the column. By this method peaks separated by the HPLC column can be verified for identity as ADDA-containing molecules. Comparison with standard, known microcystins will allow quantitation by measuring peak area. The molar absorption of microcystin variants is similar; microcystin-LR in methanol has an absorptivity of 39,800 (Harada, Matsuura et al. 1990). As a result, a reasonable approximation of the total microcystins present in a field sample can be obtained by summing the individual peak areas showing the characteristic absorption spectrum without a knowledge of which microcystins are present. An injection onto the HPLC column of 1 to10 ng of microcystin can be quantified (Meriluoto 1997); therefore if a 20-µL sample is applied, the toxin concentration for measurement must be 50 to 500 µg/L in the solution to be analyzed. To obtain this concentration then requires either extraction of a cell concentrate or concentration from a large volume of water by solid-phase extraction, as described earlier. Using the immunoaffinity column method of concentration of microcystins from water prior to HPLC separation, a limit of detection of better than 0.1 µg/L has recently been shown (Aranda-Rodriguez, Kubwabo et al. 2003).

If there is any risk of highly adsorbent compounds present in the sample to be analyzed, which is likely to be the case if a cell concentrate is extracted with an organic solvent, a guard column with C-18 packing, which can later be discarded, is recommended prior to the analytical column.

The column employed in the HPLC separation is most commonly a reversed-phase C-18 packed column; a range of these columns are commercially available. The technique was originally developed in 1985 (Krishnamurthy, Carmichael et al. 1986), and has been refined and used successfully since (Harada, Matsuura et al. 1988; Lawton, Edwards et al. 1994; Wirsing, Flury et al. 1999). After sample application on the column, a gradient of buffered methanol or acetonitrile is used to separate the constituent organic components. A discussion of methodology was compiled by Meriluoto (1997).

10.12.1 Advanced Instrument Techniques

More sensitive detection methods following liquid chromatographic separation have been developed using mass spectroscopy (Zweigenbaum, Henion et al. 2000). This can be quantitated for microcystin or nodularin variants as well as providing the identity of the peaks from the chromatographic separation. Recent use of a quadrupole mass spectrometer with an electrospray ion source demonstrated detection of 50 pg of microcystin-LR and therefore is at least 100-fold more sensitive than UV absorption. This equipment was used in a comparative study of 93 samples of cells collected in Finland in 2001. Cells from 14 water samples contained microcystins at 0.2 µg/L or above, with the highest 42 µg/L (Spoof, Vesterkvist et al. 2003). Other HPLC ionization-electrospray mass spectroscopy has, however, the same sensitivity as UV detection, with a detection limit of 20 ng per injection, which required a microcystin concentration of 1 mg/L (Ruangyuttikarn, Miksik et al. 2004).

Use of matrix-assisted laser desorption/ionization time-of-flight (MALDI-TOF) mass spectrometry also has the advantage of allowing the identification of individual microcystin variants and providing quantitation (Welker, Fastner et al. 2002). The technique is exceptionally sensitive, allowing identification of microcystins in as little as 100 cyanobacterial cells. Ability to identify variant microcystins accurately when assessing the total toxicity of a sample avoids errors due to the assumption that all microcystins are equally toxic. Provided that the individual toxicities of the identified variants is known, the sum of the toxicity of a sample can be calculated.

A wide range of other chemical techniques have been explored to obtain sensitivity and reproducibility in microcystin analysis and with the impetus of the WHO's Guideline Value determination of 1 µg microcystin per liter in drinking water, commercialization of suitable methods for automated analysis can be expected within the next few years.

One of the most innovative methods under development is the use of molecularly imprinted polymers for both concentration/extraction of microcystins from dilute solution and use in piezoelectric sensors. The principle of this technology is that a synthetic polymer can be formed that has been imprinted with the molecular configuration of the molecule to be extracted. Thus molecules of the imprinted shape will adhere in their selective pockets in the membrane, from which they can later be eluted. As a solid-phase extraction technique, a concentration of 1000-fold has been achieved. As adherence of the test molecule to the membrane alters the electrical properties, a cell can be constructed with a piezoelectric sensor recording concentration (Chianella, Lotierzo et al. 2002; Chianella, Piletsky et al. 2003). Such a technique has obvious potential for monitoring microcystin concentrations online during water treatment once the required sensitivity has been achieved. At present preconcentration is needed prior to the sensor for measurement of microcystins in bulk water.

10.13 MICROCYSTINS AND NODULARINS IN TISSUE SAMPLES

To measure microcystins or nodularins in tissues obtained from poisoned animals or people presents a different challenge. Up to the present, tissue samples have been extracted with water or methanol, cleaned by solid-phase extraction cartridges, and then assayed by ELISA or other methods. In the case of the dialysis patients from Caruaru who were poisoned by microcystin and cylindrospermopsin in the dialysis fluid, liver samples were extracted with 100% methanol, fat was removed by hexane solvent extraction, and the samples concentrated by solid phase adsorption on C-18 silica cartridges. The eluate from the washed cartridges was assayed by a polyclonal rabbit antibody raised against microcystin-LR (Carmichael, Azevedo et al. 2001). Serum samples from these patients were diluted and again concentrated on solid phase extraction cartridges and also measured by ELISA. These methods provide a measurement of the free, not covalently bound, microcystin in tissue. This can be expected to be reasonably accurate for total microcystin in serum, interstitial fluid, and contents of the gastrointestinal tract but not for intracellular toxin.

Microcystin inside liver cells has been shown to inhibit protein phosphatases (Runnegar, Kong et al. 1993). The inhibition has been identified as an association of the microcystin with the active site of the enzyme, with covalent binding of the methyl-dehydroalanine group of the microcystin peptide ring to cysteine 273 on the enzyme protein (Mackintosh, Dalby et al. 1995). This is an irreversible process, which requires the hydrolysis of the protein to liberate the microcystin-cysteine adduct. As the binding affinity of microcystin to the enzyme is very strong, intracellular microcystin can be expected to be quantitatively bound. Hence measurement of the microcystin in liver samples by water or methanol extraction will greatly underestimate the microcystin present, probably detecting less than 25% of the total.

Assessment of this difference in total and methanol-extractable microcystin has been undertaken in salmon and crab tissues. Analysis of the total microcystin present was done by oxidation of the ADDA group to a unique marker, 2-methyl-3-methoxy-4-phenylbutanioc acid. This can be quantitated by gas chromatography-MS. In salmon liver, only 24% of the microcystin was found by methanol extraction, the remainder being covalently bound. An even greater difference was found in crab tissues, in which less than 0.1% of the total microcystin was methanol-extractable (Williams, Craig et al. 1997).

An alternative approach to resolve this problem is by enzymatic hydrolysis of the tissue proteins, followed by cleanup and concentration steps. Affinity columns may be effective in this role. It can be expected that ELISA using antibodies against the ADDA portion of the molecule will detect microcystin-cysteine conjugate, whereas protein phosphatase inhibition will not. HPLC-diode array detection or liquid chromatography–tandem mass spectrometry (LC–MS/MS) will detect the molecule provided that a standard is available to set up the separation and quantification conditions.

For food safety, it is important to be able to measure microcystins and nodularins in fish, shellfish, and crustaceans harvested from contaminated waters. At present this has been done by ethanol or methanol extraction, which has demonstrated that potentially harmful concentration of free microcystins can occur in fish (de Magalhaes, Marinho et al. 2003; Mohamed, Carmichael et al. 2003). Nodularin measurement in food is likely to achieve a more accurate assessment of total toxin present than microcystin measurements, as nodularin does not form covalent adducts with protein phosphatases (Mackintosh, Dalby et al. 1995). Methanol or ethanol extraction has shown nodularin to be present in a variety of seafood from *Nodularia*-contaminated waters (Van Buynder, Oughtred et al. 2001; Kankaanpaa, Vuorinen et al. 2002).

10.14 ANALYTICAL PROBLEMS AND CHALLENGES

At present a major limitation affecting chemical analysis is the lack of analytical quality standards for most of the microcystin variants. A related problem is the presence in many samples of commercial microcystins of several variants in addition to the one required. To provide data for the total toxicity of a water sample by chemical analysis, not only does each of the significant microcystins need to be identified and quantified but the toxicity *in vivo* must be compared to microcystin-LR. Only in a proportion of cases is the LD_{50} of microcystin variants accurately known, and many more need toxicity assessment (Sivonen and Jones 1999). There is a range of about 20-fold of differences in toxicity between commonly occurring microcystins, and even larger differences when ADDA variants are included.

ELISA and protein phosphatase inhibition tests both provide an overall picture of the combined microcystins present, but neither measures toxicity to mammals. ELISA results measure the combined antigenicity of the sample to a single antibody or set of antibodies. The extent that this reflects the toxicity of the sample will vary with the antibody and with the microcystin variants present in the sample. Thus different microcystin combinations in a sample will give different results when compared to LD_{50} determination or chemical quantitation.

The phosphatase inhibition assay provides a good indication of the total inhibitory capability of the microcystin mixture but overestimates toxicity. This is because there is no correction for variants with low cell-penetrating ability, such as microcystin-RR, where there is a marked difference between *in vitro* phosphatase inhibition and *in vivo* toxicity.

The difficulties of accurate measurement of the toxicity of water samples due to the presence of microcystins or nodularins are being addressed by cross-laboratory validation studies and by multiple assay techniques carried out on the same samples. These will provide a practical guide to the most robust methods for practical use and to difficulties in interpretation of results.

A recent large-scale, international collaboration study in which 31 laboratories from 13 countries participated used the four most currently applied methods. These were HPLC with UV or photodiode-array detection, ELISA, protein phosphatase inhibition assay, and LC/HPLC coupled to mass spectrometry. It was concluded that the standard microcystin-LR solution was measured with sufficient accuracy by all laboratories irrespective of the method employed. The field sample results were considered satisfactory but required standardization between laboratories, and the HPLC results were significantly more variable than with the standard solution. Overall repeatability of 4 to 15% and reproducibility of 24 to 49% were obtained. Relatively few outlier values were recorded within or between laboratories (Fastner, Codd et al. 2002). Given the broad range of laboratories and the differences in equipment and extraction methods used on the field sample, it is notable that the results were so consistent. This gives confidence that the water utilities will be able to measure microcystins with sufficient accuracy to monitor the raw water and the "finished" drinking water and to comply with the WHO Guideline Value.

An interesting study with a similar aim was conducted by Health Canada, analyzing 100 samples of blue-green algae products sold from retail outlets across Canada. Microcystin content was assayed by each participating laboratory by ELISA, protein phosphatase inhibition, and LC–MS/MS. This was set up to detect microcystin-LA, -LR, -RR and -YR only.

In all cases the samples were extracted with 75% methanol in water and centrifuged to obtain a clear supernatant. This was then cleaned up by C-18 solid-phase extraction prior to assay. The ELISA and LC–MS/MS methods gave consistent results over a concentration range of 0.5 to 35 μg/g microcystin. Less consistency was achieved with protein phosphatase inhibition. The authors note that essentially only microcystin-LR and -LA were present in the samples; had unknown microcystins been present, the LC–MS/MS results would have been lower than the ELISA or inhibition assay results (Lawrence, Niedzwiadek et al. 2001).

With the concentrated attention that microcystin analysis is receiving at present, standardized techniques are likely to be formulated and adopted for regulatory purposes in the immediate future, as water supply utilities comply with the WHO Guideline Value or national standards.

REFERENCES

An, J. S. and W. W. Carmichael (1994). Use of a colorimetric protein phosphatase inhibition assay and enzyme linked immunosorbent assay for the study of microcystins and nodularins. *Toxicon* 32: 1495–1507.

Aranda-Rodriguez, R., C. Kubwabo, et al. (2003). Extraction of 15 microcystins and nodularin using immunoaffinity columns. *Toxicon* 46(6): 587–599.

Ash, C., C. MacKintosh, et al. (1995a). Development of a colourimetric protein phosphorylation assay for detecting cyanobacterial toxins. *Water Science and Technology* 31: 47–49.

Ash, C., C. MacKintosh, et al. (1995b). Use of a protein phosphatase inhibition test for the detection of cyanobacterial toxins in water. *Water Science and Technology* 31: 51–53.

Ashworth, C. T. and M. F. Mason (1946). Observations on the pathological changes produced by a toxic substance present in blue-green algae (*Microcystis aeruginosa*). *American Journal of Pathology* 22: 369–383.

Aune, T. and K. Berg (1986). Use of freshly prepared rat hepatocytes to study toxicty of blooms of the blue-green alga *Microcystis aeruginosa* and *Oscillatoria agardhii*. *Journal of Toxicology and Environmental Health* 19: 325–336.

Baker, P. D., D. A. Steffensen, et al. (2001). Preliminary evidence of toxicity associated with the benthic cyanobacterium *Phormidium* in South Australia. *Environmental Toxicology* 16(6, special issue): 506–511.

Botes, D. P., A. A. Tuinman, et al. (1984). The structure of cyanoginosin-LA, a cyclic heptapeptide toxin from the cyanobacterium *Microcystis aeruginosa*. *Journal of the Chemical Society, Perkin Transactions* 1: 2311–2318.

Bourne, D. G., G. J. Jones, et al. (1996). Enzymatic pathway for the bacterial degradation of the cyanobacterial cyclic peptide toxin microcystin LR. *Applied Environmental Microbiology* 62: 4086–4094.

Campbell, D. L., L. A. Lawton, et al. (1994). Comparative assessment of the specificity of the brine shrimp and microtox assays to hepatotoxic (microcystin-LR-containing) cyanobacteria. *Environmental Toxicology and Water Quality* 9: 71–77.

Carmichael, W. W. (1997). The cyanotoxins. *Advances in Botanical Research*. 27: 213–256.

Carmichael, W. W., S. M. F. O. Azevedo, et al. (2001). Human fatalities from cyanobacteria: chemical and biological evidence for cyanotoxins. *Environmental Health Perspectives* 109(7): 663–668.

Carmichael, W. W. and I. R. Falconer (1993). Diseases related to freshwater blue-green algal toxins, and control measures. *Algal Toxins in Seafood and Drinking Water*. I. R. Falconer, ed. London, Academic Press Limited: 187–209.

Chianella, I., M. Lotierzo, et al. (2002). Rotational design of a polymer specific for microcystin-LR using a computational approach. *Analytical Chemistry* 74(6): 1288–1293.

Chianella, I., S. A. Piletsky, et al. (2003). MIP-based solid phase extraction cartridges combined with MIP-based sensors for the detection of microcystin-LR. *Biosensors and Bioelectronics* 18(2–3): 119–127.

Chiswell, R. K., G. R. Shaw, et al. (1999). Stability of cylindrospermopsin, the toxin from the cyanobacterium, *Cylindrospermopsis raciborskii*: Effect of pH, temperature, and sunlight on decomposition. *Environmental Toxicology* 14(1): 155–161.

Chu, F. S., X. Huang, et al. (1989). Production and characterization of antibodies against microcystins. *Applied and Environmental Microbiology* 55(8): 1928–1933.

Chu, F. S., X. Huang, et al. (1990). Enzyme-linked immunosorbent assay for microcystins in blue-green algal blooms. *Journal of the Association of Official Analytical Chemists* 73(3): 451–456.

Dahlmann, J., A. Ruhl, et al. (2001). Different methods for toxin analysis in the cyanobacterium *Nodularia spumigena* (Cyanophyceae). *Toxicon* 39: 1183–1190.

de Magalhaes, V. F., M. M. Marinho, et al. (2003). Microcystins (cyanobacteria hepatotoxins) bioaccumulation in fish and crustaceans from Sepetiba Bay (Brasil, R J.). *Toxicon* 42(3): 289–295.

Delaney, J. M. and R. M. Wilkins (1995). Toxicity of microcystin-LR, isolated from *Microcystis aeruginosa*, against various insect species. *Toxicon* 33: 771–778.

Eaglesham, G. K., R. L. Norris, et al. (1999). Use of HPLC-MS/MS to monitor cylindrospermopsin, a blue-green algal toxin, for public health purposes. *Environmental Toxicology* 14(1): 151–154.

Falconer, I. R. (1993). Measurement of toxins from blue-green algae in water and foodstuffs. *Algal Toxins in Seafood and Drinking Water.* I. R. Falconer, ed. London, Academic Press: 165–176.

Falconer, I. R., S. J. Hardy, et al. (1999). Hepatic and renal toxicity of the blue-green alga (cyanobacterium) *Cylindrospermopsis raciborskii* in male Swiss Albino mice. *Environmental Toxicology* 14(1): 143–150.

Falconer, I. R., A. R. B. Jackson, et al. (1981). Liver pathology in mice in poisoning by the blue-green alga in *Microcystis aeruginosa. Australian Journal of Biological Science* 34: 179–187.

Falconer, I. R., M. T. C. Runnegar, et al. (1983). Effectiveness of activated carbon in the removal of algal toxin from potable water supplies: A pilot plant investigation. Technical Papers Presented at the Tenth Federal Convention, Sydney, April, 1983. Sydney, Australian Water and Wastewater Association: 26-1–26-8.

Fastner, J., G. A. Codd, et al. (2002). An international intercomparison exercise for the determination of purified microcystin-LR and microcystins in cyanobacterial field material. *Analytical and Bioanalytical Chemistry* 374(3): 437–444.

Fastner, J., I. Flieger, et al. (1998). Optimised extraction of microcystins from field samples — a comparison of different solvents and procedures. *Water Research* 32: 3177–3181.

Fischer, W. J., I. Garthwaite, et al. (2001). Congener-independent immunoassay for microcystins and nodularins. *Environmental Science and Technology* 35(24): 4849–4856.

Fitch, C. P., L. M. Bishop, et al. (1934). Water bloom as a cause of poisoning in domestic animals. *Cornell Veterinarian* 24: 30–39.

Fladmark, K. E., M. H. Serres, et al. (1998). Sensitive detection of apoptogenic toxins in suspension cultures of rat and salmon hepatocytes. *Toxicon* 36(8): 1101–1114.

Francis, G. (1978). Poisonous Australian lake. *Nature* 18(2): 11–12.

Froscio, S. M., A. R. Humpage, et al. (2001). Cell-free protein synthesis inhibition assay for the cyanobacterial toxin cylindrospermopsin. *Environmental Toxicology* 16(5): 408–412.

Froscio, S. M., A. R. Humpage, et al. (2003). Cylindrospermopsin-induced protein synthesis inhibition and its dissociation from acute toxicity in mouse hepatocytes. *Environmental Toxicology* 18(4): 243–251.

Griffiths, A. D., S. C. Williams, et al. (1994). Isolation of high affinity human antibodies directly from large synthetic repertoires. *European Molecular Biology Organisation Journal* 13: 3245–3260.

Harada, K., F. Kondo, et al. (1999). Laboratory analysis of cyanotoxins. *Toxic Cyanobacteria in Water: A Guide to Their Public Health Consequences, Monitoring and Management*. I. Chorus and J. Bartram, eds. London, E & FN Spon (on behalf of WHO): 369–405.

Harada, K., K. Matsuura, et al. (1988). Analysis and purification of toxic peptides from cyanobacteria by reversed-phase high-performance liquid chromatography. *Journal of Chromatography* 448: 275–283.

Harada, K., K. Matsuura, et al. (1990). Isolation and characterization of the minor components associated with microcystins LR and RR in the cyanobacterium (blue-green algae). *Toxicon* 28: 55–64.

Harada, K., I. Ohtani, et al. (1994). Isolation of cylindrospermopsin from a cyanobacterium *Umezakia natans* and its screening method. *Toxicon* 32: 73–84.

Hawkins, P. R., N. R. Chandrasena, et al. (1997). Isolation and toxicity of *Cylindrospermopsis raciborskii* from an ornamental lake. *Toxicon* 35(3): 341–346.

Hawkins, P. R., E. Putt, et al. (2001). Phenotypical variation in a toxic strain of the phytoplankter, *Cylindrospermopsis raciborskii* (Nostocales, Cyanophyceae) during batch culture. *Environmental Toxicology* 16(6 special issue): 460–467.

Hawkins, P. R., M. T. C. Runnegar, et al. (1985). Severe hepatotoxicity caused by the tropical cyanobacterium (blue-green alga) *Cylindrospermopsis raciborskii* (Woloszynska) Seenaya and Subba Raju isolated from a domestic supply reservoir. *Applied and Environmental Microbiology* 50(5): 1292–1295.

Heinze, R. (1996). A biotest for hepatotoxins using primary rat hepatocytes. *Phycologia* 35 (6 suppl): 89–93.

Heresztyn, T. (2001). Determination of cyanobacterial hepatotoxins directly in water using a protein phosphatase inhibition assay. *Water Research* 35(13): 3049–3056.

Heresztyn, T. and B. C. Nicholson (2001). A colorimetric protein phosphatase inhibition assay for the determination of cyanobacterial peptide hepatotoxins based on the dephosphorylation of phosvitin by recombinant protein phosphatase. *Environmental Toxicology* 16(3): 242–252.

Hiripi, L., L. Nagy, et al. (1998). Insect (Locusta Migratoria Migratorioides) test monitoring the toxicity of cyanobacteria. *Neurotoxicology* 19(4–5): 605–608.

Honkanen, R. E., B. A. Codispoti, et al. (1994). Characterization of natural toxins with inhibitory activity against serine/threonine protein phosphatases. *Toxicon* 32(3): 339–350.

Humpage, A. R. and I. R. Falconer (2003). Oral toxicity of the cyanobacterial toxin cylindrospermopsin in male Swiss albino mice: determination of No Observed Adverse Effect Level for deriving a drinking water Guideline Value. *Environmental Toxicology* 18(2): 94–103.

Humpage, A. R., M. Fenech, et al. (2000). Micronucleus induction and chromosome loss in WIL2-NS cells exposed to the cyanobacterial toxin, cylindrospermopsin. *Mutation Research* 472: 155–161.

Ito, E., A. Takai, et al. (2002). Comparison of protein phosphatase inhibitory activity and apparent toxicity of microcystins and related compounds. *Toxicon* 40(7): 1017–1025.

Kankaanpaa, H., P. J. Vuorinen, et al. (2002). Acute effects and bioaccumulation of Nodularin in sea trout (*Salmo trutta* M. *trutta* L.) exposed orally to *Nodularia spumigena* under laboratory conditions. *Aquatic Toxicology* 61(3–4): 155–168.

Kfir, R., E. Johannsen, et al. (1986a). Monoclonal antibody specific for cyanoginosin-LA: Preparation and characterization. *Toxicon* 24, 6: 543–552.

Kfir, R., E. Johannsen, et al. (1986b). Preparation of anti-cyanoginosin-LA monoclonal antibody. *Mycotoxins and Phycotoxins*. P. S. Steyn and R. Vleggaar, eds. Amsterdam, Elsevier Science Publishers: 377–385.

Kim, Y. M., S. W. Oh, et al. (2003). Development of an ultrarapid one-step fluorescence immunochromatographic assay system for the quantification of microcystins. *Environmental Science and Technology* 37: 1899–1904.

Kiviranta, J., K. Sivonen, et al. (1991). Detection of toxicity of cyanobacteria by *Artemia salina* bioassay. *Environmental Toxicology and Water Quality* 6: 423–436.

Kondo, F., Y. Ito, et al. (2002). Determination of microcystins in lake water using reusable immunoaffinity column. *Toxicon* 40(7): 893–899.

Krishnamurthy, T., W. W. Carmichael, et al. (1986). Toxic peptides from freshwater cyanobacteria (blue-green algae). I. Isolation, purification and characterization of peptides from *Microcystis aeruginosa* and *Anabaena flos-aquae*. *Toxicon* 24(9): 865–873.

Lahti, K., J. Ahtiainen, et al. (1995). Assessment of rapid bioassays for detecting cyanobacterial toxicity. *Letters in Applied Microbiology* 21: 109–114.

Lambert, T. W., M. P. Boland, et al. (1994). Quantitation of the microcystin hepatotoxins in water at environmentally relevant concentrations with the protein phosphatase bioassay. *Environmental Science and Technology* 28: 753–755.

Lawrence, J. F., B. Niedzwiadek, et al. (2001). Comparison of liquid chromatography/mass spectrometry, ELISA, and phosphatase assay for the determination of microcystins in blue-green algae products. *Journal of the Association of Official Analytical Chemists* 84: 1035–1044.

Lawton, L. A., K. A. Beattie, et al. (1994). Evaluation of assay methods for the determination of cyanobacterial hepatoxicity. *Detection Methods for Cyanobacterial Toxins*. G. A. Codd, T. M. Jefferies, C. W. Keevil, and E. Potter, eds. Cambridge, The Royal Society of Chemistry. Special publication no. 149: 111–116.

Lawton, L. A., C. Edwards, et al. (1994). Extraction and high-performance liquid chromatographic method for the determination of microcystins in raw and treated waters. *Analyst* 119: 1525–1530.

Lawton, L. A., C. Edwards, et al. (1995). Isolation and characterization of microcystins from laboratory cultures and environmental samples of *Microcystis aeruginosa* and from an associated animal toxicosis. *Natural Toxins* 3: 50–57.

Lundberg, R., L. Edler, et al. (1983). Algförgiftning hos hund. *Svensk Veterinärtidning* 35: 509.

MacKintosh, C., K. A. Beattie, et al. (1990). Cyanobacterial microcystin-LR is a potent and specific inhibitor of protein phosphatases 1 and 2A from both mammals and higher plants. *FEBS Letters* 264(2): 187–192.

Mackintosh, R. W., K. N. Dalby, et al. (1995). The cyanobacterial toxin microcystin binds covalently to cysteine-273 on protein phosphatase 1. *FEBS Letters* 371(3): 236–240.

McElhiney, J., M. Drever, et al. (2002). Rapid isolation of single-chain antibody against the cyanobacterial toxin microcystin-LR by phage display and its use in the immunoaffinity concentration of microcystins from water. *Applied Environmental Microbiology* 68(11): 5288–5295.

McElhiney, J., L. A. Lawton, et al. (2000). Detection and quantification of microcystins (cyanobacterial hepatotoxins) with recombinant antibody fragments isolated from naive human phage display library. *FEMS Microbiology Letters* 193(1): 83–88.

Mehto, P., M. Ankelo, et al. (2001). A time-resolved fluoroimmunometric assay for the detection of microcystins, cyanobacterial peptide hepatotoxins. *Toxicon* 39: 831–836.

Meriluoto, J. (1997). Chromatography of microcystins (review). *Analytica Chimica Acta* 352: 277–298.

Metcalf, J. S., K. A. Beattie, et al. (2002). Effects of organic solvents on the high performance liquid chromatographic analysis of the cyanobacterial toxin cylindrospermopsin and its recovery from environmental eutrophic waters by solid phase extraction. *FEMS Microbiology Letters* 216: 159–164.

Metcalf, J. S., J. Lindsay, et al. (2002). Toxicity of cylindrospermopsin to the brine shrimp *Artemia salina*: Comparisons with protein synthesis inhibitors and microcystins. *Toxicon* 40(8): 1115–1120.

Mikhailov, A., A. Harmala-Brasken, et al. (2001). Production and specificity of monoclonal antibodies against nodularin conjugated through N-methyldehydrobutyrine. *Toxicon* 39: 1453–1459.

Mohamed, Z. A., W. W. Carmichael, et al. (2003). Estimation of microcystins in the freshwater fish *Oreochromis niloticus* in an Egyptian fish farm containing a *Microcystis* bloom. *Environmental Toxicology* 18(2): 137–141.

Norris, R. L. G., G. K. Eaglesham, et al. (2001). Extraction and purification of the zwitterions cylindrospermopsin and deoxycylindrospermopsin from *Cylindrospermopsis raciborskii*. *Environmental Toxicology* 16: 391–396.

Ohtani, I., R. E. Moore, et al. (1992). Cylindrospermopsin: a potent hepatotoxin from the blue-green alga *Cylindrospermopsis raciborskii*. *Journal of the American Chemical Society* 114: 7941–7942.

Persoone, G., C. Janssen, et al. Eds. (2000). *New Microbiotests for Routine Toxicity Screening and Biomonitoring*. Dordrecht, Kluwer Academic Publishers.

Rinehart, K. L., K. Harada, et al. (1988). Nodularin, microcystin and the configuration of ADDA. *Journal of the American Chemical Society* 110: 8557–8558.

Rivasseau, C., P. Racaud, et al. (1999). Development of a bioanalytical phosphatase inhibition test for the monitoring of microcystins in environmental samples. *Analytica Chimica Acta* 394: 243–257.

Ruangyuttikarn, W., I. Miksik, et al. (2004). Reversed-phase liquid chromatographic-mass spectrometric determination of microcystin-LR in cyanobacteria blooms under alkaline conditions. *Journal of Chromatography B* 800(1–2): 315–319.

Runnegar, M. T., I. R. Falconer, et al. (1981). Deformation of isolated rat hepatocytes by a peptide hepatotoxin from the blue-green alga *Microcystis aeruginosa*. *Naunyn Schmiedebergs Archives of Pharmacology* 317(3): 268–272.

Runnegar, M. T., A. R. Jackson, et al. (1988). Toxicity of the cyanobacterium *Nodularia spumigena Mertens*. *Toxicon* 26(2): 143–51.

Runnegar, M. T., S. Kong, et al. (1993). Protein phosphatase inhibition and *in vivo* hepatotoxicity of microcystins. *American Journal of Physiology* 265(2 Pt 1): G224–G230.

Seawright, A. A., C. C. Nolan, et al. (1999). The oral toxicity for mice of the tropical cyanobacterium *Cylindrospermopsis raciborskii* (Woloszynska). *Environmental Toxicology* 14(1): 135–142.

Sim, A. T. R. and L. M. Mudge (1993). Protein phosphatase activity in cyanobacteria: consequences for microcystin toxicity analysis. *Toxicon* 31: 1179–1186.

Sipia, V., H. Kankaanpaa, et al. (2001). Detection of Nodularin in flounders and cod from the Baltic Sea. *Environmental Toxicology* 16(2): 121–126.

Sipia, V. O., H. T. Kankaanpaa, et al. (2002). Bioaccumulation and detoxication of nodularin in tissues of flounder (*Platichthys flesus*), mussels (*Mytilus edulis, Dreissena polymorpha*), and clams (*Macoma balthica*) from the northern Baltic Sea. *Ecotoxicology and Environmental Safety* 53(2): 305–311.

Sivonen, K. and G. Jones (1999). Cyanobacterial toxins. *Toxic Cyanobacteria in Water: A Guide to Their Public Health Consequences, Monitoring and Management.* I. Chorus and J. Bartram, eds. London, E & FN Spon (on behalf of WHO): 41–111.

Spoof, L., P. Vesterkvist, et al. (2003). Screening for cyanobacterial hepatotoxins, microcystins and nodularin in environmental water samples by reversed-phase liquid chromatography-electrospray ionisation mass spectrometry. *Journal of Chromatography A* 1020(1): 105–119.

Strachan, G., J. McElhiney, et al. (2002). Rapid selection of anti-hapten antibodies isolated from synthetic and semi-synthetic antibody phage display libraries expressed in *Escherichia coli. FEMS Microbiology Letters* 210: 257–261.

Tarczynska, M., G. Nalecz-Jawecki, et al. (2001). Tests for the toxicity assessment of cyanobacterial bloom samples. *Environmental Toxicology* 16: 383–390.

Torokne, A. (1997). Interlaboratory trial using Thamnotox kit for detecting cyanobacterial toxins. Eighth International Conference on Harmful Algal Blooms. Vigo, Xunta de Galicia and Intergovernmental Oceanographic Commission of UNESCO.

Van Buynder, P. G., T. Oughtred, et al. (2001). Nodularin uptake by seafood during a cyanobacterial bloom. *Environmental Toxicology* 16(6 special issue): 468–471.

Vasas, G., A. Gaspar, et al. (2002). Capillary electrophoretic assay and purification of cylindrospermopsin, a cyanobacterial toxin from *Aphnizomenon ovalisporum*, by plant test (blue-green Sinapis test). *Analytical Biochemistry* 302(1): 95–103.

Ward, C. J. and G. A. Codd (1999). Comparative toxicity of four microcystins of different hydrophobicities to the protozoan, *Tetrahymena pyriformis. Journal of Applied Microbiology* 86: 874–882.

Welker, M., J. Fastner, et al. (2002). Application of MALDI-TOF MS analysis in cyanobacteria research. *Environmental Toxicology* 17(4): 367–374.

Welker, M., C. Steinberg, et al. (2001). Release and persistence of microcystins in natural waters. *Cyanotoxins: Occurrence, Causes, Consequences.* I. Chorus, ed. Berlin, Springer-Verlag: 83–101.

Williams, D. E., M. Craig, et al. (1997). Evidence for a covalently bound form of microcystin-LR in salmon liver and dungeness crab larvae. *Chemical Research in Toxicology* 10: 463–469.

Wirsing, B., T. Flury, et al. (1999). Estimation of the microcystin content in cyanobacterial field samples from German lakes using the colorimetric protein-phosphatase inhibition assay and RP-HPLC. *Environmental Toxicology* 14(1): 23–29.

Yoshizawa, S., R. Matsushima, et al. (1990). Inhibition of protein phosphatases by microcystin and nodularin associated with hepatotoxicity. *Journal of Cancer Research and Clinical Oncology* 116: 609–614.

Zeck, A., M. G. Weller, et al. (2001). Generic microcystin immunoassay based on monoclonal antibodies against Adda. *Analyst* 126: 2002–2007.

Zweigenbaum, J. A., J. D. Henion, et al. (2000). Direct analysis of microcystins by microbore liquid chromatography electrospray ionization ion-trap tanden mass spectrometry. *Journal of Pharmaceutical and Biomedical Analysis* 23: 723–733.

11 Prevention, Mitigation, and Remediation of Cyanobacterial Blooms in Reservoirs

While cyanobacterial blooms are an ancient phenomenon, their frequency and extent appear to have increased in the last 50 years. The reasons behind this increase are largely anthropogenic, whether through population increase, intensification of agriculture, or global warming.

Population increase affects water quality in many ways. Increased demand for drinking water results in construction of more dams to provide reservoir capacity, leading to a decrease in river flows and also increased pumping directly or indirectly from rivers for drinking water supply. Reductions in river flow result in increases in salt and nutrient concentrations in the rivers. The new reservoirs may be located in areas draining largely agricultural land, as little untouched wilderness remains to provide clean, unpolluted catchments.

Groundwater reserves are becoming depleted in some areas — for example, in Florida — leading to saline intrusion into the groundwater and the need to use surface water as the only practical alternative. This leads to surface water abstraction from lakes, which had previously been avoided as a drinking water supply because of eutrophication and consequent problems with quality.

Urbanization contributes to nutrient enrichment of lakes and rivers, through stormwater runoff containing garden fertilizer, septic tank overflow, and domestic animal waste. By feeding persistent cyanobacterial blooms, this can make urban lakes and rivers unusable for recreation or drinking water supply. Downstream from one town, water may be pumped into off-river reservoirs for use as a drinking water supply for the next town further downstream. In other locations, weirs on rivers are used to provide a water body for drinking water supply. These relatively shallow and nutrient-enriched storages provide growth opportunities for substantial cyanobacterial blooms, causing a potential hazard to consumers and greatly increased water treatment costs.

Population growth directly contributes nutrients to the river systems through sewage discharge. Approximately half of the phosphorus in sewage comes from human waste, and the remainder from detergents and industrial products. Human waste contains 2 to 4 g of phosphorus per person per day (Siegrist and Boller 1997). For a moderate-sized town of 350,000 inhabitants, this causes 1 ton of phosphorus from human waste alone to enter the sewage treatment facility each day. This may be doubled by phosphorus from detergents in areas where the use of phosphate in washing powders is not prohibited. In the case of towns in the upper catchments of rivers, phosphate loading from urban sewage can cause widespread eutrophication. Sewage treatment ponds can carry very high concentrations of phosphorus and nitrogen, in the range of 1 to 10 mg/L or more. These nutrients can feed high concentrations of cyanobacteria in the ponds and hence large amounts of toxins. These ponds may be discharged into waterways, with potential problems for downstream users through nutrient enrichment and cyanobacterial "seeding" as well as the possibility of toxicity to livestock.

As human society becomes increasingly urban, the difficulties of providing an adequate quantity and quality of drinking water escalate. São Paulo, Brazil, a city of over 10 million people, is growing around a major water supply reservoir, leading to nutrient enrichment and cyanobacterial blooms of increasing magnitude. At a lesser scale, eutrophication threatens both drinking water supplies and the recreational waters of towns in many countries.

To enhance plant productivity, phosphatic fertilizers have been and are much used in agriculture. The most widely used form of phosphatic fertilizer is a mixture of phosphates and phosphoric acid, commonly called superphosphate. In extensive agricultural systems, this fertilizer can be distributed from aircraft; on a smaller scale, it can be spread by tractors. After rain, it partially dissolves and enters the soil. Clays in the soil adsorb phosphates, so that the groundwater below a clay soil does not contain appreciable free phosphate. However, light sandy soil has a low clay content and silica does not adsorb phosphate, so that a proportion of the applied soluble phosphate on sandy soil will wash out into rivers during heavy rain. Extensive agricultural use of fertilizers on sandy soils can lead to major eutrophication problems in the river catchments, as discussed later in the case of the Peel-Harvey estuary in Western Australia.

Soil erosion, in which clay particles carrying adsorbed phosphorus wash into rivers, provides the majority of nonsewage phosphate entering rivers. Massive soil erosion has occurred in the last 100 years in many countries, particularly those with very dry summers followed by monsoon rains. This has produced extensive lake and river sediments that carry phosphorus. Such particle-bound phosphorus can be mobilized to soluble forms under anaerobic conditions in the hypolimnion of lakes and slow-flowing rivers, thus becoming available for cyanobacterial growth.

Intensive livestock industries generate a substantial load of nitrogen and phosphorus in animal wastes. Piggery waste, dairy farm waste, intensive poultry waste, and beef feedlot waste all have the capacity to cause extensive eutrophication in lakes and rivers. One river catchment in Portugal had, at a recent count, some 30 piggeries and a highly eutrophic river. Livestock production units such as piggeries are generally subject to controls over waste discharge, aimed at reducing the entry of nutrients into river systems. In many cases these controls stipulate the spreading of waste onto land under particular conditions that minimize seepage into streams; however, heavy rain can cause substantial uncontrolled nutrient runoff into catchments, with consequent risks of eutrophication.

Global warming is now fully accepted as a real process; the remaining points of dispute are over why and how fast it is occurring and what should be done to minimize warming. The climatic effects are being modeled to assist in the prediction of future rainfalls and storm patterns. It can be expected that lake and river temperatures will rise as atmospheric and sea temperatures increase. The distribution of the formerly tropical species *Cylindrospermopsis raciborskii* into the Northern Hemisphere has been suggested to be a response to global warming (Padisak 1997; Briand, Leboulanger et al. 2004). It is difficult to predict which other tropical species will similarly spread as water temperature rises, since the ecology of tropical phytoplankton has not been as extensively studied as has temperate ecology. One species commonly found in tropical lakes is *Microcystis aeruginosa* (Ganf 1974; Oliver and Ganf 2000), which is abundant in eutrophic water supplies worldwide. The intensity of water blooms of this organism may rise with a longer "growing season" in temperate climates and greater stratification of water bodies due to higher ambient temperatures.

11.1 NUTRIENT REDUCTION

The ultimate approach to reduction in nutrient concentration in reservoirs, lakes, and rivers must be integrated catchment management. This is receiving considerable attention worldwide, requiring multiple approaches that include the effects of economic and social issues as well as scientific solutions. The United Nations Environment Programme has recently released its *Guidelines for Integrated Watershed Management*, focusing on phytotechnology and ecohydrology (Zalewski 2002).

As discussed earlier in Chapters 4 and 9, the magnitude of cyanobacterial biomass that can grow in a reservoir or lake is determined by the combination of light availability, phosphorus, nitrogen, and the hydrophysical characteristics of the water body. The component that has received most attention for the prevention and mitigation of cyanobacterial blooms is phosphorus (Chorus and Mur 1999). The established relationship between the maximum bloom potential and total water phosphorus concentration for a large number of temperate lakes shows the possible benefit of phosphorus reduction (Vollenweider and Kerekes 1982; Reynolds 1997). Examination of the data of Vollenweider (1982) shows a wide scatter of points, with about a 20-fold range in maximum phytoplankton content at any given phosphorus concentration. This scatter reflects the influence of other factors, especially depth. Calculation of the maximum phytoplankton content at different mixing depths in the water body shows the impact of light availability. This decreases exponentially with depth and results in greatly reduced growth through shading at increased mixing depths (Chorus and Mur 1999). In this way the hydrophysical character of the reservoir will substantially affect the response to phosphorus reduction, as phosphorus concentration may not be the factor limiting cyanobacterial growth. In deep-mixing lakes, the total phosphorus concentration may have to be reduced below 40 μg/L before substantial reduction in cyanobacterial growth occurs. In shallow lakes or lakes that have shallow mixing depths, progressive reductions in phytoplankton may occur with reduced phosphate concentrations, which commence at much higher initial phosphorus loadings.

Nitrogen availability may be limiting, rather than phosphate, at high phytoplankton densities in eutrophic waters. Eukaryotic phytoplankton, macrophytes growing in the water, and non-nitrogen-fixing cyanobacteria require inorganic nitrogen for growth. Under conditions of nitrogen limitation through competition, reductions in phosphorus will not be reflected in reduced cyanobacteria until phosphorus becomes limiting. The nitrogen-fixing genera of cyanobacteria may have a competitive advantage under conditions of nitrogen limitation, but they are at an energetic disadvantage compared with eukaryotic phytoplankton or non-nitrogen-fixing genera because of the high energy requirement for nitrogen fixation. They are thus more likely to compete successfully in clear, less eutrophic waters lacking inorganic nitrogen. Fortunately most strategies for nutrient reduction in water bodies will reduce both nitrogen and phosphorus, thus resulting in more effective outcomes than reduction of only one nutrient.

Thus effective measures for reduction of cyanobacterial concentrations in reservoirs are best undertaken when the limiting factors for cyanobacterial growth have been identified.

11.2 PHOSPHORUS REDUCTION

To minimize the biomass of cyanobacteria in a reservoir on a long-term basis, phosphorus reduction will ultimately be the most successful approach. Because of the complex interactions between the potentially limiting factors in biomass development, initial reduction in phosphorus input may have no observable effect; however, as the total phosphorus available for biomass falls, so eventually will the biomass decrease.

11.2.1 Reduction to Inflow

Identification of the main sources of phosphorus to the water body is a necessary first step. Some rural industrial sources are readily identified, such as meat- and wool-processing plants or intensive livestock industries. Nutrient discharge from these sources can be regulated by legislation, such as environmental protection acts, which can specify the allowable discharge of phosphorus by a license to the company. For agricultural and food industries, alternative discharge mechanisms, such as holding ponds from which the high-nutrient water is used for crop or pasture irrigation, are preferable to discharge into streams. A reduction in phosphorus load from detergents can be achieved by regulating discharge from industries using phosphates in cleaning processes and by either a public campaign against phosphates in household washing powders or a ban on the sale of products containing phosphorus. The ban on the sale was successfully undertaken in Canada but strongly opposed in other countries by commercial interests.

The human poisoning caused by a toxic *Microcystis* bloom in a drinking water supply reservoir, described in Chapter 5, occurred in a catchment in which a meat-processing plant discharged wastewater directly into a watercourse leading to the reservoir. The effluent stream is now diverted into irrigation use. This has reduced nutrient loading of the reservoir but has not, to the present date, reduced cyanobacterial bloom formation.

11.2.2 Phosphorus Stripping

Sewage treatment plants have the potential to discharge substantial quantities of phosphorus into watercourses, however this can be greatly reduced by phosphorus reduction built into the operation of the plant. Use of precipitants for phosphorus in the later part of the treatment train — for example, ferric salts — can remove 99.9% of the incoming phosphorus load in state-of-the-art wastewater treatment facilities. The plant processing wastewater from Canberra, Australia, discharges into the headwaters of a major river. After phosphate precipitation, the content of the discharge is reduced from 8.7 mg/L of phosphorus to 0.07 mg/L in effluent (ACTEW 2000). The phosphorus is trapped as insoluble ferric phosphate in the sludge from the plant, which can be processed for use as a fertilizer or, more commonly, dried and put into landfill or incinerated. Aluminum precipitants are also effective. In both cases the precipitants assist in clarification of the final product, carried out by centrifugal separation and finally filtration, leading to a low turbidity of the effluent discharge.

Biological phosphorus removal is based on microbial uptake of phosphorus from the effluent and uses a combination of anaerobic and aerobic digestion to facilitate microbial growth in an activated sludge. Settling reduces phosphorus through transfer into the sludge at each stage. Final clarification removes remaining particulate phosphorus; however, soluble phosphorus will not be reduced as effectively as with chemical precipitation. Initial phosphorus concentration in urban sewage may contain more than 10 mg/L, which can be reduced to 0.2 mg/L with microbial phosphorus removal (Harremoes 1997). This concentration is more than sufficient to support cyanobacterial growth in the effluent.

Both chemical and biological phosphorus stripping techniques can be used for improving the quality of inflow water to reservoirs and recreational lakes. The use of such methods is costly, and the benefits of reduction or prevention of cyanobacterial blooms have to be set against the cost of the facility and ongoing operating costs. In Germany, phosphate stripping of inflow water to lakes and reservoirs has had beneficial results (Sas 1989). The response to reduction of phosphorus entry into a reservoir may be very slow, due to the low turnover time of the water in large reservoirs and the large pool of phosphorus retained by the sediments. The biomass itself also acts as a phosphorus reservoir, with phosphorus moving between algae, diatoms, and cyanobacteria. Reduction in nutrient inflow to a water body is considered to be the key to long-term control of eutrophication, with other measures secondary (Chorus and Mur 1999).

11.2.3 Wetlands

Artificial ponds, lagoons, and wetlands are widely used for nutrient reduction in wastewater treatment, in urban runoff, and in rivers draining intensively used agricultural land (Greenway and Simpson 1996; Williams, Pettman et al. 1998). These can be effective for phosphorus and nitrogen reduction in sewage plant effluent and other nutrient-enriched waters. They rely on aquatic macrophytes and sediment microorganisms, with clay and organic debris in the sediment acting as nutrient sinks. The wetlands may be a series of simple shallow lagoons planted with rushes or more sophisticated systems in which the treated sewage effluent is fed into the wetland from subsurface pipes and moves through the plant layer. These wetland systems are very sensitive to variations in load, flooding and drying, and operating temperature, as they depend on biological activity. If they are overloaded, they may become anaerobic through the excess oxygen demand of the water entering the lagoons and release high concentrations of nutrients. Long-term maintenance of these systems is required to ensure that the lagoons remain aerobic, as even temporary anoxic conditions will remobilize phosphorus, and lead to a pulse of eutrophication downstream. Flooding will also upset the operation of these systems, washing out sediments rich in phosphorus as well as possible pathogenic microorganisms.

The characteristics of the lagoons and ponds for optimal nutrient reduction should include retention times in days, so that suspended particulate material and eukaryotic algae can sediment out. Total phosphorus input to a reservoir can be reduced by 50 to 65% by this approach (Klapper 1992; Chorus and Mur 1999) and a reduction of greater than 70% in phosphorus has been set for urban pond/wetland design discharging into a river (ACT Department of Urban Services 2001).

A large-scale example of this approach to nutrient reduction is the construction of the Kis-Balaton reservoir and wetlands in Hungary. The Zala River flowing into Lake Balaton drains an agricultural and urban area, which resulted in substantial nutrient loads entering the lake, phosphorus in the amount of 2.47 g/m^2/year entering the western part of the lake (Padisak and Istvanovics 1997). Cyanobacterial blooms resulted, commencing in the 1970s (Voros, Hiripi et al. 1975). Lake Balaton is a very important ecological and recreational resource in Hungary, so a major phosphorus reduction strategy was implemented, which reduced the overall phosphorus loading of the lake from 0.5 to 0.3 g/m^2/year (Istvanovics and Somlyody 2001). This was carried out by construction, within natural wetlands, of a large, shallow reservoir followed by a series of small dams and reed-bed wetlands, giving an average retention time of 1 month (Chorus and Mur 1999). A progressive reduction in the cyanobacterial population resulted, though blooms of *C. raciborskii* still occurred during warm summers (Padisak and Istvanovics 1997).

11.2.4 Low-Flow Effects

A factor that exacerbates the eutrophication arising from sewage discharge is the continual flow from sewage plant outfalls in summer, when the natural flow in the river system may be minimal. This problem is very apparent in Mediterranean climates when summer rainfall is negligible and river flows are affected both by low inputs from the catchment and use of the water for crop irrigation. In areas of low population, the rivers cease to flow in dry summers. In Europe, in areas with some rainfall in summer but with high population density, the more northern rivers — such as the Thames in England and the Havel in Germany — have almost all the flow arising from wastewater in a dry summer (Gray 1994; Kohler and Klein 1997). The same result is seen in the Sydney area in Australia, with the low summer flow in the Hawkesbury River arising from wastewater coming from 100 licensed sewage treatment plants (SPCC 1983). In this tidal river, cyanobacterial blooms form in the lower section above the brackish water zone, which move up and down the river with the tide. This results in the intermittent intake of high cyanobacterial loads at a drinking water supply facility drawing water from the river.

11.2.5 Agricultural Land

Phosphorus from agricultural sources can be from livestock waste or from soil erosion. Livestock waste inputs from intensive production units can be controlled by suitable engineering design for holding tanks and ponds, with either land disposal under controlled conditions or waste treatment. High-nutrient wastewater after treatment is normally used for irrigation of pasture or crops, limiting the contamination of waterways. Environmental protection agencies regulate large-scale intensive animal production facilities, often by a system of licensing that allows only specified discharge into waterways. Any water discharge to the catchment can be monitored for nutrients. The allowable discharge into a waterway should not permit a phosphorus concentration that would result in eutrophication. The actual concentration limits would depend on the volume of flow in the waterway into which the discharge occurs, particularly considering the minimum flows in dry periods and the downstream use of the water. In the case of waterways supplying drinking water reservoirs, a complete prohibition of intensive livestock production in the supply catchment is desirable. This applies to piggeries, intensive poultry units, beef feedlots, and large dairy units. Where this is not possible, considerable control over the design and operation of the production units is essential, for health reasons as well as preventing eutrophication.

Phosphorus arising from diffuse sources of soil erosion is more difficult to control, as heavy rain will wash suspended clays from soil, as well as any soluble phosphorus from recently applied fertilizer, into watercourses. The clays carry an adsorbed phosphorus load. This source of phosphorus can be minimized by a zone of riparian vegetation (vegetation on the banks of watercourses), which will intercept nutrients and sediment from surface runoff. Where reservoirs are supplied from watercourses running through agricultural land, management of the riparian vegetation is an important element of preventing eutrophication. A zone of 100-m width of mature native vegetation each side of a watercourse will reduce phosphorus load greatly (Hairsine 1997; Zalewski 2002). The load of phosphorus carried into a reservoir by a single flood can be greater than the total phosphorus content of the water body plus the phosphorus inflow for the remainder of the year, so that upstream erosion control and flood mitigation will have major effects on phosphorus reduction (Jones and Poplawski 1998).

In grazing land, preventing cattle and sheep from entering watercourses to drink will reduce nutrient input, both directly from feces falling into the water and by stopping channeling from the higher grazing land down to the water (Robertson 1997). These livestock paths become waterways during heavy rain, carrying soil and fecal material into the water without any interception by vegetation. Both restoring the natural vegetation along waterways and prevention of livestock entry are expensive. Fencing is necessary as well as construction of stock watering points that are away from the natural watercourse. Maintenance of a riparian zone in arable farming is relatively straightforward, requiring only the conservation of a band of trees, shrubs, and grasses along the watercourse edges. As a considerable proportion of soil erosion arises from arable farmland from which phosphorus is carried into rivers, the benefits from riparian zone management are apparent and of relatively low cost.

11.3 CATCHMENT MANAGEMENT

There is increasing emphasis on whole-catchment management for the reduction of nutrient inputs into rivers and reservoirs. Since the catchment area may contain land with a wide range of ownership, including land totally owned and managed by the water utility, national parks, leasehold land, freehold land used for farming, small urban areas, and isolated housing, considerable cooperation is required. Catchment groups with a coordination and education role have been established in several countries, with representation from the landholders, the water supply agency, local councils, and other concerned groups. However, unless there is direct political or financial motivation for change, these groups have difficulty in achieving progress. Government financial assistance for riparian zone management can be targeted through catchment management groups, thus using local knowledge and government money for rectifying areas of major erosion. Changing land-use practices is more difficult, and education through catchment groups or provision of expert help is most likely to succeed. Agricultural practices can also be improved, especially if it involves minimal cost; the suggestion of expensive, unsustainable practices may result only in opposition and noncooperation from landholders. Chorus and Mur (1999) indicate that cooperation has been most effective in Germany when the land is owned by the water supply agency and the landholders are leasehold.

Fertilizer use directly contributes to the phosphorus load entering watercourses, especially in sandy soils with low phosphorus-retention capacity. The arable farming, primarily wheat, in southwestern Australia uses phosphatic fertilizers to enhance crop yields in otherwise relatively infertile soils. The area has winter rainfall, and dry summers, with the growing period over winter. Soluble phosphorus from fertilizer application on the sandy soils washes down into watercourses in winter, leading to massive blooms of *Nodularia spumigena* in the Peel-Harvey estuary. Changes in fertilizer type, to lower-solubility slow release forms, has reduced the magnitude of the blooms (Lukatelich and McComb 1981). Unfortunately the reduction was insufficient to prevent the blooms recurring, which required the dredging of a channel into the sea; this was 2.5 km long, 200 m wide, and approximately 2 m deep and cost more than $20 million (Hosja, Grigo et al. 2000). The increased tidal flushing of the estuary removed phosphorus and also increased the salinity of the water during winter. No further *Nodularia* blooms have occurred in the estuary, although the rivers entering the estuary are still badly affected by cyanobacterial blooms.

Subtropical reservoirs have enormously variable nutrient input due to the rainfall pattern. Monsoonal summer rains bring down huge quantities of phosphorus-containing sediments in floodwaters, followed by 10 to 11 months of the year of reduced or no inflow. The high energy input from the sun results in highly stratified water in the reservoirs. The hypolimnion is anoxic for most of the year, and nutrient supply from soluble phosphorus and ammonia nitrogen arising from the sediments is the main factor influencing cyanobacterial growth. Catchment management for nutrient control is relatively ineffective in these circumstances (Jones 1997). Management of the reservoir hydrology, discussed later, can be effective in reducing cyanobacterial blooms.

11.4 NITROGEN REDUCTION

Inorganic nitrogen in water arises naturally from plant and microbial decomposition in soil and nitrogen fixation. It also results from lightning, industrial waste gases, fossil fuel burning in power generation, and burning of biomass in forest and grassland fires. Most significantly, inorganic nitrogen in water arises from treated sewage discharge, animal wastes, and nitrogenous fertilizer application. Urea, ammonia, ammonium sulfate, and ammonium nitrate are widely used in agriculture to enhance plant production. In the U.K. alone, more than 2 million tons of nitrogenous fertilizer are applied each year (Gray 1994). Because they are soluble and if not incorporated into biomass, wash out of the soil, they are applied annually. Oxidation within the soil will convert ammonium ions, if not taken up by biota, into nitrate. As nitrate is highly soluble and not substantially removed from solution by adsorption to clays, it can enter watercourses from groundwater as well as surface runoff. Estimated release of fertilizer nitrogen applied to crops into groundwater and runoff, under ideal conditions for crop use, ranges from 2 to 10%, and will be appreciably greater with heavy rainfall or low soil temperatures (Gray 1994). The accumulation of nitrate and nitrite in groundwater used for public water supply is an issue of medical significance, with a Maximum Acceptable Concentration of nitrate in drinking water in the European Economic Community of 50 mg/L, which is frequently exceeded (Gray 1994).

Animal wastes high in nitrogen are spread onto agricultural land to enhance fertility and also to dispose of them at low cost. In intensive animal production, large volumes of waste in the form of manure or slurry are spread on land, usually under the regulation of the local authority, river board, or environmental protection agency to minimize losses into watercourses. Surplus nitrogen from wastes moves down into groundwater or washes out of soil as nitrate and into watercourses.

Normal sewage treatment will not remove nitrogen, which is discharged as nitrate after aerobic decomposition of fecal material. Combined nitrification/denitrification processes in sewage treatment will reduce nitrogen discharge, as the final anerobic denitrification stage results in microbial nitrate conversion to nitrogen gas.

The overall consequence of agricultural use of fertilizers and manures has been a dramatic rise in nitrate in rivers, reaching over 100 mg/L in many European rivers in winter (Gray 1994). From the basis that phytoplankton have a nitrogen:phosphorus ratio in their biomass of 7:1, it is apparent that nitrogen will not be the limiting nutrient for cyanobacterial growth in such waters, which may have a phosphorus concentration 1000 times lower (Chorus and Mur 1999). As a result, reduction in nitrogen in surface waters may not have any effect on the extent of eutrophication, though it will affect which cyanobacterial species is dominant. *M. aeruginosa* does not fix atmospheric nitrogen and is frequently the dominant species in eutrophic lakes and rivers, in which nitrogen is in large excess. Under circumstances of intensive land use, as occurs in Europe and parts of North and South America and Asia, reduction in nitrate in surface waters is relevant to meet drinking water standards but not in control of cyanobacterial eutrophication.

Nitrogen limitation of cyanobacterial growth does, however, occur in lake and river systems in semiarid areas in other parts of the world, in which industrial activity is minimal and nitrogenous fertilizers are applied to only a small proportion of the land area. The large Murray-Darling River basin in southeastern Australia is an example; there, nitrogen availability limits cyanobacterial growth in the river (Brookes, Baker et al. 2002). In these circumstances the nitrogen-fixing *A. circinalis* is the dominant bloom-forming species. This organism causes taste and odor problems in drinking water at relatively low cell concentrations (5000 cells/mL), and livestock poisoning at high concentrations due to the presence of neurotoxic saxitoxin derivatives (Humpage, Rositano et al. 1994). Reduction of available nitrogen in these circumstances can be expected to reduce overall cyanobacterial biomass, as the energetic efficiency of nitrogen fixation is low compared with cyanobacterial use of nitrate or ammonia. Control of sewage treatment plant discharge and runoff from feedlots and irrigated agriculture will be effective in reducing cyanobacterial populations in nitrogen-limited rivers and reservoirs.

Nitrogen in organic debris in sediments in reservoirs, lakes, ponds, and rivers decomposes aerobically to release nitrate or nitrite and anaerobically to release ammonia. The anaerobic release of ammonia during prolonged periods of stratification of a water body is of considerable significance in subtropical reservoirs. A nitrogen gradient of 1.6 mg/L was observed from the surface to 20-m depth during the summer stratification of a deep reservoir. A *Microcystis* bloom of 10,000 cells per milliliter occurred simultaneously, lasting 6 to 9 months in 2 successive years (Jones 1997).

Weir pools on subtropical rivers are also susceptible to cyanobacterial blooms at times of high temperatures and low water flows (Bormans, Ford et al. 2000). Under these circumstances, stratification occurs in shallower systems, with the sediments becoming anaerobic, which releases phosphorus and ammonia into the hypolimnion. Mixing by wind or river flow as well as vertical migration of cyanobacteria provide access to increased nutrients, resulting in cyanobacterial blooms (Fabbro and Duivenvoorden 1996).

The "capping" of sediments to reduce nutrient availability is discussed in the next section.

11.5 RESERVOIR REMEDIATION

In circumstances when major reductions of nutrient supply to a reservoir are impractical or the reservoir sediments contain a massive nutrient store, a range of hydrophysical and chemical approaches to cyanobacterial control may be possible. The hydrophysical methods rely on mixing techniques, which range from aeration to flow control, whereas the chemical techniques rely on algicides, precipitants, and sealants added to the water in the reservoir. Because of the difficulty of reducing nutrient inputs to reservoirs, which requires cooperation over a wide sector of the upstream population, many water supply agencies have opted for engineering solutions.

11.6 DESTRATIFICATION

Stratification of reservoirs and weir pools in summer, due to solar input of thermal energy, greatly enhances the risk of cyanobacterial blooms. The stratified warmer upper layer (the epilimnion) provides a stable environment with greatly reduced vertical mixing. This favors cyanobacterial growth through clarification of the epilimnion as heavy algae, diatoms and other particles settle downward from the surface, allowing greater light penetration. These phytoplankton rely on mixing in the water column to gain access to light; in a stable, stratified water column, they sink to lower levels and lose competitive advantage. The cyanobacterial species causing water blooms have variable buoyancy, as discussed in Chapter 4, and can move up and down in the water column, thus optimizing their access to light and ability to remain in the upper layers.

The variable buoyancy of cyanobacteria also allows downward movement into nutrient-enriched zones deeper in the water, where ammonia and phosphorus from the sediments are diffusing upward.

Organic decomposition in the sediments depletes the oxygen in the adjoining water, and as the stratification continues, a deeper and deeper layer of anoxic water forms above the sediments. Experimental measurement of the solubilization of phosphorus in upper sediments under anaerobic conditions showed that about 80% of the phosphorus content could be made available for cyanobacterial growth, representing a massive reserve of nutrient (Chambers, Olley et al. 1997). Ammonia is also liberated, and is a preferred nutrient for cyanobacterial growth.

Destratification aims to combat all of these processes that enhance cyanobacterial growth. By providing artificial mixing, cyanobacteria are carried down below the euphotic zone, reducing energy availability. Diatoms and eukaryotic algae are carried upward into the light, allowing competition for nutrients. In particular, the deoxygenated water above the sediments becomes oxygenated, and nutrient enrichment is greatly reduced as a result.

In practice it has not been uniformly possible to achieve these desired ends despite considerable efforts. One of the most significant problems is the energy input needed to destratify the water body. Partial destratification may result in mixing high-nutrient water into the lower layers of the epilimnion above the area where the destratification equipment is operating, with the shallower parts of the reservoir still highly stratified. This can exacerbate the bloom biomass. It has been suggested that at least 80% of the water volume should be destratified under northern European conditions of relatively low solar energy input (Visser, Ibelings et al. 1996; Chorus and Mur 1999). With higher solar energy inputs in Mediterranean or subtropical climates, the epilimnion–hypolimnion temperature gradient will be higher, requiring both more energy input for destratifying the surface layer and a greater proportion of the total volume to be destratified.

If the destratification is sufficient to render the surface of the sediment oxygenated, the nutrient release will be substantially reduced, hence reducing the potential cyanobacterial biomass (Sherman, Whittington et al. 2000). The extent of the sediment surface oxygenated will depend on the method employed and the hydrology of the reservoir.

Bubble plume destratification has been widely used to increase sediment oxygenation, with compressed air lines across the deepest portions of the reservoir and air nozzles at intervals. In the Myponga reservoir in South Australia, a water body of maximum depth of 36 m and mean depth of 15 m with a 27-GL (10^9 L) capacity, an aerator 200 m in length with 160 outlets at a depth of 30 m delivering 120 L/s of air was installed in 1994. This proved inadequate to destratify the surface layer. In 1999, two surface-mounted mechanical mixers driving water downward through a delivery tube 15 m in length at 3.5 m^3/s were installed to take surface water and cyanobacteria down to depths at which light intensity was inadequate for growth and to reinforce the surface destratification and oxygenation from the bubble plume aerators.

The reservoir has annual summer blooms of *A. circinalis*, and modeling studies have indicated that the combined effect of the two types of destratifiers has significantly reduced the period over which the cyanobacteria can grow, thus enhancing the growth of diatoms and green algae. However in midsummer blooms of *Anabaena* still occur, causing taste and odor problems; these are treated by application of copper sulfate (Brookes, Baker et al. 2002).

The hydrology of a reservoir has a major impact on the potential success of destratification. Narrow, deep reservoirs with little shallow water are more likely to be successfully destratified as long as the energy input through the destratification is sufficient to overcome the surface heating from solar energy. In reservoirs with extensive water of shallow depth, local stratification with enhanced cyanobacterial growth will occur due to the low rate of horizontal mixing, notwithstanding effective destratification of the deep water. In reservoirs in which cyanobacterial biomass is determined by light availability, mixing to sufficient depth to inhibit growth through lack of light is required. This may be as deep as 20 m, indicating that mixing of shallower water will not inhibit cyanobacterial growth (Chorus and Mur 1999).

In one small compact reservoir, destratification by bubble plume aerators has had substantial effects on the cyanobacterial bloom species and numbers. Solomon Dam on Palm Island in tropical Australia formerly had large blooms of the toxic *C. raciborskii* in summer. After installation and operation of the aerators these blooms did not recur, but winter blooms of *Anabaena* and *Microcystis* continued. These blooms were due to the reduction in water level in the dry season, rendering the destratification ineffective, and the hypolimnion became deoxygenated. When winter turnover occurred, the nutrients released from the anoxic sediments triggered major cyanobacterial growth (Griffiths, Saker et al. 1998).

Considerable site-specific development is needed for successful destratification of reservoirs, with ensuing capital and recurrent costs. If the main aim of the destratification is to maintain oxygen availability at the sediment surface, rather than total reservoir mixing, more cost-effective strategies may be possible. Injection of pure oxygen or of nitrate may be more effective in oxygenating deep water than the use of mixing techniques (Gemza 1997; Prepas, Murphy et al. 1997). Addition of nitrate must be sufficient to maintain oxygenation of the sediment surface, as any reversion to anaerobic conditions will cause increased ammonia release and enhancement of cyanobacterial blooms.

11.7 FLOW

Flow regulation and flushing have been shown to be effective in cyanobacterial control in river systems, lakes, and weir pools used as sources of drinking water (Sas 1989). In slow-flowing rivers and pools, diurnal or persistent stratification occurs in regions with high day temperatures in the summer months (Webster, Sherman et al. 2000). In deeper pools, locally anoxic sediments release phosphorus and ammonia and the stratification enhances the sedimentation of particles and heavier phytoplankton. Cyanobacteria proliferate, the species depending on the nitrogen availability, with *Microcystis* often predominant in nitrogen-rich waters and *Anabaena* in nitrogen-limited waters.

The massive 1000-km bloom of *A. circinalis* in the Darling River in Australia in 1991 was brought about by drought and irrigation demand for water from the river. Flow was almost stopped, and the deeper pools stratified with an anoxic hypolimnion. Phosphorus liberated from the sediments triggered the water bloom in a river system generally deficient in dissolved inorganic nitrogen. The bloom was terminated by natural flushing caused by rain in the upper catchment, which effectively eliminated the bloom in 2 weeks (Bowling 1992; NWSBGA Task Force 1992).

Changes in water turbidity and local destratification can by brought about by increases in water flow and controlled discharges from weir pools. Stratification was identified as the main factor in cyanobacterial growth in weir pools, though in the weir investigated the deeper layers of water remained oxygenated even when the upper layers were stratified. This was brought about by colder oxygenated water flowing into the weir pool along the bottom. Pulsed discharge was identified as the most effective approach to destratification and bloom control (Webster, Sherman et al. 2000).

Flushing flows of water from upriver water storages can substantially reduce blooms in rivers used as drinking water sources. In the Murray River in South Australia, *Anabaena* blooms occur in summer near drinking water intakes supplying a number of towns. To minimize the frequency of these blooms, a study was undertaken of the factors influencing stratification. Wind velocity was identified as a major influence, which, in conjunction with river flow, reduced stratification through mixing. A modeling study determined the flows necessary to destratify the river under a range of wind conditions and the optimum location for the source of the water (Maier, Burch et al. 2001).

Release of hypolimnetic water from a dam or weir in summer will, however, have multiple effects. The removal of dissolved phosphorus from the reservoir will assist in reducing phosphorus load in the water body and hence reduce the potential biomass of cyanobacteria (Nurnberg 1997). The overall effect, however, may not be uniformly beneficial, as the deoxygenated cold water will have strong local effects on the ecology. The increased downstream phosphorus concentration may result in later eutrophication of the river if the water flow then ceases or the hypolimnetic water is insufficiently diluted by other water sources.

11.8 PHOSPHORUS PRECIPITATION, SEDIMENT CAPPING, AND DREDGING

Whole reservoirs and lakes can also be treated to reduce phosphorus concentrations in the water and suppress cyanobacterial blooms. Ferric salts, when added to reservoirs, precipitate phosphorus as insoluble ferric phosphate, and ferric alum (ferric aluminum sulfate) has been used as a flocculent for both cells and phosphorus (May 1974). After dog and sheep deaths had occurred, ferric salts for phosphorus control were applied to Rutland Water, a major drinking water storage in the U.K., to prevent the growth of toxic *Microcystis* blooms (Corkill, Smith et al. 1989; Matthews 1990). This method requires the sediment surface to remain oxygenated, as under anaerobic conditions ferrous iron salts become soluble. Aeration may be needed to prevent stratification and development of anaerobic conditions, adding to the cost of this approach. Unless the entry of phosphorus to the water body is restricted, repeated application of ferric salts is required (Chorus and Mur 1999).

Application of lime as calcium hydroxide or calcium carbonate has been used successfully to flocculate cyanobacterial cells and phosphate in small reservoirs and also to form a cap on the sediments (Murphy, Prepas et al. 1990). Flocculation with lime does not appear to result in cell lysis, so that toxins are likely to be trapped in the sediments and to undergo microbial breakdown (Lam, Prepas et al. 1995). The rate of lime application is high, with calcium hydroxide applications of 50 to 250 mg/L (Zhang and Prepas 1996). This is likely to affect water pH and to be less effective in acidic waters. The success of in-lake phosphorus removal is variable and likely to depend on pH, hardness, dosage, and frequency of treatment (Cooke, Welch et al. 1993).

The capping of the sediment surface with clay has been used to suppress phosphorus solubilization in small recreational lakes and will additionally adsorb microcystins from solution (Morris, Williams et al. 2000). Dredging has been undertaken to remove phosphorus-rich sediments, since they provide a major source of soluble phosphorus in stratified lakes in summer. This process requires considerable preconditions to be met if it is to be effective. Sediment depth may be great, and dredging down to low phosphorus or low-phosphorus solubility layers will be needed to provide any benefit. Soluble interstitial phosphorus contained within the sediment will be released during the dredging and contribute greatly to the availability of phosphorus in the year in which the work is undertaken. The sediment removed must be taken out of the upstream catchment or it will provide a supply of phosphorus in subsequent years (Chorus and Mur 1999). The year following a systematic dredging operation in the recreational Torrens Lake, Adelaide, South Australia, a massive bloom of highly toxic *M. aeruginosa* occurred, causing closure of the lake. In the 4 years following that event, only one substantial bloom has occurred in the lake. There had not been any major toxic blooms before dredging.

11.9 ALGICIDES

11.9.1 COPPER

Control of cyanobacterial blooms in reservoirs by copper dosing is the oldest and most commonly applied method in large areas of the world. It is also controversial, as copper is a nonspecific poison for aquatic organisms and accumulates in sediments. As a result of the ecological harm likely to occur from repeated applications of copper to water bodies, its use is subject to regulation by state and local authorities and is totally banned in some countries. While application of copper to dedicated drinking water supply reservoirs is permitted in other countries, use of copper in flowing water and in natural lakes may be prohibited. Use of copper as an algicide is relatively simple to undertake, and while the quantity (and hence cost) of material used can be considerable, it may be cheaper than alternative control measures or upgrading water treatment.

The main motivation for the water supply industry to use algicides in drinking water reservoirs is the need to improve the taste and odor of the final drinking water. Blooms of some diatoms and cyanobacteria release geosmin and methylisoborneol into the water, causing musty off-flavors and consequent complaints from consumers. These off-flavors may be from diatoms that are not toxic or that occur at cyanobacterial cell concentrations below those likely to cause toxicity problems. Hence poor taste and odor cannot be directly related to toxicity. Unfortunately for plant operators, the absence of a smell does not imply absence of toxic organisms, as some toxic cyanobacterial species produce little or no odor. Copper dosing of a reservoir at appropriate concentration will kill both diatoms and cyanobacteria and thus remove the problems of taste and odor (Burch 2002).

11.9.2 PROBLEMS WITH THE USE OF COPPER

The toxic component of copper in aqueous solution is the cupric ion Cu^{2+}. The active ion is toxic to phytoplankton at concentrations of 10^{-11} to 10^{-6} M, depending on the organisms involved (McKnight, Chisholm et al. 1983). This corresponds to 0.00063 to 63 μg/L of Cu^{2+}, a wide range but at very low concentrations. Field data for effective cyanobacterial control by copper in reservoirs used 0.5 to 2.0 mg/L of copper sulfate; assuming the application was crystalline $CuSO_4(5H_2O)$, this is 0.125 to 0.5 mg Cu^{2+}/L (Burch 2002).

The very large increase between the effective concentration of toxic ion and the field concentrations of applied copper are due to a series of reactions occurring in the water as the copper sulfate dissolves. In acidic waters low in organic content, the concentration of cupric ion in solution remains high. By contrast, in alkaline waters, particularly those with a high content of dissolved Ca^{2+} and Mg^{2+} carbonates, copper will rapidly form carbonates and related compounds, which are insoluble. In addition, suspended clays, organic material containing acidic groups, and other colloidal agents will adsorb or complex with copper, removing the cupric ions from solution and rendering them nontoxic (Hart 1981). The consequence of the combined effects of these reactions is that the cupric ion has a short lifetime once dissolved and rapidly becomes nontoxic to phytoplankton. The lifetime of the ion will vary between water bodies, and the concentration of copper needed to kill cyanobacterial cells will be accordingly different. In one reservoir, negligible free cupric ion was present 60 to 120 min after copper addition (Burch 2002). Modeling can provide a guide to the concentration to be used, incorporating the parameters for pH, hardness, and dissolved organic matter for the water body (Sunda and Hanson 1979).

The adsorbed and complexed copper will slowly sediment, thus accumulating and having toxic effects on sediment-dwelling species (Hanson and Stefan 1984). Under anoxic conditions, some solubilization occurs, which may also have ecological consequences (Prepas and Murphy 1988).

To avoid the rapid loss of effectiveness of dissolved copper sulfate when added to reservoirs and therefore reduce the quantity of copper to be used, a number of chelated forms of copper have been marketed commercially. These are based on a slower release of the active cupric ion and therefore maintenance of toxic concentrations of copper for an extended period. One of these copper compounds is the citrate salt, which appears to provide an extended availability of cupric ions in alkaline waters (Hrudey, Burch et al. 1999; Burch 2002). Part of the mechanism of this effect will be dependent on the dissociation characteristics of the salt, but it is likely that microorganisms in the water will metabolize the citrate, also releasing the copper.

Calculation of the dosage of copper to apply can be based on the volume of the top 1 or 2 m of depth in the reservoir. At 1 mg/L of final concentration and 2 m of depth, 2 tons of copper will be required per square kilometer of area (8 tons of copper sulfate crystals). Application methods include towing woven sacks of crystals behind a boat, spreading crystals from a boat, dropping them from a helicopter, spraying solution from a boat, and spreading crystals from fertilizer-spreading aircraft. In one location known to the author, copper sulfate crystals have been seen adjacent to superphosphate granules on the edge of a drinking water reservoir, both delivered by air (Figure 11.1).

A separate problem arises when the application of copper has been successful. Copper causes the lysis of phytoplankton cells, liberating the contents into the water. If the intention was to remove a diatom bloom causing taste and odor problems, the result of copper application would be a sudden increase in bad taste and odor. Similarly, cyanobacteria lyse after copper treatment, releasing toxins and organic load into the water. Both of the documented instances of human poisoning from cyanobacterial toxins in tap water arose from bloom lysis following copper sulfate treatment of the reservoir, after complaints of bad taste and odor (see Chapter 5). Water treatment is strongly affected by a sudden increase in organic load, and conventional treatment plants do not have specific capacity to remove dissolved cyanobacterial toxins. This is discussed further in Chapter 12.

To minimize the problems of increased organic load, bad taste, odor, and possible cyanobacterial toxins in drinking water following copper treatment of reservoirs, the treatment should be applied very early in bloom development. This requires careful and frequent monitoring of the cyanobacterial population so that major increases in cell numbers are not allowed to occur. In water bodies with recurrent problems with cyanobacterial blooms in summer, it may be necessary to treat a storage reservoir with copper at 2-week intervals for several months. Some small water-holding

FIGURE 11.1 **(See color insert following page 146.)** Distribution of copper sulfate to a drinking water reservoir. An airplane, normally used to distribute superphosphate fertilizer, is dropping copper sulfate on a reservoir with a *Microcystis* bloom. (Photograph courtesy of M. Choice, Armidale, Australia. With permission.)

reservoirs have been treated up to three times a week with copper to prevent formation of cyanobacterial tastes and odors.

As a precaution against increased copper in tap water, and cyanobacterial toxins released into the water by copper use, discontinuing use of the treated reservoir for up to 2 weeks is advisable. Cyanobacterial toxins decompose slowly when released into water by copper dosing, and the removal of these toxins is particularly important if only conventional water treatment is available and the cyanobacterial cell population has risen to bloom concentrations prior to dosing (Jones, Bourne et al. 1994).

11.9.3 Oxidants and Herbicides

Oxidants are routinely used in drinking water treatment, as discussed in Chapter 12, but less often in reservoir management. Permanganate at concentrations of 1 to 5 mg/L is toxic to cyanobacteria and has been used by some drinking water utilities in the U.S. for control of blooms in reservoirs (Fitzgerald 1966; District 1987).

The herbicides diquat and Simazine have been tested for use in cyanobacterial control in reservoirs, but their use has not been adopted (Blackburn and Taylor 1976; Lam, Prepas et al. 1995). Some herbicides are highly persistent, with broad ecological impacts, including adverse effects on fish. Application to drinking water would not be desirable from the perspective of possible human health effects (WHO/IPCS 2002).

11.10 BIOLOGICAL REMEDIATION

The concept of modifying the ecology of a reservoir or lake to prevent cyanobacterial blooms without use of chemicals, the cost of destratification, or the complexity of changing land use in the catchment is very attractive. Previously in this chapter, the influence of light and nutrient supply on cyanobacterial growth has been highlighted, as have the approaches and difficulties experienced in bringing about major changes through technological means. If biological manipulation can be used to prevent cyanobacterial blooms, it may be able to become a cost-effective means of improving the quality of water.

There are two potential ways in which the biological control of cyanobacteria can be brought about. One is through competition for nutrients, thus depriving cyanobacteria of means of growth. This forms part of the "bottom-up" control discussed by Reynolds (1997). In oligotrophic lakes, the submerged macrophytes and shallow-water reed beds utilize and retain the majority of the low concentrations of nutrients available. In eutrophic lakes, the turbidity of the water through phytoplankton growth may shade out the submerged macrophytes, removing this competitive nutrient sink. Thus the competition for nutrients is reduced and the eutrophic status maintained.

If the growth of submerged macrophytes can be enhanced, then there is the possibility that a new equilibrium with clear water above the macrophytes can be established (Scheffer 1990). This is likely to be successful if there is a substantial area of shallow water in which macrophytes can be introduced or encouraged and if the phosphorus status of the lake is marginal for eutrophication (Chorus and Mur 1999). It is also more likely to be successful if combined with reduction in nutrient inputs and in the population of bottom-feeding fish.

Where the water level in the reservoir or lake is highly variable, as is the case with drinking water reservoirs, it is difficult to maintain a stable population of submerged macrophytes due to repeated drying of the shoreline. This approach is unlikely to be of value in reservoirs located in regions with marked differences in summer and winter rainfall.

11.10.1 Fish Population

Bottom-feeding fish such as carp will consume macrophytes, plankton, and filamentous algae and also generate substantial turbidity as they scavenge the sediment for food. Nutrient release from the sediment will be higher if such benthic fish are present. Planktivorous fish will consume zooplankton, which themselves consume phytoplankton. Lakes with high populations of carp are frequently eutrophic, with seasonal cyanobacterial blooms; this does not appear to disadvantage the fish, as they feed on the cyanobacteria with minimal ill effects.

Hence an approach to remediation of lakes subject to cyanobacterial blooms through modification of the fish population has been tried, with some success (Hrbacek, Desortova et al. 1978; Scheffer 1990). The method employed is to reduce the population of herbivorous and planktivorus fish by a program of netting. Predatory fish are returned to the lake, to continue the reduction of the other species. Predatory fish population may be increased by introduction of fingerlings of predatory species naturally present or enhancement of the lake environment for predators by providing cover, such as sunken trees (Kitchell 1992). As the competitive pressure among bottom-feeding fish is reduced, their reproductive success will be enhanced, so that continued netting, or the introduction of predators, is likely to be required to maintain any improvements in water quality. These biological approaches are more appropriate for recreational lakes, especially if fishing if permitted, generating an income that can be used to maintain the program. This approach is currently achieving success in South Africa and Europe and has the considerable advantage of ecological soundness.

11.10.2 Straw

In small ponds and lakes, water quality may be improved by the addition of bales of barley or similar straw to the incoming stream (Harriman, Adamson et al. 1997). This has also been applied to drinking water reservoirs (Barrett, Curnow et al. 1996); however, the data are not consistent (Cheng, Jose et al. 1995). The addition of a substantial organic load to a reservoir can be expected to increase oxygen demand considerably, augmenting deoxygenation of the hypolimnion and nutrient release from the sediments. Considerable further work, using controlled matched water bodies, will be required before any clear understanding of the benefit from straw is available.

11.10.3 Phage

Natural cyanobacterial blooms can rapidly disappear, and among the possible reasons for this is infection by cyanophages. Interest in these organisms for biological control of cyanobacterial blooms arose in the late 1960s and attracted attention in Europe and the USSR, with headlines of "Soviet Virus Claimed to Clear 'Green Water'"(Anon 1967) and "Swedes Find a Virus that Attacks Algae" (Stubbs and Wick 1969). The mechanism of phage attack on *Anabaena* was described in 1969 (Granhall and Hofsten 1969). The research on cyanobacterial phages was reviewed by Martin and Benson (1982) and by Gromov (1983) but has received little attention since that time as a possible means of biological control. Use of species-specific phage for the control of major toxic cyanobacteria may have potential for the future, where diversion of the species composition of eutrophic reservoirs from *Microcystis*-dominated into green algae–dominated water would have a health benefit and could reduce treatment costs.

REFERENCES

ACT Department of Urban Services (2001). *Canberra's Urban Lakes and Ponds*. Canberra, ACT, Australia: 83.

ACTEW (2000). *Lower Molongolo Water Quality Control Centre*. Canberra, Australian Capital Territory Electricity and Water.

Anon (1967). Soviet virus claimed to clear "green water." *New Scientist* 36(14): 677.

Barrett, P. F. R., J. C. Curnow, et al. (1996). The control of diatom and cyanobacterial blooms in reservoirs using barley straw. *Hydrobiologia* 340: 307–311.

Blackburn, R. D. and J. B. Taylor (1976). Aquazine™, a promising algicide for the use in southeastern waters. *Proceedings of the Soil and Weed Science Society* 29: 365–373.

Bormans, M., P. W. Ford, et al. (2000). Temporal changes in nutrients and cyanobacterial populations in a dammed, stratified tropical river. *Verhandlungen der Internationalen Vereinigung für Theoretische und Angewandte Limnologie* 27: 3239–3242.

Bowling, L. (1992). Th*e Cyanobacterial (Blue-Green Algal) Bloom in the Darling/Barwon River System, November–December, 1991*. Sydney, Technical Services Division, New South Wales Department of Water Resources, Australia.

Briand, J.-F., C. Leboulanger, et al. (2004). *Cylindrospermopsis raciborskii* (cyanobacteria) invasion at mid-latitudes: Selection, wide physiological tolerance, or global warming. *Journal of Phycology* 40: 231–238.

Brookes, J. D., P. D. Baker, et al. (2002). Ecology and management of cyanobacteria in rivers and reservoirs. *Blue-Green Algae; Their Significance and Management within Water Supplies*. Salisbury, South Australia, Cooperative Research Centre for Water Quality and Treatment. Occasional paper 4: 33–42.

Burch, M. D. (2002). Algicides for control of toxic cyanobacteria. *Blue-Green Algae: Their Significance and Management within Water Supplies*. Salisbury, South Australia, Cooperative Research Centre for Water Quality and Treatment. Occasional paper 4: 23–32.

Chambers, L., J. Olley, et al. (1997). Conditions affecting the availability and release of phosphorus from sediments in the Maude Weir Pool on the Murrumbidgee River, New South Wales. *Managing Algal Blooms*. J. R. Davis, ed. Canberra, CSIRO Land and Water: 41–50.

Cheng, D., S. Jose, et al. (1995). *Assessment of the Possible Algicidal and Algistatic Properties of Barley Straw in Experimental Ponds: Confirmatory Trial*. Report prepared for the State Algal Coordinating Committee. Sydney, University of Technology: 21.

Chorus, I. and L. Mur (1999). Preventative measures. *Toxic Cyanobacteria in Water: A Guide to Their Public Health Consequences, Monitoring and Management*. I. Chorus and J. Bartram, eds. London, E & FN Spon (on behalf of WHO): 236–273.

Cooke, G. D., E. B. Welch, et al. (1993). *Restoration and Management of Lakes and Reservoirs*. Boca Raton, FL, Lewis Publishers.

Corkill, N., R. Smith, et al. (1989). Poisoning at Rutland Water. *The Veterinary Record* 125: 356.

District, C. M. W. (1987). *Current Methodology for the Control of Algae in Surface Water Reservoirs*. Denver, American Water Works Association Research Foundation.

Fabbro, L. D. and L. J. Duivenvoorden (1996). Profile of a bloom of the cyanobacterium *Cylindrospermopsis raciborskii* (Woloszynska) Seenaya and Subba Raju in the Fitzroy River in tropical central Queensland. *Marine and Freshwater Research* 47(5): 685–694.

Fitzgerald, G. P. (1966). Use of potassium permanganate for control of problem algae. *Journal of the American Water Works Association* 58: 609–614.

Ganf, G. G. (1974). Phytoplankton biomass and distribution in a shallow eutrophic lake (Lake George, Uganda). *Oecologia (Berlin)* 16: 9–29.

Gemza, A. F. (1997). Water quality improvements during hypolimnetic oxygenation in two Ontario lakes. *Water Quality Research Journal of Canada* 32: 365–390.

Granhall, U. and A. V. Hofsten (1969). The ultrastructure of a cyanophage attack on *Anabaena variabilis*. *Physiologia Plantarum* 22: 713–722.

Gray, N. F. (1994). *Drinking Water Quality: Problems and Solutions*. Chichester, U.K., John Wiley & Sons.

Greenway, M. and S. Simpson (1996). Artificial wetlands for wastewater treatment, water reuse and wildlife in Queensland, Australia. *Water Science and Technology* 33: 221–229.

Griffiths, D. J., M. L. Saker, et al. (1998). Cyanobacteria in a small tropical reservoir. *Water* 25(1): 14–19.

Gromov, B. V. (1983). Cyanophages. *Annals of Microbiology (Institute Pasteur)* 134B: 43–59.

Hairsine, P. (1997). *Controlling Sediment and Nutrient Movement within Catchments.* Clayton, Victoria, Cooperative Research Centre for Catchment Hydrology: 22.

Hanson, M. J. and H. G. Stefan (1984). Side effects of 58 years of copper sulfate treatment of the Fairmont Lakes, Minnesota. *Water Resources Bulletin of the American Water Resources Association* 20(6): 889–900.

Harremoes, P. (1997). The challenge of managing water and material balances in relation to eutrophication. *Eutrophication Research, State-of-the-Art.* R. Riojackers, R. H. Aalderink and G. Blorn, eds. Wageningen, the Netherlands, Department of Water Quality Management and Aquatic Ecology, Wageningen Agricultural University: 3–12.

Harriman, R., E. A. Adamson, et al. (1997). An assessment of the effectiveness of straw as an algal inhibitor in an upland Scottish loch. *Bioconservation Science and Technology* 7: 287–296.

Hart, B. T. (1981). Trace metal complexing capacity of natural waters: A review. *Environmental Technology Letters* 2: 95–110.

Hosja, W., S. Grigo, et al. (2000). The effect of the Dawesville Channel on cyanobacterial blooms and associated microalgae in the Peel-Harvey estuarine system, Western Australia. *Ninth International Conference on Harmful Algal Blooms.* Hobart, Tasmania, United Nations Educational, Scientific and Cultural Organization.

Hrbácek, J., B. Desortová, et al. (1978). Influence of fish stock on the phosphorus-chlorophyll ratio. *Verhandlungen der Internationalen Vereinigung für Theoretische und Angwandte Limnologie* 20: 1624–1628.

Hrudey, S., M. Burch, et al. (1999). Remedial measures. *Toxic Cyanobacteria in Water: A Guide to Their Public Health Consequences, Monitoring and Management.* I. Chorus and J. Bartram, eds. London, E & FN Spon (on behalf of WHO): 275–312.

Humpage, A. R., J. Rositano, et al. (1994). Paralytic shellfish poisons from Australian cyanobacterial blooms. *Australian Journal of Marine and Freshwater Research* 45: 761–771.

Istvanovics, V. and L. Somlyody (2001). Factors influencing lake recovery from eutrophication: The case of basin 1 of Lake Balaton. *Water Research* 35(3): 729–735.

Jones, G. J. (1997). Limnological study of cyanobacterial growth in three South-East Queensland reservoirs. *Managing Algal Blooms.* J. R. Davis, ed. Canberra, CSIRO Land and Water: 51–66.

Jones, G. J., D. G. Bourne, et al. (1994). Degradation of the cyanobacterial hepatotoxin microcystin by aquatic bacteria. *Natural Toxins* 2: 228–235.

Jones, G. J. and W. Poplawski (1998). Understanding and management of cyanobacterial blooms in sub-tropical reservoirs of Queensland, Australia. *Water Science and Technology* 37(2): 161–168.

Kitchell, J. F., Ed. (1992). *Food Web Management: A Case Study of Lake Mendota.* New York, Springer-Verlag.

Klapper, H. (1992). *Eutrophierung und Gewasserschutz.* Stuttgart, Gustav Fischer Verlag.

Kohler, A. and M. Klein (1997). Cyanobakterien und die Nutzung der Berliner Gewasser. *Toxische Cyanobakterien in Deutschen Gewassern.* I. Chorus, ed. Berlin, WaBoLu Hefte: 58–66.

Lam, A. K. Y., E. E. Prepas, et al. (1995). Chemical control of hepatotoxic phytoplankton blooms: Implications for human health. *Water Research* 29: 1845–1854.

Lukatelich, R. J. and A. J. McComb (1981). *The Control of Phytoplankton Populations in the Peel-Harvey Estuarine System*. Perth, Department of Conservation and the Environment, Western Australia.

Maier, H. R., M. D. Burch, et al. (2001). Flow management strategies to control blooms of the cyanobacterium, *Anabaena circinalis*, in the River Murray at Morgan, South Australia. *Regulated Rivers: Research and Management* 17: 637–650.

Martin, E. L. and R. L. Benson (1982). Algal viruses, pathogenic bacteria and fungi: Introduction and bibliography. *Selected Papers in Phycology II*. J. R. Rosowski and B. C. Parker, eds. Lawrence, Kansas, Phycological Society of America: 793–798.

Matthews, P. (1990). *Algal Bloom at Rutland Water in 1989. Part I: The Reservoir*. Cambridge, U.K., Anglian Water Services.

May, V. (1974). Suppression of blue-green algal blooms in Braidwood Lagoon with alum. *Journal of the Australian Institute of Agricultural Science* 40: 54–57.

McKnight, D. M., S. W. Chisholm, et al. (1983). $CuSO_4$ treatment of nuisance algal blooms in drinking water reservoirs. *Environmental Management* 7: 311–320.

Morris, R. J., D. E. Williams, et al. (2000). The adsorption of microcystin-LR by natural clay particles. *Toxicon* 38(2): 303–308.

Murphy, T. P., E. E. Prepas, et al. (1990). Evaluation of calcium carbonate and calcium hydroxide treatments of prairie drinking water dugouts. *Lake and Reservoir Management* 6: 106–108.

NSWBGA Task Force (1992). Blue-green algae. *Final Report of the New South Wales Blue-Green Algae Task Force*. Parramatta, New South Wales, Department of Water Resources.

Nurnberg, G. K. (1997). Coping with water quality problems due to hypolimnetic anoxia in Central Ontario lakes. *Water Quality Research Journal of Canada* 32: 391–405.

Oliver, R. L. and G. G. Ganf (2000). Freshwater blooms. *The Ecology of Cyanobacteria. Their Diversity in Time and Space*. B. A. Whitton and M. Potts, eds. Dordrecht, Kluwer Academic Publishers: 150–194.

Padisak, J. (1997). *Cylindrospermopsis raciborskii* (Woloszynska) Seenayya et Subba Raju, an expanding, highly adaptive cyanobacterium: worldwide distribution and review of its ecology. *Archiv für Hydrobiologie* 107 (suppl): 563–593.

Padisak, J. and V. Istvanovics (1997). Differential response of blue-green algal groups to phosphorus load reduction in a large shallow lake: Balaton, Hungary. *Verhandlungen der Internationalen Vereinigung für Theoretische und Angewandte Limnologie* 26: 574–580.

Prepas, E. E. and T. P. Murphy (1988). Sediment-water interactions in farm dugouts treated with copper sulphate. *Lake and Reservoir Management* 4: 161–168.

Prepas, E. E., T. P. Murphy, et al. (1997). Lake management based on lime application and hypolimnetic oxygenation: The experience in eutrophic hardwater lakes in Alberta. *Water Quality Research Journal of Canada* 32: 273–293.

Reynolds, C. S. (1997). *Vegetation Processes in the Pelagic: A Model for Ecosystem Theory*. Ohlendorf/Luhe, Germany, Excellence in Ecology, Ecology Institute.

Robertson, A. I. (1997). Land-water linkages in floodplain river systems: the influence of domestic stock. *Frontiers in Ecology: Building the Links*. N. Klomp and I. Lunt, eds. Oxford, Elsevier Scientific: 207–218.

Sas, H. (1989). *Lake Restoration by Reduction of Nutrient Loading: Expectations, Experiences, Extrapolations*. St. Augustin, Germany, Academia Verlag Richarz.

Scheffer, M. (1990). Multiplicity of stable states in freshwater systems. *Hydrobiologia* 200/201: 475–486.

Sherman, B. S., J. Whittington, et al. (2000). The impact of destratification on water quality in Chaffey Dam. *Archiv für Hydrobiologie* 55 (Special Issues in Advanced Limnology): 15–29.

Siegrist, H. and M. Boller (1997). Effects of the phosphate ban on sewage treatment. *EAWAG News:* 42 E.

SPCC (1983). *Water Quality in the Hawkesbury-Nepaen River: A Study and Recommendations*. Sydney, New South Wales, State Pollution Control Commission.

Stubbs, P. and G. Wick (1969). Swedes find a virus that attacks algae. *New Scientist* November 6: 278.

Sunda, W. G. and P. J. Hanson (1979). Chemical speciation of copper in river water: Effect of total copper, pH, carbonate and dissolved organic matter. *Chemical Modeling in Aqueous Systems: Speciation, Sorption, Solubility and Kinetics*. E. A. Jenne, ed. Washington, D.C., American Chemical Society: 147–180.

Visser, P. M., B. W. Ibelings, et al. (1996). Artificial mixing prevents nuisance blooms of the cyanobacterium *Microcystis* in Lake Nieuw Meer, the Netherlands. *Freshwater Biology* 36: 435–450.

Vollenweider, R. and J. Kerekes (1982). *Eutrophication of Waters: Monitoring, Assessment, Control*. Paris, Organization for Economic Cooperation and Development.

Voros, L., L. Hiripi, et al. (1975). Cyanobacteria and water quality of Lake Balaton. *Hidrologiai Kozlony* 79(6): 343–344.

Webster, I. T., B. S. Sherman, et al. (2000). Management strategies for cyanobacterial blooms in an impounded lowland river. *Regulated Rivers: Research and Management* 16(5): 513–525.

WHO/IPCS (2002). *Global Assessment of the State-of-the-Science of Endocrine Disruptors*. Geneva, World Health Organization/International Program on Chemical Safety: 180.

Williams, W. D., I. Pettman, et al. (1998). Management issues associated with constructed wetlands in urban areas of Australia. *Wetlands in a Dry Land: Understanding for Management*. W. D. Williams, ed. Canberra, Environment Australia, Biodiversity Group: 177–186.

Zalewski, M., Ed. (2002). *Guidelines for Integrated Management of the Watershed: Phytotechnology and Ecohydrology*. Osaka, United Nations Environment Programme.

Zhang, Y. and E. E. Prepas (1996). Short-term effects of $Ca(OH)_2$ additions on phytoplankton biomass: A comparison between laboratory and *in situ* experiments. *Water Research* 30: 1285–1294.

12 Water Treatment

Drinking water treatment began in response to high levels of waterborne diseases — such as dysentery, typhoid, and cholera — transmitted through fecal contamination of food and water in urban populations. Cities arose close to freshwater sources, often rivers, which became progressively polluted. One of the earliest and still most effective treatments for river water is slow sand filtration. In this process, incoming raw water is passed slowly through a bed of sand, which builds up a layer of microorganisms on the surface. As the water passes down through this surface layer, pathogens are removed and organic molecules oxidized. For over 150 years, this process has been in use in treating water for the city of London. About 100 years ago, this water filtration process was reinforced by the addition of chlorine as a sterilizing agent to destroy pathogens that might escape removal by filtration and also pathogens that enter the drinking water supply after treatment.

From this relatively unsophisticated beginning, modern drinking water treatment has developed into a well-researched area of chemical engineering. As the demand for potable water has increased, treatment processes that will handle larger throughputs of water of varying quality have been required. Because of the low rate of flow through the beds of slow sand filters, many hectares of filter beds would be needed to filter sufficient water for average modern cities. Some of the filter beds will be out of action at any given time, as slow sand filters gradually clog up with incoming detritus and grow algal mats. They then have to be cleaned, and after cleaning their filtration effectiveness is much lower. Hence more compact systems were devised that can handle large flow rates in a more controllable manner. Rapid sand filters were developed that will remove flocculated particles but lack the fine filtering capability of slow sand filters and also lack the biological processes that decontaminate water. Thus their filtering effectiveness had to be enhanced by prior flocculation and sedimentation/clarification; in addition, the chemical contamination of the raw water had to be reduced by oxidants and adsorbents.

Figure 12.1 illustrates two alternative treatment sequences for drinking water, one a basic and conventional system and the other more advanced, including ozone and activated carbon filtration.

The most common and older water treatment plants follow the flow diagram in Figure 12.1 from left to right, across the top of the figure, and down the right-hand side. To keep the plant clean, especially from algal and bacterial growths on the tanks, a dose of chlorine is added at the beginning of the process. As well as killing some of the microorganisms, this preoxidation dose may assist in later flocculation of suspended material. To obtain optimal performance, the pH of the incoming water is adjusted; with rapid mixing, a coagulant is added, most frequently aluminum

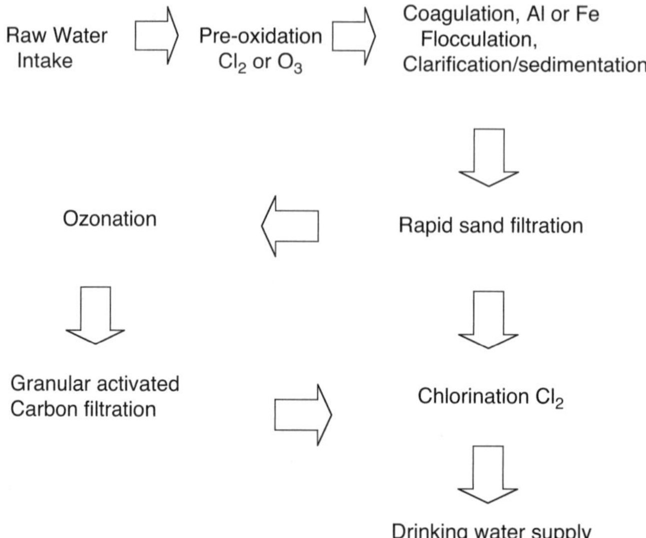

FIGURE 12.1 Simplified diagram of a drinking water treatment plant.

sulfate, but ferric and ferrous salts are also in use. A polyelectrolyte may be added to help flocculation. The water then forms soft flocs of metal hydroxide with entrapped organisms and organic debris, which provide the main element in water purification (Gray 1994).

The flocs then require removal, as they carry live pathogens, plankton, and other organic material, including cyanobacterial cells. Three methods are widely used for clarification:

1. Sedimentation by gravity in slow flowing tanks after initial stirring.
2. The "sludge blanket," in which the water and new flocs are forced upward through a layer of old flocs which acts as a filter.
3. Dissolved air flotation, in which compressed air is dissolved in water and then released from pressure at the base of the tank holding the flocs. This forms fine bubbles, which trap the flocs and carry them up and out of the tank.

Each of these processes will remove the great majority of particulate contaminants but still leave enough suspended material to need further filtration so as to provide a bright, clear drinking water. Rapid sand filtration is used for this step, often with a mix of material such as crushed anthracite (hard coal) and sand in the filter bed (Gray 1994).

Prior to supply into the distribution system, chlorine is added (in most countries) at a concentration that will leave a minimal amount of residual free chlorine in the water as it leaves the household tap. This ensures that pathogens sensitive to chlorine are killed within the pipelines, even if there are breaks in the pipes allowing pathogen entry.

This relatively simple treatment process will provide bacteriologically safe water to consumers and, if operated well, will prevent almost all pathogens from being distributed in the drinking water. The limitation of this process is that harmful organic compounds — such as pesticides, some industrial chemicals, and pharmaceuticals and heavy metals — will not be removed, as they are in solution. This limitation applies equally to cyanobacterial toxins in solution, which have been shown to pass through into tap water and cause adverse human health effects (Byth 1980; Falconer, Beresford et al. 1983).

The additional processes shown on the left side of the flow diagram in Figure 12.1 were designed to remove harmful organic compounds from drinking water and to reduce unpleasant tastes and odors. Many surface waters worldwide have detectable concentrations of pesticides, surfactants, plasticizers, and pharmaceuticals. In a recent survey in the U.S., organic wastewater contaminants were found in 80% of 139 streams in 30 states across the nation (Kolpin, Furlong et al. 2002). In many major European rivers, indirect reuse of treated wastewater (sewage) discharged into rivers is common practice, with measurable concentrations of pesticides and pharmaceuticals in the water. In the U.K., 324 organic compounds were detected in drinking water samples, many of them toxic (Fielding, Gibson et al. 1981).

Most of these potentially harmful compounds found in surface water are included in the World Health Organization's *Guidelines for Drinking Water Quality* (WHO 1996) and Guideline Values for their safe concentration in drinking water have been determined. When these values are adopted in drinking water regulations by state and national legislatures, the compounds specified should be monitored by drinking water suppliers. If the treated water contains compounds that frequently exceed the concentrations specified in the Guideline Values, especially if they exceed the guidelines by a factor of 10 or more, additional water treatment may be legally required to meet the concentrations of organic compounds that are specified.

Persistent organic chemicals, including some pharmaceuticals and hormones, will pass through simple flocculation/filtration plants; as a result, specific removal techniques must be implemented. The most widely used are activated carbon adsorption and ozone treatment, often combined. Ozone is a very powerful oxidant and will degrade most organic compounds, hence it is a broadly effective method for removing anthropogenic chemicals from drinking water. Newer drinking water treatment plants in developed countries, which rely on river water for supply, increasingly use ozone as an oxidant for unwanted organic compounds and also as a preoxidant to enhance flocculation.

As there is a possibility of ozone producing toxic degradation products from oxidation of organic material, oxidation is followed by filtration through granular activated carbon to ensure that these oxidized organic compounds do not remain in the water. An alternative approach for the removal of organic compounds from water (discussed later), is incorporating powdered activated carbon into the filtration process. The powdered activated carbon adsorbs organic compounds, which are then removed together with the carbon during filtration.

An example of a water treatment plant using the ozone/granular activated carbon sequence is the Feltham Works of Thames Water in the U.K., which supplies water for London. Water from the Thames River provides the bulk supply, which has passed through industrial areas and many wastewater treatment plants prior to reaching Feltham. It is interesting that this plant, which was originally commissioned during thc 1800s, still has slow sand filters, which are used at the end of the treatment sequence prior to final chlorination and supply.

12.1 PROCESSES FOR REMOVING CYANOBACTERIAL TOXINS FROM DRINKING WATER SUPPLIES

Since more and more of the surface freshwaters of the world are becoming eutrophic because of raised nutrient inputs, the abundance of toxic cyanobacterial blooms is increasing. Reservoirs with relatively clean immediate surroundings are becoming affected by more distant agricultural intensification or by urban encroachment within the catchment. As groundwater resources as depleted, communities that previously relied on groundwater are forced to change to surface water, especially in locations such as Florida, where the population is steadily increasing.

The resulting deterioration in the quality of source water for the drinking water supply through eutrophication, and use of surface water, necessitates improvement of the water treatment facilities. These are driven in the first instance by increased off-flavor and odors generated by cyanobacterial blooms. As these decreases in water quality are quickly noticed by consumers, water supply utilities have responded by improving treatment. Only in the last 20 years, through increased awareness of adverse human health effects and of livestock poisoning at water supply sources, has the existence of cyanobacterial toxins in water supplies been recognized.

In the U.K. in 1989, a well-publicized example of a eutrophic, poisonous drinking water reservoir was Rutland Water, a major supply reservoir of 1260 hectares for the East Anglian area of 1.5 million people. Here the deaths of 20 sheep and 15 dogs that had drunk or played in the water were reported in the national press. A massive water bloom of *Microcystis aeruginosa* had formed, shown to be toxic by mouse testing and containing microcystin-LR by high-performance liquid chromatography detection. The reservoir is a pumped storage site, water being taken from two rivers flowing through agricultural areas and supplemented by treated effluent from the local sewage treatment works. The two water treatment plants supplying drinking water sourced from this reservoir had previously been fitted with granular activated carbon filtration to minimize unpleasant taste and odor and remove harmful organic contaminants.

12.1.1 Control of Abstraction

One of the most applicable methods of reducing the intake of cyanobacterial cells into a drinking water supply is to regulate the depth at which the water is taken. As discussed in Chapter 4, cyanobacterial blooms are rarely at uniform cell density down the depth profile of a reservoir. Overnight, under calm conditions, *Microcystis* blooms tend to float to the surface, resulting in very high local concentrations in the top few meters of water. Intakes well below the surface will lessen the possibility of drawing these cell concentrates into the treatment plant.

Location of the intake is also a major factor determining the immediate cyanobacterial concentration, as intakes downwind of the prevailing air movement across the lake surface, and especially in sheltered bays, will accumulate cyanobacterial scums. Figure 9.1 illustrates an intake tower in such a location, which has repeatedly resulted in problems of *Microcystis* entering the treatment plant.

Other cyanobacterial species form bands of peak cell concentration deeper in the water. In stable, stratified lakes, *Cylindrospermopsis raciborskii* and *Planktothrix rubescens* will grow to maximum concentrations deep in the metalimnion, taking advantage of higher nutrient concentrations and their ability to photosynthesize at low light levels. When these species are known to be present, depth profiles of cyanobacterial concentration are required to determine the best level for water abstraction. Diurnal variations occur in the depth at which peak cell concentrations accumulate during cyanobacterial blooms; this occurs because the cells become heavier during the day, as a result of the accumulation of photosynthetic products, and then sink. As a consequence, a single determination of the depth profile at a particular time of day will not necessarily provide useful information for abstraction depth for the whole day and could be misleading. A series of depth profiles for cyanobacterial cell concentration spaced over 24 h are needed for optimal water abstraction. These profiles are then used to adjust the depth of abstraction so as to minimize the intake of cyanobacteria.

Floating barriers have been tested to prevent cyanobacterial scums from accumulating around water intakes. They may have a place in small systems with a single depth of intake, mainly to reduce the quantity of toxin released in the vicinity of the intake when lysis of a bloom is occurring in an adjacent scum. They are unlikely to be of assistance in a large reservoir, especially if wind action is redistributing the scum.

12.1.2 Bank Filtration

A widely used method of obtaining good-quality water for a drinking water treatment facility is to draw water from shallow wells along the banks of a reservoir or river or from beneath a riverbed. In these cases the geology of the river or lake bed and adjacent floodplain is crucial for success. If these wells are located in fine alluvial gravel deposits, there is likely to be a good horizontal flow of water and effective filtration of the river or lake water. Depending on the distance between the water's edge and the wells, the transit time of the water will vary, so that more or less filtration is available by adjusting the well location.

A study of the removal of taste and odor from cyanobacterial blooms by filtration of water through the river bank provided promising results; it was therefore expected that cyanobacterial toxins would also decrease (Chorus, Klein et al. 1993). Laboratory-based experiments have demonstrated that both adsorption and degradation of microcystins can occur in lake sediments and soils. Lake water to which both microcystin and *Microcystis* cultures were added showed over 90% removal of free toxin when passed down sediment and soil columns over a week (Lahti, Kilponen et al. 1996). More recently, nodularin adsorption onto five different soils was examined. The soils with the highest clay or organic carbon content showed the highest adsorption, which increased as the ionic strength of the solution increased and pH decreased (Miller, Critchley et al. 2001). Studies of the degradation of microcystin and nodularin in soils showed that within 10 to 16 days at room temperature and in the dark, complete toxin removal occurred on two of three soils. The 98.5% sand soil showed no degradation (Miller and Fallowfield 2001). For effective long-term removal of microcystins by bank filtration, degradation as well as adsorption must take place.

A field study of removal of microcystin by riverbank filtration showed that careful selection of the well sites could result in adequate removal of microcystins. In this example, a complicating factor was the salinity of the groundwater in some locations, which restricted well location (Dillon, Miller et al. 2002).

A potential problem with riverbed abstraction and bank filtration is due to channeling, whereby water moves directly from the lake or river into the wells without any filtration or delay. Two past outbreaks of human gastroenteritis may be partially attributed to the channeling of river water containing cyanobacteria into the drinking water supply, along with other deficiencies in the supply system. Impervious substrata, such as clay, prevent the use of bank filtration, as does a highly saline groundwater, which is unsuitable for drinking.

12.2 WATER FILTRATION, COAGULATION, AND CLARIFICATION

On the basis that the majority of microcystins and nodularins and a substantial proportion of cylindrospermopsins are contained within the live cyanobacterial cells, the first priority in the removal of cyanobacterial toxins is the removal of live cells from the drinking water stream (Drikas, Chow et al. 2001). This is more easily achieved in water treatment plants located adjacent to the supply reservoir than in plants at some distance. Cyanobacteria may lyse during transit through pipelines, particularly if there is a substantial drop in height between the reservoir outlet and the treatment works. Under these circumstances, pressure-reduction valves, which have the ability to burst cyanobacterial cells, are fitted to the pipelines. The extent of lysis will depend on the degree of pressure reduction and also on the genus of cyanobacteria. *Microcystis* is relatively robust, whereas *Anabaena* is easily disrupted. *Planktothrix* and *Cylindrospermopsis* are intermediate. Even *Microcystis* will lyse if the pipeline is long and the climate hot (Dickens and Graham 1995).

On entry into the treatment plant, pH adjustment and mixing do not appear to injure cyanobacterial cells within the operational range (WRc 1996). However, preoxidation by chlorine dosing of the incoming water will lyse the cyanobacterial cells, liberating the toxins into solution in the water. In an experimental evaluation of the effect of prechlorination, a 64% release of intracellular microcystin on chlorine dosing was found (Lam, Prepas et al. 1995). An additional disadvantage of prechlorination is the formation of chlorinated organic molecules early in the treatment sequence, when a maximum amount of organic material is present. These chlorinated molecules are normally described as disinfection by-products. They include a range of chloro- and bromoorganic molecules, many of which are harmful (Gray 1994).

Prechlorination of incoming water will cause effectively all of the applied chlorine to react with organic molecules in the water, whether in solution or as particles, giving chloroorganic products. Thus prechlorination is to be avoided when cyanobacteria are present in the raw water, as the consequence will be the release of toxins into the water solution and the production of chlorinated by-products, both with potentially adverse effects on the health of consumers.

Ozone as a preoxidant appears to be less active in cell lysis; at low doses of 1 mg/L, little cyanobacterial lysis was reported (Mouchet and Bonnelye 1998). As discussed later, ozone rapidly destroys microcystins in solution, and hence preozonation may reduce soluble microcystins while not liberating cell-bound microcystins. Ozone also assists in the coagulation and removal of cells. On this basis the use of ozone as an initial step in the treatment of water containing live toxic cyanobacteria and soluble cyanobacterial toxins appears to have significant benefits.

Potassium permanganate pretreatment appears to reduce total microcystin in raw water by about 50%, which is likely to include some cell-bound toxin as well as dissolved microcystin (Karner 2001).

If this preoxidation step with chlorine, ozone, or permanganate is omitted, the majority of the live cyanobacterial cells in the raw water will enter the coagulation and clarification process intact. The proportion of cyanobacterial toxin contained in the cells at this step will depend on the species and the growth phase of the bloom. Senescent blooms leak toxin into the water solution, and natural lysis will allow the majority of toxin to enter the water (NRA 1990). In the case of toxic *Anabaena* and *Cylindrospermopsis*, a larger proportion of the toxin will be in the free water phase. Hence it cannot be assumed that cell removal at the flocculation/clarification step of drinking water purification will provide a total answer to extraction of cyanobacterial toxins from raw water; however, in most circumstances, it will have a markedly beneficial effect.

Coagulation and clarification/filtration remove the particulate content of raw water with efficiency, including protozoa, bacteria, cyanobacteria, viruses, and general debris. Alum coagulation removes about 90% of fecal indicator bacteria and up to 99% of viruses present in raw water (Gray 1994). Applied to water containing cyanobacterial cells, a progressive reduction in intracellular microcystin concentration was seen with increased alum dosage, reaching a plateau at which about 90% of the original content had been removed, while the free soluble microcystin concentration remained unchanged (Hart, Fawell et al. 1997). The alum dosage to be applied for optimal removal of cyanobacterial cells is determined by the alkalinity of the water and the cyanobacterial cell concentration (Mouchet and Bonnelye 1998). Alum and ferric salt flocculation do not lyse *Microcystis* cells, which are effectively removed intact, or cause toxin leakage (Drikas, Newcombe et al. 2002).

The soluble toxins present in raw water are essentially unaffected by flocculation/coagulation processes with aluminum or ferric salts. Alum, polyaluminum chloride, and ferric sulfate were used as flocculants in the presence of purified microcystins. There was negligible toxin removal (Rositano and Nicholson 1994).

Measurement of total toxin removal within a small water treatment plant across the alum flocculation/sedimentation stage showed removal of 0 to 39% of microcystins present in the natural raw water. This plant received water with a high total organic carbon load of 25 to 30 mg/L and microcystin concentrations from 0.27 to 2.9 μg/L during the sampling period. Measurement of microcystin removal between the stage of raw water intake and that following sand/anthracite filtration ranged from 48 to 60%, indicating that some flocculated material containing intact cyanobacterial cells was not sedimented during the first part of the clarification process but was removed by the dual-medium filters (Lambert, Holmes et al. 1996). In general, the conventional drinking water treatment plant, as illustrated in the top and right side of the flow diagram in Figure 12.1, will remove intact cyanobacterial cells at moderate concentrations with reasonable effectiveness, but it will not remove toxin in solution (Hoffmann 1976; Keijola, Himberg et al. 1988; Himberg, Keijola et al. 1989; Tarczynska, Romanowska-Duda et al. 2000; Lahti, Rapala et al. 2001). If the load of cyanobacterial cells entering a commercial water treatment plant becomes very large, for example 10^5 to 10^6 cells per milliliter, flocculation/clarification processes become less effective and whole cells can appear in the final treated water (Falconer, unpublished data).

The process used for the removal of flocculated material containing cyanobacterial cells may also affect the outcome of this stage of water treatment. Three alternative systems are in widespread use. Sedimentation tanks followed by rapid sand or dual-medium filters are the most common. When these are employed in clarification of water containing a cyanobacterial bloom, the fluctuating operational demands for optimization of settling and the effectiveness of the filters pose a considerable problem due to the high variability of the organic load entering the plant. As the cyanobacterial cell concentration in the raw water can vary perhaps 100-fold during one daily cycle as vertical movement of cells occurs, the settling rate of floc and frequency of filter backwashing change. Filter clogging is often observed under these conditions, particularly with filamentous cyanobacteria.

Dissolved air flotation followed by rapid dual-medium filtration can be highly effective in removal of *Microcystis* and *Anabaena* from raw water and has been applied in both high- and moderate-capacity water treatment plants that regularly have to deal with fluctuating cyanobacterial bloom conditions. In one such treatment plant, the incoming raw water is drawn from a river in which overall flow changes markedly during the year, with almost no net flow in summer in dry years (Falconer 1994). As this river is also tidal, in summer blooms of toxic cyanobacteria move up and down with the tides, generating large changes in the organic load at the drinking water offtake. Dissolved air flotation has proved a satisfactory technique to accommodate these demands.

Sludge blanket filtration with upward flow has been suggested as the most effective system of water clarification and is capable of use at a range of rates of water flow (Gray 1994). It is adaptable for use in small treatment plants with high demands from fluctuating cyanobacterial concentrations in raw water. In one plant in Australia drawing water from a river with frequent cyanobacterial blooms, in a semi-desert region (Hay Plains, New South Wales), this method has been used for clarification. The plant provides a town population with high-quality drinking water via a dual-reticulation system. Bulk water supply of variable quality is supplied as chlorinated river water, whereas the drinking water supply is treated by successive flocculation, sludge blanket clarification, rapid filtration, and chlorination, with powdered activated carbon available when toxic cyanobacteria are present in the raw water.

In all of the clarification process, a sludge of flocculent and contained particles is produced, which is collected in tanks; water extracted from this material is added back to the raw water intake. In the case of trapped cyanobacterial cells, the death of the cells in the sludge will release toxin back into solution. This has recently been investigated and has shown an almost quantitative release of microcystin from cells in sludge within 2 days of separation (Drikas, Newcombe et al. 2002). The soluble toxin concentration in sludge remained high for a further 4 days, when microbial degradation reduced the concentration over a subsequent 10 days.

If toxic cells are present, the return of water from the sludge to the supply after sludge separation can be expected to return soluble microcystin back into the drinking water. As discussed above, soluble microcystin passes through conventional treatment, so that the benefits of toxic cell removal in flocculation/clarification will be nullified if extracted sludge water containing toxins is added back to the system.

There is no direct evidence at present of the effect of coagulation and clarification on removal of cylindrospermopsin from water during treatment. Cylindrospermopsin appears to leak from growing *Cylindrospermopsis* cells, so that there is an appreciable content in solution in water during a bloom of the organism. The cells also appear easier to disrupt than are *Microcystis* cells, so that it can be expected that the majority of toxin will be in solution during the clarification stage of water treatment. In the Palm Island poisoning episode, discussed in Chapter 5, the drinking water causing the poisoning had been treated in a conventional water treatment plant. Recent monitoring of treated drinking water in Florida also showed substantial toxin loads in some cases (Burns, Chapman et al. 2000).

12.3 ACTIVATED CARBON

To reduce the potential risks to health from drinking water containing cyanobacterial toxins, an additional treatment step has been investigated in which activated carbon is included in the process. This material has been used extensively to adsorb organic pollutants, such as industrial chemicals and pesticides, in water treatment, as discussed earlier.

The earliest research into the use of activated carbon for the removal of cyanobacterial toxins from water supplies was done in South Africa, as a result of substantial water contamination by toxic *Microcystis* in a major supply reservoir near Pretoria, the national capital. This problem prompted the South African Council for Scientific and Industrial Research to initiate an investigation of water treatment. The normal sequence of prechlorination, flocculation, sedimentation, and sand filtration had no observed effect on toxicity in laboratory-scale jar tests. However, the addition of granular activated carbon filtration or powdered activated carbon resulted in the removal of toxins (Hoffmann 1976).

A comparable requirement in Australia to improve the quality of tap water drawn from a reservoir annually carrying massive *Microcystis* blooms resulted in the construction of a pilot plant to test activated carbon at the existing water treatment works at Armidale, New South Wales. Toxic *Microcystis* was collected from natural blooms in 1980, frozen and thawed to cause lysis, and centrifuged for removal of cell debris. The clear blue supernatant was diluted for use. The pilot plant was connected to raw and flocculation tank water supplies and included a flash mixer with pH adjustment, alum and polyelectrolyte addition, a 4000-L holding tank, a 6-m-high rapid sand/activated carbon filter with a cross-sectional area of 0.3 m^2, and necessary pumps, valves, and metering equipment. Water toxicity was assessed in mice after concentrating 20-L samples by boiling at pH 7. In all, 14 different samples of powdered activated carbon and 11 samples of granulated activated carbon were assessed, both in the laboratory and in the pilot plant. In the pilot plant, granular activated carbon was shown to be effective in toxin removal at a bed depth of only 7 to 8 cm, an empty-bed contact time of 0.9 min, and a flow rate of 5.4 m/h. However, marked differences in detoxification capability were demonstrated between the carbon samples tested over a greater than twofold range in adsorption capacity. In all cases the carbon became progressively saturated and less adsorbent as the volume filtered increased (Falconer, Runnegar et al. 1983, 1989). The progressive saturation of granular activated carbon has been observed in other pilot-plant studies, probably due to the combination of adsorption of general organic material on the surface of the carbon and the occupancy of adsorption sites for the toxins themselves (Jones, Minato et al. 1993).

Other investigations at pilot-plant level with granular activated carbon have shown high performance in toxin removal, with greater than 90% removal of microcystins from an initial concentration of 30 to 50 μg/L. In this study, 7,000 to 10,000 volumes of water were treated via the activated carbon bed before its adsorption efficiency dropped to less than 63% (Bernezeau 1994). In the small full-scale treatment plant monitored by Lambert, Holmes et al. (1996), granular activated carbon achieved 43 and 60% reductions in microcystin-LR, in one step, at water concentrations of less than 1 μg/L.

Under large-scale commercial water treatment conditions, a granular activated carbon filtration system may have a bed depth of 1 to 3 m with a contact time of 10 to 15 min and a flow rate of 12 m/h, which would provide for an extended working life of the filter. It is likely that water reaching the filter with a high content of organic matter would reduce the working life of a granular activated carbon filter; therefore the effectiveness of the prior coagulation/clarification system will substantially determine the life of the filter.

12.3.1 Biological Activated Carbon

Under extended use, granular activated carbon develops a biofilm on the surface, which was clearly shown by Lambert and coworkers (1996) by electron microscopy. In this study, activated carbon that had had 5 months of continuous use was compared with unused carbon from the same supplier. A reduction in adsorption capacity in the used carbon was observed, which could be eliminated by crushing, exposing fresh adsorption sites. Other results indicated that, in addition to adsorption, toxin degradation in the filter may play a part in the removal of microcystins.

In the pilot-plant study by Falconer and colleagues (1989), a sample of granular carbon obtained from a commercial water treatment plant where it had been in use for an extended period showed almost equal capacity for toxin removal to the same material that had not been used (Falconer, Runnegar et al. 1989). In a more definitive study, Carlile (1994) compared a new sample of activated carbon with the same material that had been used in a previous unrelated trial for removal of microcystin-LR. Empty-bed contact times of 7.5 and 15 min were tested. With a microcystin concentration of about 3 μg/L, excellent removal efficiency was reported at both contact times and with both samples of carbon (Carlile 1994). Modeling of this system, as discussed by Hrudey, Burch et al. (1999), indicated that the microcystin removal was substantially better than could be expected from adsorption alone, which was interpreted as reflecting biological degradation in addition to adsorption.

A further study of the removal of microcystins with granular activated carbon over a period of 85 days at lab scale showed clearly the biological degradation of microcystins, which were totally removed prior to sterilization of the filter. After sterilization, 25% of microcystin-LR and 100% of microcystin-LA appeared in the filtrate (Drikas, Newcombe et al. 2002).

Water treatment plants constructed to use granular activated carbon can have large fixed filters operating in parallel, allowing the filter medium to be replaced when necessary. This is the normal process when pesticides, industrial organics, and other persistent pollutants are in the raw water. An alternative emergency arrangement is to modify the rapid sand filter tanks to hold an upper bed of granular activated carbon. As a temporary measure when an unexpected toxic cyanobacterial bloom occurs or there is a toxic spill in the catchment, this has some limited use. The operational difficulty is that the normal filter backwashing practices will displace the carbon into the backwash water, at considerable cost to the treatment plant. More gentle backwashing is required to retain the carbon in the filters.

Irrespective of the technique or mechanism by which microcystins are removed during filtration through granular activated carbon, the outstanding feature of all the comparative experiments is the large differences in adsorption capability between the carbons of different types and manufacturers. Experimentation has shown that chemically activated wood-based carbons are more effective in microcystin adsorption (Drikas, Newcombe et al. 2002). It is recommended that measurement of toxin adsorption capacity be carried out prior to purchase of large quantities of granular activated carbon.

12.3.2 Powdered Activated Carbon

Water treatment plants that are subject to intermittent contamination by cyanobacterial blooms — for example, during only one or two summer months each year — can employ powdered activated carbon for toxin removal on a temporary basis. This technique is widely adopted when taste and odor occurs in the treated supply, which may be at cyanobacterial cell concentrations too low to be a health hazard. Activated carbon can be incorporated at several points of a standard water treatment system. The effectiveness of the activated carbon is affected by competition between organic material in the water and toxins for adsorption sites and also by the presence of ionic compounds and chlorine and its products.

In the two water treatment plants examined by Lambert, Boland et al. (1994), one had no prechlorination and granular activated carbon filtration was located at the postclarification step, by which stage almost all the organic material would have been removed from the water stream. This can be expected to assist in extending the life and effectiveness of the carbon filter. By contrast, the other, larger plant introduced powdered activated carbon immediately prior to the beginning of treatment, which was followed by alum, polymer, lime, and sulfuric acid prior to clarification and final anthracite/sand filtration. Under these conditions the adsorption capacity of the carbon would have been substantially reduced through attachment of organic material as well as charge alteration during the large pH shifts. In this larger plant, powdered activated carbon at 30 mg/L was added to the water stream, which resulted in an overall removal of about 50% of microcystin (Lambert et al. 1994; Lambert, Holmes et al. 1996). This level of microcystin removal by activated carbon has also been recorded in Finnish water treatment (Lepisto, Lahti et al. 1994). Other plants incorporate powdered activated carbon after the addition of flocculent (Karner 2001).

Because of these different approaches and the varying quality of the water to be treated, use of powdered activated carbon has been disappointing in many cases. The dosage rates of 20 to 30 mg of carbon per liter, used in some plants, may be appreciably below the level needed for effective microcystin removal. An interesting modeling study was undertaken by Drikas et al. (2002), in which water containing 2 to 10 μg/L of different microcystin variants was exposed to powdered activated carbon for 60 min and the microcystin concentration measured. The resulting data were then modeled to calculate the carbon concentration required to reduce microcystin to 1 μg/L, the WHO Guideline Value. For microcystin-LR at a concentration of 10 μg/L in the input water, 38 mg/L of activated carbon was required; whereas for microcystin-LA, over 100 mg/L of carbon was the calculated requirement (Drikas, Newcombe et al. 2002).

Microcystin concentrations of above 10 μg/L have been observed in drinking water reservoirs, though concentrations of 1 to 5 μg/L of microcystins are more common. In natural German lakes, a median concentration of 21 μg/L of microcystins was measured when water blooms of *Planktothrix agardhii* predominated, compared with a median of 3.1 μg/L when *Microcystis* was the dominant cyanobacterium (Fastner, Wirsing et al. 2001).

Cylindrospermopsin removal by activated carbon awaits thorough investigation. The accumulation of cylindrospermopsin on the activated carbon filters at the dialysis treatment clinic at Caruaru, Brazil, at which the cyanobacterial toxin poisoning occurred, indicates that adsorption on carbon may be an effective mechanism for removal (Carmichael, Azevedo et al. 2001).

12.4 OZONATION AND CHLORINATION

Ozone is an exceptionally powerful oxidizing agent and is widely used in water treatment in developed countries to remove potentially harmful organic pollutants and for disinfection. It has therefore been extensively examined as a technique for the removal of a range of cyanobacterial toxins.

Studies of the effect of ozone on removal of microcystins and nodularin under laboratory conditions showed complete removal of toxin, which could essentially be titrated out with increasing ozone concentrations (Rositano, Nicholson et al. 1998; Shawwa and Smith 2001; Drikas, Newcombe et al. 2002). As ozone is an unselective oxidizing agent, any organic material present will be attacked, so that oxidation of toxins occurs in competition with overall organic oxidation. Hence the concentration of ozone that must be applied to provide confident removal of microcystin has to be above the ozone demand of the water for oxidation of organic material.

This cause of complexity in the successful use of ozone has been extensively reviewed by Hitzfeld et al. (2000). A summary of microcystin removal by ozone shows that only partial destruction of microcystins has resulted when organic matter was present in the water (Hitzfeld, Hoger et al. 2000; Drikas, Newcombe et al. 2002). Drikas et al. (2002) recommend that an ozone residual should be maintained after oxidation for over 1 min to ensure that all microcystins have been oxidized.

When ozone is used as a preoxidant in water treatment plants at doses of 0.5 mg/L, for example, partial oxidation of organic matter occurs at this step. If cyanobacterial cells are present, continuing a low preoxidant dose is preferable to increasing the dose so as to avoid cell lysis, which will increase the soluble organic load passing through the coagulation/clarification step. This, in turn, increases the ozone demand at the postclarification step, when under normal conditions the organic content of the water stream has been greatly reduced. A commonly used postclarification dose of 1.0 mg/L of ozone, however, may be insufficient to maintain an ozone residual under cyanobacterial bloom conditions due to the increased soluble organic load associated with the biomass. Ozonation results in by-products that themselves may have adverse health effects and is followed in some large treatment plants by granular activated carbon adsorption filters. The products from ozonation of microcystins are discussed by Hitzfeld, Hoger et al. (2000), and it can be expected that the ADDA side chain of the microcystin molecule will be cleaved at C-6 to C-7. This approach has been used in the analysis of ADDA (Harada 1996). Further research is required to clarify the safety or otherwise of ozonation products of microcystins or nodularins. In the absence of this information, after oxidation with ozone, filtration through granular activated carbon is advisable.

Comparison of ozone, potassium permanganate, and powdered activated carbon added at the preflocculation step in a pilot plant indicated that the lowest dissolved microcystin concentration in the final water was achieved by activated carbon. Both the permanganate and the ozone appeared to cause leakage of microcystin from the cells at the flocculation/filtration stage (Schmidt, Willmitzer et al. 2002).

Ozone has also been tested as a treatment for water supplies containing cylindrospermopsin. Preliminary results are available and appear promising, but no published results are available at present from laboratory, pilot, or full-scale treatment. Since cylindrospermopsin has several locations in the molecule at which ozone could act, it is very likely that ozone will be a most effective means of eliminating cylindrospermopsin from drinking water. The effects of overall organic load on ozone demand will be equally relevant when cylindrospermopsin destruction is necessary.

12.4.1 Chlorine

Chlorination alone has been shown in jar tests to remove microcystins provided that the pH is 8.0 or below and there is a chlorine residual of at least 0.5 mg/L for a period of 30 min (Nicholson, Rositano et al. 1994; Drikas, Newcombe et al. 2002). However, in several operating treatment plants in which chlorine was used as a postclarification disinfectant, microcystins were not effectively removed, probably because the total chlorine demand was not being met under conditions of raised organic load (Keijola, Himberg et al. 1988; Himberg, Keijola et al. 1989). In both of the cases of human injury from cyanotoxins in the drinking water supply (described in Chapter 5), the treated drinking water had been subjected to postclarification chlorination.

Chloramination has not been shown to be effective in cyanobacterial toxin removal, probably because it is not as powerful an oxidant as chlorine (Nicholson, Rositano et al. 1994).

Chlorine has been shown to destroy cylindrospermopsin under conditions that can be met in water treatment plants at pH 7.5 with a residual chlorine of 0.5 to 0.7 mg/L. Trihalomethanes were produced at concentrations of 23 μg/L when an extract of *C. raciborskii* was chlorinated, with negligible cylindrospermopsin remaining. This concentration of trihalomethanes is well below the WHO guideline of 200 μg/L.

The chlorinated cell extract was supplied to male mice in their drinking water, and 40% showed fatty vacuolation of the liver. Further investigation of the biological effects of chlorinated water containing cyanobacterial extracts was recommended (Senogles-Derham, Seawright et al. 2003).

12.5 TITANIUM DIOXIDE PHOTOCATALYSIS

Oxidative degradation of organic pollutants by semiconductor photocatalysis has shown that this is an effective method of wastewater treatment. The pollutants are substantially degraded to carbon dioxide, inorganic salts, and water (Hoffmann, Martin et al. 1995). Application to the degradation of microcystins in water treatment appears highly promising.

Robertson, Lawton, and coworkers used a suspension of TiO_2 irradiated with a xenon UV lamp (330- to 450-nm wavelength) to demonstrate the destruction of microcystin-LR at concentrations from 15 to 80 μM within 10 to 40 min (Robertson, Lawton et al. 1997a). Toxicity studies on the decomposition products using *Artemia* assay (brine shrimps) indicated that the products had minimal toxicity. Flow reactors in which the TiO_2 was coated onto glass demonstrated that the process had potential for water treatment at plant scale (Robertson, Lawton et al. 1997b).

The process required an oxygen atmosphere; no degradation occurred under nitrogen. The initial reaction products were hydroxylated derivatives of microcystin (Robertson, Lawton et al. 1998). The influence of other organic contaminants on microcystin degradation in this system largely awaits clarification, though the abundant cyanobacterial pigment phycocyanin was shown to reduce the effectiveness of the TiO_2, while causing oxidative degradation itself (Robertson, Lawton et al. 1999).

Addition of low concentrations of hydrogen peroxide greatly enhanced the photocatalytic oxidation, though determination of the extent of conversion to inorganic products showed only 18% could be accounted for as CO_2 (Cornish, Lawton et al. 2000). From the low conversion of microcystin to CO_2, it is apparent that the majority of degradation products are organic derivatives. These have been extensively investigated, and identified. The oxidation process occurs at the conjugated diene bonds of the ADDA component of the molecule, with the insertion of two hydroxyl groups, either at 4 and 5 or 6 and 7 carbon atoms. The next step is the cleavage of the ADDA component, with further oxidation leaving a carboxylic acid connected to the intact peptide ring. As the ADDA diene is essential for the activity of microcystins, these oxidations result in molecules that are nontoxic; hence this degradation is effective for water treatment (Liu, Lawton et al. 2003).

The microcystin variants have a range of levels of hydrophobicity, which affect the surface adsorption onto titanium dioxide and subsequent oxidation. They also have a range of isoelectric points, which affect surface properties (Lawton, Robertson et al. 2003). The optimization of water treatment by titanium dioxide, UV light, and hydrogen peroxide will require extensive technical development. This approach may, however, provide an economical and effective mechanism for the treatment of drinking water from sources that require removal of cyanobacterial toxins and other organic pollutants.

This process has not yet been tested on cylindrospermopsin; however, as the oxidation potential of photocatalysis is appreciably greater than that of chlorine or ozone, there is little doubt that it will be effective provided that the surface properties of the titanium dioxide are suitable for reversible adsorption of cylindrospermopsin.

12.6 SLOW SAND FILTRATION

This old-established method of water purification relies on the formation of a layer of microorganisms forming on the surface of the sand, which acts both as a particulate filter and as an active metabolic degradation system for organic molecules (Ellis 1985). The water flow through the filter is slow, allowing uptake and metabolism of xenobiotics and natural toxins by bacteria and fungi. In a full-scale test filter, flow rates of 0.8 and 0.2 m/day were used (Grutzmacher, Bottcher et al. 2002). Studies of the degradation of microcystin by bacteria from lake sediments have demonstrated the ability of natural systems to remove toxin, with evidence of adaptation to the presence of microcystin (Bourne, Jones et al. 1996). Similarly, soil and sediment samples have shown capacity to degrade microcystins (Lahti, Kilponen et al. 1996; Miller, Critchley et al. 2001; Miller and Fallowfield 2001).

Experimental verification of the effectiveness of slow sand filtration as a method for microcystin removal from drinking water was carried out by Grutzmacher et al. (2002). In this study, a full-scale sand filter system was used, employing a holding reservoir supplying two slow sand filter beds, each with 0.5 m of water above 0.8 m of sand. The filters had earlier been exposed to water containing microcystins, so that the need to precondition the microflora to metabolize microcystin was reduced. In one experiment, dissolved microcystin at a concentration of about 8 μg/L was passed through the filter at 0.8 m/day with an average contact time of 4.5 h on the filter. Ambient temperature was not reported, but the experiment was conducted in summer. More than 94% removal of toxin was recorded.

In another experiment, live filaments of *P. agardhii* were supplied to the filter under conditions in which cyanobacterial growth was occurring. The filter was operated at a flow rate of 0.2 m/day with a contact time of 18 h. The filter was run for 25 days. During the first half of this period, the ambient daytime temperatures were 15 to 20°C, and about 90% of the microcystin was contained within the cells. Toxin removal was better than 85%. This was followed by a rapid drop in daytime temperature to less than 10°C, with a night temperature of 0°C causing cell lysis and a reduction of intracellular toxin to only 16% of the total. Removal of toxin across the filter dropped to below 60%, presumably due to the combination of increased dissolved toxin and reduced metabolic capability of the microflora at low ambient temperatures (Grutzmacher, Bottcher et al. 2002).

The practical implication of these data is that slow sand filtration operating at warm ambient temperatures can be expected to remove microcystins effectively, so that the technique has application in warm climates where low operating costs and low maintenance are major considerations for drinking water treatment. The large land area needed for these filters is a disadvantage in locations with high land values. Due to evaporative water loss, they are inappropriate for arid areas with high daytime temperatures.

There is no information about the effectiveness of cylindrospermopsin removal in slow sand filters; however, the natural breakdown of this toxin by microorganisms is likely to be faster than that of microcystin, as the former is a more reactive molecule. Laboratory-scale investigation should be able to clarify this easily without the need for pilot or full-scale trials.

12.7 MEMBRANE FILTRATION

Membrane filtration technology is in a rapid stage of commercialization for water treatment. Full-scale production plants using hollow-fiber technology are in current operation. Domestic membrane filters for water purification are widely on sale. However, the experimental validation of membrane filtration for the removal of cyanobacterial cells or dissolved toxins is in an early stage. Small-scale experimental evaluation of both microfiltration and ultrafiltration membranes has shown effective removal of *Microcystis* cells (Chow, Panglisch et al. 1997; Drikas, Newcombe et al. 2002). The cells formed a layer on the membrane, requiring frequent backwashing, with more difficulty in restoring flow rate to the rougher microfiltration membrane than to the smooth-surfaced ultrafiltration membrane. A reduction of 20 to 30% of dissolved microcystin was seen across the ultrafiltration membrane.

Nanofiltration membranes can be obtained with a range of cutoffs for exclusion of molecules of different molecular weights. With a cyclic structure and molecular weight of about 1000 Da, microcystin should be relatively easy to exclude in an appropriate filter. Trials of this technique have appeared promising, with effective removal of microcystin (Hart and Stott 1993; Muntisov and Trimboli 1996; unpublished personal data).

With all membrane filtration systems, the particulate and organic content of the raw water entering the filtration system substantially affects the filtration rate. High-quality raw water from oligotrophic sources can be processed more effectively and with less backwashing and membrane cleaning than more eutrophic water, especially if cyanobacterial blooms are present. Multistep membrane filters or installation of coagulation/clarification processes prior to membrane filtration are likely to be required for the removal of cyanobacterial cells and toxins. These technologies require considerable further development for removal of cyanobacterial toxins at the scale of drinking water supply.

An alternative approach to the use of membranes is the possibility of degrading microcystins on a biofilm of organisms, using a membrane surface as a support film (Saito, Sugiura et al. 2003).

Cylindrospermopsin has a molecular weight about half that of microcystin and would require a nanofiltration membrane with an appropriately lower cutoff, approaching that of a reverse osmosis membrane. Reverse osmosis units in current use for water purification would be expected to provide water free of both microcystins and cylindrospermopsin, though this remains to be verified experimentally.

12.8 CONCLUSIONS

A wide range of technologies are available for treatment of drinking water sources contaminated with cyanobacterial toxins. Studies undertaken in operating commercial plants have shown progressive reductions in microcystin concentrations along the treatment sequence. Raw water containing low toxin concentrations has been successfully treated in plants using powdered activated carbon (Karner 2001) and ozone (Tarczynska, Romanowska-Duda et al. 2000). Conventional coagulation/clarification treatment is only partially effective and becomes less so if prechlorination is used, as this redistributes cyanobacterial toxins into the free water solution. When high concentrations of cyanobacterial colonies enter the plant, the particulate load on the flocculation and clarification system will block filters, causing whole cells and free toxin to pass through the system. The chlorine demand of the "finished" water rises under these circumstances, so that the normal residual concentration of chlorine may decrease to zero, allowing dissolved toxin to enter the distribution system. Conventional treatment is therefore effective in reducing cyanobacterial toxins when the problem is minor but increasingly ineffective as the problem escalates.

The most cost-effective solutions to the removal of microcystins and cylindrospermopsin from drinking water depend very much on the other requirements of the system, in particular the need to remove other harmful organic compounds. The highest level of protection currently operating for removal of unwanted organic chemicals from drinking water, which includes the cyanobacterial toxins, is post-clarification ozonation followed by granulated activated carbon filtration. However, if the major problem is intermittent, with seasonal cyanobacterial blooms, other systems with lower operating costs may be applicable. As long as prechlorination is avoided and the coagulation/clarification system can handle the fluctuating organic load, the addition of powdered activated carbon may be sufficient to remove dissolved toxins and cyanobacterial tastes and odors. In examples known to the author, dissolved air-flotation technology has proved capable of handling variable cyanobacterial loads, as has (on a smaller scale) upflow sludge blanket clarification. In the future, titanium oxide/UV oxidation and membrane filtration may prove to be cost-effective options.

Posttreatment chlorination can provide a last defense against microcystins and cylindrospermopsin in the drinking water supply, but only if a residual chlorine level of at least 0.5 mg/L is maintained in the finished water for more than 30 min at neutral pH. While this is commonly achieved in normal water treatment, the circumstances of a major cyanobacterial bloom in the raw water may override the chlorination, allowing harmful toxins to reach the consumer. If postchlorination or postozonation are not practiced, there is an increased possibility of cyanobacterial toxins reaching the consumer, which must be recognized in assessing finished water safety.

REFERENCES

Bernezeau, F. (1994). Can microcystins enter drinking water distribution systems? *Toxic Cyanobacteria: Current Status of Research and Management.* Adelaide, Australia, Australian Centre for Water Quality Research, American Water Works Association Research Foundation, Centre for Water Research, Belgium.

Bourne, D. G., G. J. Jones, et al. (1996). Enzymatic pathway for the bacterial degradation of the cyanobacterial cyclic peptide toxin microcystin-LR. *Applied Environmental Microbiology* 62: 4086–4094.

Burns, J., A. Chapman, et al. (2000). Cyanotoxic blooms in Florida's lakes, rivers and tidal river estuaries: The recent invasion of toxigenic *Cylindrospermopsis raciborskii* and consequences for Florida's drinking water supplies. *Ninth International Conference on Harmful Algal Blooms.* Hobart, Tasmania.

Byth, S. (1980). Palm Island Mystery disease. *Medical Journal of Australia* 2: 40–42.

Carlile, P. R. (1994). *Further Studies to Investigate Microcystin-LR and Anatoxin-a Removal from Water.* Medmenham, U.K., Foundation for Water Research.

Carmichael, W. W., S. M. F. O. Azevedo, et al. (2001). Human fatalities from cyanobacteria: Chemical and biological evidence for cyanotoxins. *Environmental Health Perspectives* 109(7): 663–668.

Chorus, I., G. Klein, et al. (1993). Off-flavours in surface waters, how efficient is bank filtration in their abatement in drinking water? *Water Science and Technology* 25: 251–258.

Chow, C. W. K., S. Panglisch, et al. (1997). A study of membrane filtration for the removal of cyanobacterial cells. *Aqua (Oxford)* 46(6): 324–334.

Cornish, B. J. P. A., L. A. Lawton, et al. (2000). Hydrogen peroxide enhanced photocatalytic oxidation of microcystin-LR using titanium dioxide. *Applied Catalysis B: Environmental* 25: 59–67.

Dickens, C. and P. Graham (1995). The rupture of algae during abstraction from a reservoir and the effects on water quality. *Journal of Water Supply: Research and Technology — Aqua* 44(1): 29–37.

Dillon, P. J., M. J. Miller, et al. (2002). The potential of riverbank filtration for drinking water supplies in relation to microcystin removal in brackish aquifers. *Journal of Hydrology* 266: 209–221.

Drikas, M., C. W. K. Chow, et al. (2001). Using coagulation, flocculation, and settling to remove toxic cyanobacteria. *Journal of the American Water Works Association* 193: 100–111.

Drikas, M., G. Newcombe, et al. (2002). *Water Treatment Options for Cyanobacteria and Their Toxins*. Salisbury, South Australia, Cooperative Research Centre for Water Quality and Treatment: 75–92.

Ellis, K. V. (1985). Slow sand filtration. *CRC Critical Reviews in Environmental Control* 15: 315–354.

Falconer, I. R. (1994). Blue-green algae in the Hawkesbury, what are the health risks? *National Parks Journal* 38: 13–14.

Falconer, I. R., A. M. Beresford, et al. (1983). Evidence of liver damage by toxin from a bloom of the blue-green alga, *Microcystis aeruginosa. Medical Journal of Australia* 1(11): 511–514.

Falconer, I. R., M. T. C. Runnegar, et al. (1983). Effectiveness of activated carbon in the removal of algal toxin from potable water supplies: A pilot plant investigation. *Technical Papers Presented at the Tenth Federal Convention*. Sydney, Australian Water and Wastewater Association: 26-1–26-8.

Falconer, I. R., M. T. C. Runnegar, et al. (1989). Use of powdered and granular activated carbon to remove toxicity from drinking water containing cyanobacterial toxins. *Journal of the American Water Works Association* 18: 102–105.

Fastner, J., B. Wirsing, et al. (2001). Microcystins and hepatocyte toxicity. *Cyanotoxins: Occurrence, Causes, Consequences*. I. Chorus, ed. Berlin, Springer-Verlag: 22–45.

Fielding, M., T. M. Gibson, et al. (1981). *Organic Micropollutants in Drinking Water*. Medmenham, U.K., Water Research Centre.

Gray, N. F. (1994). *Drinking Water Quality: Problems and Solutions*. Chichester, U.K., John Wiley and Sons.

Grutzmacher, G., G. Bottcher, et al. (2002). Removal of microcystins by slow sand filtration. *Environmental Toxicology* 17(4): 386–394.

Harada, K.-I. (1996). Chemistry and detection of microcystins. *Toxic Microcystis*. M. F. Watanabe, K.-I. Harada, W. W. Carmichael, and H. Fujiki, eds. Boca Raton, FL, CRC Press: 103–148.

Hart, J., J. Fawell, et al. (1997). *The Fate of Both Intra- and Extracellular Toxins during Drinking Water Treatment*. IWSA World Congress, Oxford, Blackwell Science.

Hart, J. and P. Stott (1993). *Microcystin-LR Removal from Water*. Marlow, England, Foundation for Water Research.

Himberg, K., A. M. Keijola, et al. (1989). The effect of water treatment processes on the removal of hepatotoxins from *Microcystis* and *Oscillatoria* cyanobacteria: A laboratory study. *Water Resources Bulletin of the American Water Resources Association* 23(8): 979–984.

Hitzfeld, B. C., S. J. Hoger, et al. (2000). Cyanobacterial toxins: Removal during drinking water treatment and human risk assessment. *Environmental Health Perspectives* 108: 113–122.

Hoffmann, J. R. H. (1976). Removal of *Microcystis* toxins in water purification processes. *Water SA* 2(2): 58–60.

Hoffmann, M. R., S. T. Martin, et al. (1995). Environmental applications of semiconductor photocatalysis. *Chemical Reviews* 95: 69–96.

Hrudey, S., M. Burch, et al. (1999). Remedial measures. *Toxic Cyanobacteria in Water: A Guide to Their Public Health Consequences, Monitoring and Management*. I. Chorus and J. Bartram, eds. London, E & FN Spon (on behalf of WHO): 275–312.

Jones, G. J., W. Minato, et al. (1993). Removal of low level cyanobacterial peptide toxins from drinking water using powdered and activated granular carbon and chlorination: Results of laboratory and pilot plant studies. *Proceedings of the Australian Water and Wastewater Association*. Australian Water and Wastewater Association.

Karner, D. A., J. H. Standridge, et al. (2001). Microcystin algal toxins in source and finished drinking water. *Journal of the American Water Works Association* 93(8): 72–81.

Keijola, A. M., K. Himberg, et al. (1988). Removal of cyanobacterial toxins in water treatment processes: Laboratory and pilot-scale experiments. *Toxicity Assessment* 3: 643–656.

Kolpin, D. W., E. T. Furlong, et al. (2002). Pharmaceuticals, hormones, and other wastewater contaminents in U.S. streams, 1999–2000: A national reconnaissance. *Environmental Science and Technology* 36: 1202–1211.

Lahti, K., J. Kilponen, et al. (1996). Removal of cyanobacteria and their hepatotoxins from raw water in soil and sediment columns. *Artificial Recharge of Groundwater*. A.-L. Kivimaki and T. Suokko. Helsinki, Nordic Hydrological Report No. 38: 187–195.

Lahti, K., J. Rapala, et al. (2001). Occurrence of microcystins in raw water sources and treated drinking water of Finnish waterworks. *Water Science and Technology* 43: 225–228.

Lam, A. K. Y., E. E. Prepas, et al. (1995). Chemical control of hepatotoxic phytoplankton blooms: Implications for human health. *Water Research* 29: 1845–1854.

Lambert, T., C. Holmes, et al. (1996). Adsorption of microcystin-LR by activated carbon in full-scale water treatment. *Water Research* 30: 1411–1422.

Lambert, T. W., M. P. Boland, et al. (1994). Quantitation of the microcystin hepatotoxins in water at environmentally relevant concentrations with the protein phosphatase bioassay. *Environmental Science and Technology* 28: 753–755.

Lawton, L. A., P. K. J. Robertson, et al. (2003). Process influencing surface interaction and photocatalytic destruction of microcystins on titanium dioxide photocatalysts. *Journal of Catalysts* 213(1): 109–113.

Lepisto, L., K. Lahti, et al. (1994). Removal of cyanobacteria and other phytoplankton in four Finnish waterworks. *Algological Studies* 75: 167–181.

Liu, I., L. A. Lawton, et al. (2003). Mechanistic studies of the photocatalytic oxidation of microcystin-LR: An investigation of byproducts of the decomposition process. *Environmental Science and Technology* 37(14): 3214–3219.

Miller, M. J., M. M. Critchley, et al. (2001). The adsorption of cyanobacterial hepatotoxins from water onto soil during batch experiments. *Water Research* 35: 1461–1468.

Miller, M. J. and H. J. Fallowfield (2001). Degradation of cyanobacterial hepatotoxins in batch experiments. *Water Science and Technology* 43: 229–232.

Mouchet, P. and V. Bonnelye (1998). Solving algae problems: French expertise and worldwide applications. *Journal of Water Supply: Research and Technology — Aqua* 47: 125–141.

Muntisov, M. and P. Trimboli (1996). Removal of algal toxins using membrane technology. *Water* 23(3): 34.

Nicholson, B. C., J. Rositano, et al. (1994). Destruction of cyanobacterial peptide hepatotoxins by chlorine and chloramine. *Water Research* 28: 1297–1303.

NRA (1990). *Toxic Blue-Green Algae. A Report by the National Rivers Authority*. London, National Rivers Authority.

Robertson, P. K. J., L. A. Lawton, et al. (1997a). Destruction of cyanobacterial toxins by semiconductor photocatalysis. *Chemical Communications* 4: 393–394.

Robertson, P. K. J., L. A. Lawton, et al. (1997b). The destruction of cyanobacterial toxins by titanium dioxide photocatalysis. *Journal of Advanced Oxidation Technology* 4: 20–26.

Robertson, P. K. J., L. A. Lawton, et al. (1998). Processes influencing the destruction of microcystin-LR by TiO_2 photocatalysis. *Journal of Photochemistry and Photobiology A: Chemistry* 116: 215–219.

Robertson, P. K. J., L. A. Lawton, et al. (1999). The involvement of phycocyanin pigment in the photodecomposition of the cyanobacterial toxin, microcystin-LR. *Journal of Porphyrins and Phthalocyanines* 3: 544–551.

Rositano, J. and B. C. Nicholson (1994). *Water Treatment Techniques for the Removal of Cyanobacterial Toxins from Water.* Salisbury, South Australia, Australian Centre for Water Quality Research.

Rositano, J., B. C. Nicholson, et al. (1998). Destruction of cyanobacterial toxins by ozone. *Ozone Science and Engineering* 20: 223–238.

Saito, T., N. Sugiura, et al. (2003). Biodegradation of microcystis and microcystins by indigenous nanoflagellates on biofilm in a practical treatment facility. *Environmental Technology* 24(2): 143–145.

Schmidt, W., H. Willmitzer, et al. (2002). Production of drinking water from raw water containing cyanobacteria: Pilot plant studies for assessing the risk of microcystin breakthrough. *Environmental Toxicology* 17(4): 375–385.

Senogles-Derham, P.-J., A. Seawright, et al. (2003). Toxicological aspects of treatment to remove cyanobacterial toxins from drinking water determined using the heterozygous p53 transgenic mouse model. *Toxicon* 41(8): 979–988.

Shawwa, A. R. and D. W. Smith (2001). Kinetics of microcystin-LR oxidation by ozone. *Ozone Science and Engineering* 23(2): 161–170.

Tarczynska, M., Z. Romanowska-Duda, et al. (2000). Removal of cyanobacterial toxins in water treatment processes: IV. *International Conference Water Supply and Water Quality.* Krakow, Poland.

WHO (1996). *Guidelines for Drinking Water Quality.* Second Edition, Volume 2. Geneva, World Health Organization.

WRc (1996). The fate of intracellular microcystin-LR during water treatment. Report No. 96/DW/07/4. London, U.K. Water Industry Research Ltd.

13 Emerging Issues

The whole research field of toxic cyanobacteria and associated hazard to human health is relatively new, with the first international meeting on the topic at Wright State University, Dayton, Ohio, in 1980 (Carmichael 1981). Since that time there have been very considerable developments in understanding of the organisms, toxins, health impacts, and processes for the removal of unwanted substances in the treatment of drinking water. A landmark development in public safety was the determination in 1998 by the World Health Organization (WHO) of a provisional Guideline Value for microcystin-LR in drinking water. This has been followed by individual national and regional governments adopting a regulatory standard for microcystins in drinking water, which has been closely related to the WHO guideline of 1.0 μg/L. The regulations have varied from the WHO guideline only by specifying toxicity equivalent to microcystin-LR, thus including all microcystins, and through different population body weights and assumed proportion of the toxin in drinking water. In addition, New Zealand and Brazil have adopted Maximum Acceptable Values or Guideline Values for cylindrospermopsin of 3.0 and 15 μg/L, respectively, in drinking water.

The increasing acceptance that cyanobacterial toxins in drinking water may be a hazard to human health has raised a series of currently unanswered questions, which fall roughly into three areas: ecological, health, and water treatment. In the ecological area, the dominant question is how to respond to the changing world, in which population pressure and global warming are together intensifying the likelihood of the appearance of toxic cyanobacteria in water sources. In the health field, the largest unresolved issue is the relationship between cyanobacterial toxins and human cancer. In water treatment, the main question is the most cost-effective way of providing safe water for the population, including safety from cyanobacterial toxins.

13.1 ECOLOGICAL ISSUES

The continuing growth of the overall world population, together with the need for additional agricultural production, is placing increased demands on water supply. Intensification of agriculture has resulted in greater use of water for irrigation as well as greater nutrient loads within water catchments. In more arid areas, including those with Mediterranean climates, groundwater has often provided the bulk of the supply for drinking and for irrigation. In parts of the U.S. and Australia, groundwater depletion and deterioration in quality are forcing a reexamination of the use of surface water, which was earlier rejected because of eutrophication or contamination.

In circumstances where demand for irrigation water has much depleted water flow in rivers, the flow reduction is frequently accompanied by eutrophication in the summer months. This may be exacerbated by discharge of treated or untreated sewage into the rivers. Use of such supplies as sources of drinking water increasingly leads to problems with toxic cyanobacteria in the drinking water.

Use of eutrophic water for drinking water supply raises the need for fast and clear identification of toxic organisms. The recent development of genetic approaches offers the most promising advance, since the presence or absence of toxin-coding genes provides the key information for potential hazard. With the use of genetic identification of toxic potential, there is no need for species identification or toxin measurement in the first instance. If a potential hazard requires quantitation, which is secondary to the initial identification, then cell numbers or biovolume and toxin concentration are required.

The genetic methods lend themselves to development of automated techniques capable of handling multiple samples rapidly. They can also be developed as simple color test kits, capable of use in basic laboratories or even at the water's edge.

Global warming has implications for the world distribution of cyanobacterial species. In particular, the organisms now mainly restricted to subtropical and tropical regions will be enabled to grow in cooler latitudes. This is already evident from the increased range of *Cylindrospermopsis raciborskii*, which was first characterized as a tropical species in Indonesia and has since been found in Australia, Thailand, Brazil, the U.S., Canada, France, Germany, and Hungary. The other widely distributed species producing cylindrospermopsin is *Aphanizomenon ovalisporum*, reported in the Mediterranean region and Australia in the subtropics. This too can be expected to increase its range with global warming.

C. raciborskii is capable of producing a divergent range of toxins, including cylindrospermopsins, saxitoxins, and so far unidentified neurotoxins and hepatotoxins. As a species that can form dense concentrations in the lower epilimnion, out of sight of surface observation, it poses a problem for reservoir managers, who may not detect its presence until the cyanobacterial filaments block the filters at a treatment plant. As the filaments leak toxin into the water, potentially high concentrations of dissolved cylindrospermopsin may reach consumers.

The need for effective reservoir and catchment management is becoming more important as eutrophication increases. In both areas of management the problems are intractable, with reservoir management often the only option. The most common technique for controlling cyanobacterial blooms (in countries where it is legal to do so) is to add herbicide to the water. Copper sulfate is the most widely used herbicide, added in some reservoirs every 14 days in summer. This has highly adverse impacts on the local ecology and can lead to the appearance of copper-resistant organisms. It also releases toxin when the cyanobacterial bloom lyses. The other technique in wide use is compressed-air destratification, in which diffusers on the reservoir bottom send streams of air upward, mixing the deeper and surface layers. This has had variable success, leaving the issue of a cost-effective method of reservoir management open for further investigation.

The need for catchment management on a long-term basis to reduce nutrient inputs to water storage sites is becoming more important and also more difficult.

Intensity of land use in catchments inevitably rises, with increases in population leading to pressure for residential development, agricultural intensification, and access to recreational facilities. Control of catchments is made more difficult by private land ownership, increasing land value, and — in less developed countries — occupation by squatters. In many instances effective catchment management is a political impossibility. Some of the most polluting activities can be controlled by regulations, such as discharge of high-nutrient waste by agroindustry and wastewater treatment plants into drinking water storage catchments, but lower-intensity discharges — such as unsewered residences, high-density but small-scale livestock farming, and the use of fertilizer — are practically uncontrollable. Studies on optimal management strategies for partially polluted drinking water catchments are required to provide a factual basis for political decision making.

13.2 HEALTH ISSUES

Although there is much basic information on the acute toxicity of microcystins and cylindrospermopsins, there are very significant gaps in our knowledge of chronic effects. In particular little is known at present on possible actions of cylindrospermopsin on teratogenesis, reproductive health, and carcinogenesis in humans or experimental mammals. The increasing threat posed by cylindrospermopsin in drinking water as a result of global warming makes the urgency of research more acute. The present preliminary data raise the possibility of cylindrospermopsin being a potent human carcinogen. This requires investigation by carcinogenicity trials using formal, established protocols. These trials must be supported by studies of the molecular mechanisms of cylindrospermopsin carcinogenicity as well as genotoxicity studies. Where possible, epidemiological studies of cancer rates in human populations known to be exposed to cyanobacterial toxins in drinking water should be undertaken.

Because of the urgent need for the WHO to set a provisional Guideline Value for cylindrospermopsin in drinking water, the available subchronic toxicity data for experimental animals should be used in the absence of carcinogenicity data. There are several relevant studies, in particular one following the protocols of the Organization for Economic Cooperation and Development (OECD) for such trials, which can provide a sufficient basis for a provisional determination.

The carcinogenicity of microcystins is still an unresolved issue. The evidence for tumor promotion in experimental animals is very strong, but the relevance of this to human cancer has not yet been clearly shown. The most informative data come from southern China, where the rate of human hepatocellular carcinoma has been related to hepatitis, aflatoxin in the diet, and the drinking of surface water. The component of surface water that is proposed as the active agent in the cancer rate is microcystin. Because microcystin is taken up and excreted through the gut, there is also the possibility of increased rates of gastrointestinal tumors in the human population caused by microcystin. This awaits further exploration. In the mouse colon, tumor precursors were shown to grow faster with microcystin in the drinking water.

Until an exposure marker for cyanobacterial toxin intake in the human population is demonstrated, it is very difficult to carry out effective human epidemiology. This requires the comparison of known rates of intake of the postulated carcinogen with cancer rates, so that a dose–response relationship can be demonstrated. Without accurate measures of exposure, only indicative information can be acquired. An exposure marker that directly relates to toxin intake would remove the need to measure the actual intake of individuals, as it would give a cumulative intake measurement. Research in this area on both cylindrospermopsin and microcystin is needed.

Since the recognition that endocrine-disrupting compounds in drinking water may harm human health, the number of chemical compounds that may require to be monitored in drinking water has become very large. Together with the present chemical substances for which Guideline Values or Maximum Contaminant Levels have been set, water utilities may in the future be expected to measure very large numbers of substances that are at concentrations below detection limits in most instances.

To alleviate the necessity of carrying out uninformative analyses for compounds that are not likely to be present at detectable levels in any particular water supply, and to provide a proactive approach to safe drinking water, the WHO is developing a quite different methodology for drinking water safety. It is entitled *Water Safety Plan*, which would be required to be set up individually for each drinking water supply utility (WHO 2003). The basic approach is that of risk assessment at each stage of the drinking water process by evaluating the potential risks of harmful contaminants from the catchment and source water and their reduction, the risks and risk reduction within the treatment system, and risks of contamination and risk reduction in the distribution system. This would be done on an individual plant basis, so that the critical risks are assessed and minimized, and monitoring is based on the risk assessment. It is intended to include a process of continual revision and improvement, with legislative backing for the whole process and formal approval of the plan.

The New Zealand government is implementing such an approach in 2005 on the basis of a structure set up and publicized in 2001 with the title *How to Prepare and Develop Public Health Risk Management Plans for Drinking Water Supplies* (New Zealand Ministry of Health 2001). The supporting information includes risk management and alert levels for cyanobacterial contamination of drinking water reservoirs, which are ongoing management problems for New Zealand drinking water utilities.

The Australian National Health and Medical Research Council is developing a comparable approach to safety in drinking water based on planning of multiple barrier risk reduction and assessment of critical control points (Australian Department of Health 2003).

When the WHO *Water Safety Plan* final document is available it is likely that a range of other countries will evaluate this new approach to drinking water safety.

13.3 WATER TREATMENT

There are several emerging issues related to water treatment that arise through legislative and technical development. Now that countries are adopting as regulations the WHO provisional Guideline Value for microcystin in drinking water, the need for standardized test protocols for these toxins becomes acute. The increased interest in applying Hazard Analysis at Critical Control Points will reduce water industry requirements for analysis in circumstances where toxins are unlikely, but will not remove the need for accurate and straightforward analytical techniques that can be applied when needed. Cross-laboratory validation of methods is essential and has commenced.

Similarly, when a WHO Guideline Value for cylindrospermopsins is determined, standardized analytical methods will be required, which can be undertaken with normal laboratory facilities and at moderate cost. An enzyme-linked immunosorbent assay may be the most feasible.

Treatment technologies are in a process of continual development, with the newer approaches of titanium oxide/UV oxidation and membrane filtration offering the possibility of lower operating costs than the present highly effective ozone/activated carbon treatment. Conventional coagulation/sedimentation offers some protection from cyanobacterial toxins, but only with focused operational provisions in place. As toxic blooms become more common and occur to a greater extent in drinking water storage sites, upgrading of water treatment practices will be increasingly necessary.

Because a substantial proportion of the population in the developing world does not have access to treated water, consumption of cyanobacterial toxins is more likely among them. Often surface water is the only supply, with seasonal eutrophication. Very low-cost methods of purifying water from pathogens and toxins are needed. Boiling is common practice to reduce enteric disease transmission, but it does not remove microcystins or cylindrospermopsins. It is possible that low-cost slow sand filters can be developed, based on 44-gal drums half filled with sand, for remote and rural families. These would reduce hazards from protozoa, bacteria, and metabolizable organic contaminants as well as to cyanobacterial toxins.

REFERENCES

Australian Department of Health (2003). Public consultation draft framework for *Management of Drinking Water Quality: A Preventive Strategy from Catchment to Consumer.* www.health.gov.au/nhmrc/publications/synopses/eh19syn.htm.

Carmichael, W. W. (1981). *The Water Environment.* New York, Plenum Press.

New Zealand Ministry of Health (2001). *How to Prepare and Develop "Public Health Risk Management Plans for Drinking Water Supplies."* www.moh.govt.nz/water.

WHO (2003). *Guidelines for Drinking Water Quality.* Third Edition. Draft Chapter 4, "Water Safety Plans." www.who.int/docstore/water_sanitation_health/GDWQ/Updating/draftguide1/draftchap4.htm.

Index

A

B

D

E

F

G

H

I

O

P

Q

R

S

T

U

V

W